WILD PIGS:
ENVIRONMENTAL PEST OR ECONOMIC RESOURCE?

WILD PIGS:
ENVIRONMENTAL PEST
OR
ECONOMIC RESOURCE?

C.A. Tisdell
Professor of Economics
University of Newcastle
New South Wales
Australia

Pergamon Press
Sydney • Oxford • New York • Toronto • Paris • Frankfurt

Pergamon Press (Australia) Pty Ltd,
19a Boundary Street, Rushcutters Bay, N.S.W. 2011, Australia.

Pergamon Press Ltd,
Headington Hill Hall, Oxford OX3 OBW, England.

Pergamon Press Inc.,
Maxwell House, Fairview Park Elmsford, N.Y. 10523, U.S.A.

Pergamon Press Canada Ltd,
Suite 104, 150 Consumers Rd, Willowdale, Ontario M2J 1P9, Canada.

Pergamon Press GmbH,
6242 Kronberg-Taunus, Hammerweg 6, Postfach 1305
Federal Republic of Germany.

Pergamon Press SARL,
24 rue des Ecoles, 75240 Paris, Cedex 05, France.

First published 1982

Cover design by Deidre Huxley
Printed in Singapore by Singapore National Printers (Pte) Ltd

National Library of Australia Cataloguing in Publication Data:

Tisdell, C.A. (Clement Allan)
Wild pigs.

Bibliography.
Includes index.
ISBN 0 08 024820 9.
ISBN 0 08 024821 7 (soft cover).

1. Wild boar. 2. Feral livestock - Control - Australia. I. Title.

632'.69734

Contents

Also published by Pergamon Press

Economics of Fibre Markets

C. Tisdell and P.W. McDonald

Preface

This book represents the results of my research over the past few years on wild pigs and began after a conversation with an agricultural economist, Mr Cyril Catt. I had previously done some research on the economics of the conservation of wildlife and I remarked to him that I would like to study a vertebrate pest. He suggested that feral pigs would make a worthwhile study. After considering this, I applied to the Rural Credits Development Department of the Reserve Bank of Australia for a small research grant with the intention of studying some limited special aspects of the problem and then moving on to a new one. But it soon became clear that this case study in pest control involved many unresolved issues, that the problem was important in the Australian context and that overseas experience with wild pigs needed consideration. Furthermore, the grantees of my research funds encouraged me to complete this research before taking up a new topic. So instead of a one-year project this became a five-year study taking most of my 'spare time'. I hope that the readers of this book will find that the results justify the extra time spent on this research.

This study is a general rather than narrow disciplinary one. I found it quite impracticable to proceed otherwise. Much of the basic ecological and environmental data that I needed for the study was not available so I had little option but to collect it myself. Economic evaluation depends on the availability of such material, and I could not sensibly proceed to the latter without the other data. Hence the study is a broad one in which I have exercised my right as a citizen to consider matters additional to those in which I have a professional training. To go beyond the bounds of one's discipline is of course to court sniping from specialists, including those from one's own field of specialization. However, in many instances the integration of knowledge is a prerequisite for rational policy decisions. The formulation of policies for the management or control of wild pigs requires this integration.

Many individuals have assisted and encouraged me with this study and it is impossible to list all their names in this preface. Many are mentioned in the text but others are not. I am grateful for all the help and response that I have had. This includes individuals who have spent time to complete my survey forms and have sent me personal correspondence and material on wild pigs. This type of help came from within Australia and from several overseas countries, and I am especially grateful to those embassies in Australia that showed goodwill by aiding me in making such contacts.

There are, however, some individuals that I should mention since they helped me with specific queries or with material at crucial points in the study. In Australia, these include Cyril Catt, Jack Giles, Jim Hone, Bunny Fennessey, Will Andrew, Professor Jack Golson and Dr Peter Jarman; in the United States, Professor Reginald Barrett of the University of California; and in the United Kingdom, Dr Juliet Clutton-Brock of the British Museum (Natural History). The study also benefitted from a number of points put to me by scholars at the University of York (namely Dr John Beddington of the Department of Biology and Professor Jack Wiseman of I.S.E.R.) during my visit in 1979 and from comments, especially by Professor David Pearce, at the University of Aberdeen, where I gave a seminar in 1979. Other seminars such as those given at the School of Economics, University of New South Wales, and at annual conferences of the Australian Agricultural Economics Society also yielded useful feedback.

I wish to thank Paul Ross, Ross Turvey and Peter Jackson for their enthusiasm and research assistance, the Rural Credits Development Department of the Reserve Bank for making financial assistance available for this *independent* study, the University of Newcastle for providing excellent overhead facilities for this research, my colleagues for tolerating my single-minded concentration on this research, and especially my former colleague, Dr Roley Piggott for his useful comments on aspects of it. I am grateful for permission to use the photographs taken and supplied by Mr Terry Pichler, Mr Bob Penfold, Mr Rup Hanna, the National Parks and Wildlife Service of N.S.W., and the R.S.P.C.A. These are included along with those taken by my research assistants and myself.

Beverley Barnes deserves credit for editing the manuscript so carefully, as does Gillian Hewitt (and others at Pergamon Press) for guiding this publication through to completion.

Elizabeth Williams typed the draft of the manuscript and Lorraine King the final copy. Both these demanding tasks were done competently and with good humour. Joan Allridge assisted with some typing and kindly translated some Spanish material.

My immediate family (Mariel, Ann-Marie and Christopher) took an active interest in the project and my wife Dr Mariel Tisdell translated a considerable amount of literature from German and Swedish for the project. My father-in-law, Dr Richard Eckermann, collected relevant material and did further translations, thereby acting as an unofficial research assistant. He was ably assisted by my mother-in-law, Mrs A. Eckermann. All this support is sincerely appreciated.

January, 1982

Clem Tisdell

CHAPTER 1

Status of Wild Pigs: General Features

While the range of many animal species has been reduced by man's activities, and indeed many have been driven to extinction, this has not been the fate of the wild pig species. *Sus scrofa* (apart from the domestic pig) is comprised of Eurasian wild pigs and feral pigs. Feral pigs (wild populations derived from domestic pigs that have returned to the wild)[1] can trace their ancestry to Eurasian wild pigs because all domestic pigs have been developed from this source.[2] Indeed (despite their disappearance in historic times from some areas such as the British Isles) wild pigs have extended their range in the world as a result of man's activities. The (wild) pig has colonized or is successfully colonizing new continents and islands to which it has been transported in comparatively recent times by man. Australia is one of the island/continents colonized in less than the last 200 years by feral pigs.

Man's interests in wild pigs are many-sided: To agriculturalist and pastoralists, they are often a pest requiring control. They can have beneficial as well as adverse affects on forests. Thus foresters may need to manage the species. Sporting hunters find wild pigs to be an important hunting asset, and the sale of hunting rights to wild pigs can provide landholders with supplementary income. Hunters, up to a point, have an interest in conserving wild pigs as game. Wild pigs can also be used commercially for food and other products, and in some countries make a significant contribution to the game-meat industry. Conservationists consider pigs to be a threat to other wildlife in a number of situations; they pose such a threat when they are introduced to countries and areas where the flora and fauna has developed in the absence of pigs, and they can also become a menace to other wildlife when their population expands beyond its usual ecological limits. Wild pigs also play a significant role in the spread of diseases and parasites infectious to livestock and man. These aspects and the methods and economics of controlling populations of wild pigs are all discussed in this book.

Although this study focuses in depth on feral pigs in Australia, it does so in a worldwide context and in relation to studies elsewhere. Thus the book reviews and provides new information about feral pigs in Australia as well as giving a general overview of relevant worldwide aspects of wild pigs.

This chapter is designed to provide general background about wild pigs, especially *Sus scrofa*. It looks at the wild pig family; its geographical distribution in the world; the geographical distribution of feral pigs in Australia and nearby, their origin and broad consequences; and the general experiences of other countries having wild pigs. Relevant aspects of the biology, habits and ecological requirements of *Sus scrofa* are also outlined.

THE WILD PIG FAMILY (SUIDAE)

Although the Suidae (pig family)[3] belongs to the order of Artiodactyla (ungulates) - that is, cloven-hooved or even-toed animals - they do not chew their cud and so are included under a sub-order Suiformes which also includes hippopotamuses and peccaries (family Tayassuidae).

There are five genera of the Suidae family, the original distribution of which was confined to Europe, Asia and Africa. These genera are set out in Table 1.1 together with some information about their geographical distribution.

Authorities differ about the number of species of *Sus* that exist. Some recognize three distinct[4] species and others four.[5] If four are accepted, they are *Sus scrofa* (including the wild boar, feral pigs, etc.), *Sus barbatus* (the bearded pig, which occurs in Kalimantan (Borneo)), *Sus verrucosus* (the Javan Pig, which occurs in Java and Sulawesi), and *Sus salvanius* (the Pygmy hog, which occurs in Asia). Each has a number of subspecies and varieties. *Sus scrofa* has the largest number of subspecies and varieties and by far the largest population of all these species. It is this species that is the focus of attention in this book, especially in its feral form.

TABLE 1.1 Genera of the Pig Family (Suidae), Common Names, and Distribution in the Wild

Genus	Common names for some members or species of the genus	Distribution
Sus	Wild boar, Eurasian wild pig, wild pig, feral pig	Europe, Asia, Northern Africa (has spread to Americas, Oceania and other areas)
Potamochoerus	Bushpig	Africa (south of the Sahara)
Phacochoerus	Warthog	Africa
Hylochoerus	Giant forest pig	Central Africa
Babyrousa	Babirussa (deer-pig)	Celebes (Sulawesi)

Fig. 1.1 European wild boar Sus scrofa. *There are a number of varieties of this species, and it is from this species that the domestic pig has been derived.*

Bushpigs are included in one species, *Potamochoerus porcus*, but have many subspecies. They occur in Africa south of the Sahara and on Madagascar. Their habits are very similar to those of *Sus* and indeed some authorities doubt whether they should be included in a separate genus to *Sus* because of their resemblance to the wild boar is so close.[6]

Warthogs are included in one species, *Phacochoerus aethiopicus*, but it has seven subspecies. Unlike members of the *Sus* and *Potamochoerus* genera, warthogs tend to be active by day, spending the night underground (usually in the abandoned den of an aardvark).[7] Warthogs are considered to be less of a pest to farmers than bushpigs. By all reports their flesh is tastier than that of the bushpig, the meat of which is considered to be rather dry and inferior to that of the European wild boar. Some commercial use is made of their meat in South Africa (see Chapter 4) and Mr C. K. Manu, Chief Game and Wildlife Officer of Ghana informed me in July 1981 that serious attention is being given to the inclusion of warthog in a game ranching project in Ghana. They considerably supplement the animal protein supply in Ghana.

The giant forest pig, *Hylochoerus meinertzhageni*, was not discovered until the beginning of this century in Central Africa. Adults weigh about 250 kg and are covered by long hair. This pig appears to prefer heavily forested or jungle areas and little is known about its habits.

The only species of the *Babyrousa* genus appears to be *Babyrousa babyrussa*. It has canine teeth that grow like horns and hence the allusion to Rusa deer in its name. Babirussa occurs only in Sulawesi (Indonesia) and seems to be an endangered species because of land-clearing and the hunting of it for meat.

A pig-like animal belonging to a separate family to the Suidae is the peccary. It belongs to the family Tayassuidae, of which there is only one existing genus, *Tayassu*, but three species. Tayassuidae now only occur in the New World from the south-west U.S.A. southwards, but in the geological past also occurred in Eurasia. They weigh 18–30 kg and their habits and external appearance are not unlike that of the European wild pig. They have a strong pungent smell from a scent gland in front of their tail. They can be a pest to agriculturalists but are sought after for their meat and skins, and are claimed to have become rare as a result of this.[8] They are not nearly as dependent on water as wild pigs.

Fig. 1.2 Collared peccary Tayassu tajacu, *photographed at the* Taronga Zoological Park. *This species ranges from Arizona, New Mexico and Texas in North America, through Central America, and then south into Argentina. While its external appearance is like a pig except that it has smaller ears, various physiological differences result in its being classified in a different family to the Suidae.*

GEOGRAPHICAL DISTRIBUTION AND SPREAD OF THE WILD PIG *SUS SCROFA*

According to Heptner, Nasimovic and Bannikov the original geographical distribution of the wild boar *Sus scrofa* was in the area in Fig. 1.3 enclosed by the two heavy lines.[9] In the north its distribution appeared to be limited by the depth of snow (a depth of 30-40 cm)[10] and it was absent in the Gobi Desert area because of the aridity of the region.

Heptner *et al.* point out that its original natural distribution in South-east Asia is uncertain. It *may* have been present in the Philippines area - that is, within the area enclosed by the dashed line and the solid line and it *may not* have extended to the lesser south-eastern Indonesian islands and the New Guinea area. Translating their statement, they point out:

> 'It is very difficult to work out the actual natural (original) area of distribution of the wild pig in the very far south-eastern region and perhaps already impossible. This is so because the inhabitants of these areas, who were excellent sailors, brought already in early times the domestic pig with them when they colonized the islands ... These domestic pigs became wild and changed into genuine wild pigs. This is particularly true of New Guinea and the adjacent group of islands which are populated by *Sus scrofa papueusis*. However, wild pigs of other parts on the islands between Australia and Asia also have similar features to domestic pigs. Even the wild pigs on the Celebes Is. (i.e. the complete group of *Sus scrofa celebeusis*) have these "domestic" characteristics. Apparently interbreeding took place in some areas between genuine wild pigs native to these areas and introduced pigs who became wild. It is quite possible that the original area to which genuine wild pigs were native may not have extended as far to the south east as indicated on the map.'[11]

It must be regarded as extremely doubtful whether New Guinea and neighbouring Melanesian Islands were a part of the original range of the wild pig. Its original distribution might have extended only to the islands of Indonesia and been stopped by the deep-water straits to their east - at divisions between the regions that are known as Wallace's Line and Weber's Line. In other words it might *not* have spread by natural means beyond the areas linked by land bridges to Asia approximately 20 000 years ago.[12] At that time Java, Borneo and Sumatra were linked to the Asian mainland, but not Sulawesi, the Mollucas and other islands in that area. New Guinea was certainly not linked to Asia by a land bridge but was linked to Australia. Other islands where the wild pig was already present in earlier times (such as Japan and Hainan) were also linked to the Asian mainland.

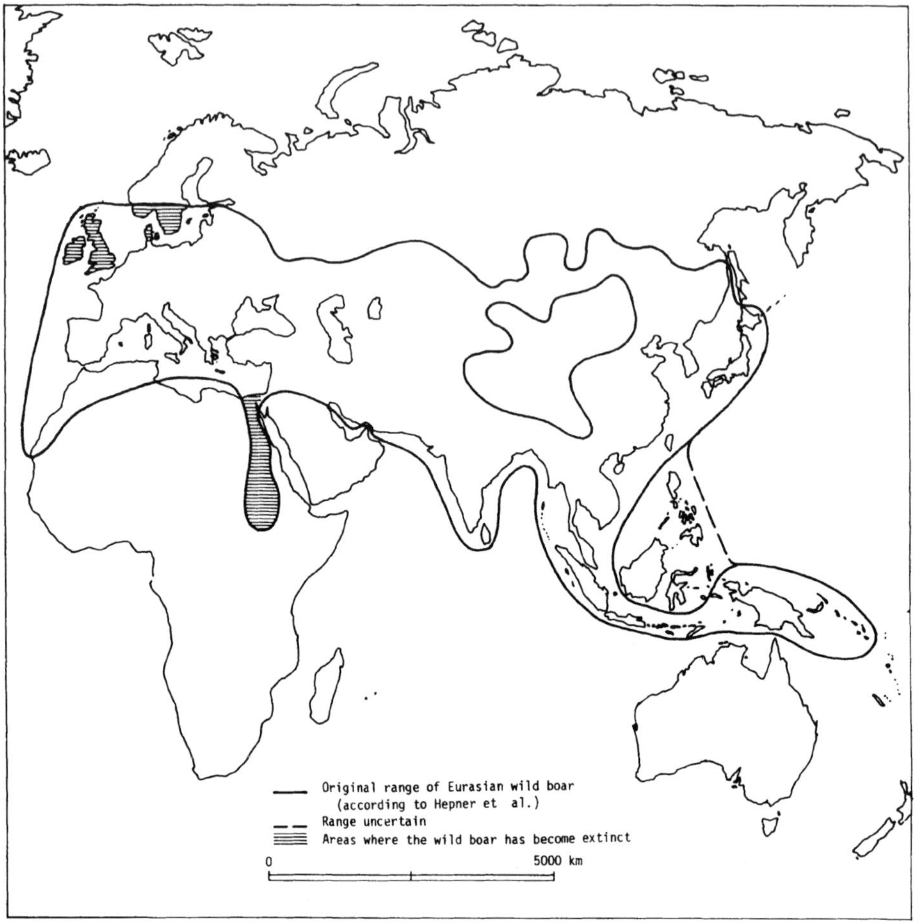

Fig. 1.3 Original native range of Sus scrofa *(after Heptner* et al.). *As explained in the text, it is most unlikely that this range extended to New Guinea. The south-eastern limit of the original range is uncertain.*

Indeed, it appears unlikely that New Guinea and neighbouring islands were inhabited by wild pigs in early times. If they had been present, say, 20 000 years ago, these pigs would have presumably crossed by the land bridge to Australia and would have already successfully established themselves here. No evidence has been found of the existence of such animals in Australia prior to the coming of European man.[13]

Golson suggests that on the basis of available evidence pigs were introduced to New Guinea about 6000 years ago[14] and were brought there, it seems, as domestic animals from Asia, possibly by migrating groups having Lapita pottery in common.

New Guinea itself already had inhabitants at least 26 000 years ago and possibly (as for Australia) around 50 000 years ago and appears to have been well settled by 8000 B.C.[15] However, not all the islands to the east were settled.

It has been claimed that there is evidence of the presence of pigs in New Guinea 10 000 years ago but this evidence has yet to be evaluated.[16] Many of the introduced domesticated pigs would have undoubtedly returned to a wild or semi-feral state. The pig spread with human migration eastwards into Melanesia and thence to most of Polynesia. Most of Melanesia was settled (for the first time) by the first and second millenia B.C.[17] For example, human settlement had spread as far as Fiji by 1300 B.C. and brought with it the pig. Settlement expanded into Polynesia, most of it being settled by the end of the first millenium B.C. However, pigs did not make it to all parts of the Polynesian area. New Zealand was settled by Polynesians in about 850 A.D. but their pigs perished on the long journey from the area of the Marqesas Islands or Society Islands. Pigs were not introduced into New Zealand until the coming of European man.[18]

Figure 1.4 indicates the current distribution of the Eurasian wild boar and of feral pigs.[19] It shows by hatching the regions in which the wild boar is extinct. The distribution of the bushpig is also indicated, since as mentioned earlier it is closely related to *Sus scrofa* and also has similar habits. Between them *Sus scrofa* and *Potamochoerus porcus* are present in most countries of the world. Notable exceptions are the absence of wild pigs in the British Isles and Canada. It should be pointed out that the distinction between feral pigs and wild boar population is not clearcut in many parts of the world.

In the United States, for example, the population is principally of feral origin but there has been much interbreeding with wild boar which were introduced to North Carolina in 1912 and subsequently escaped. Wood and Barrett point out that except for a small population of wild boar in New Hampshire, few, if any, European wild boar exist in the U.S. today because of interbreeding with feral stock.[20] Interbreeding has also presumably occurred in South America, where wild boar were introduced to Argentina in this century.[21] In Eurasia itself, escaped domestic pigs soon merge, it seems, with the population of wild boar. Their intermixing has been facilitated by the free-roaming of domestic pigs in forests and elsewhere in Asia and in Europe in the past.

Let us now consider the spread of *Sus scrofa* in the world and some of the reasons for its extinction in particular regions. As mentioned earlier, pigs spread from Asia to Melanesia and then subsequently to most of Polynesia as a result of human migration from south-east Asia and settlement commencing approximately 6000 years ago. By 1000 A.D. pigs had spread to most of Oceania, Australia and New Zealand excepted. They reached Hawaii about 1000 A.D.

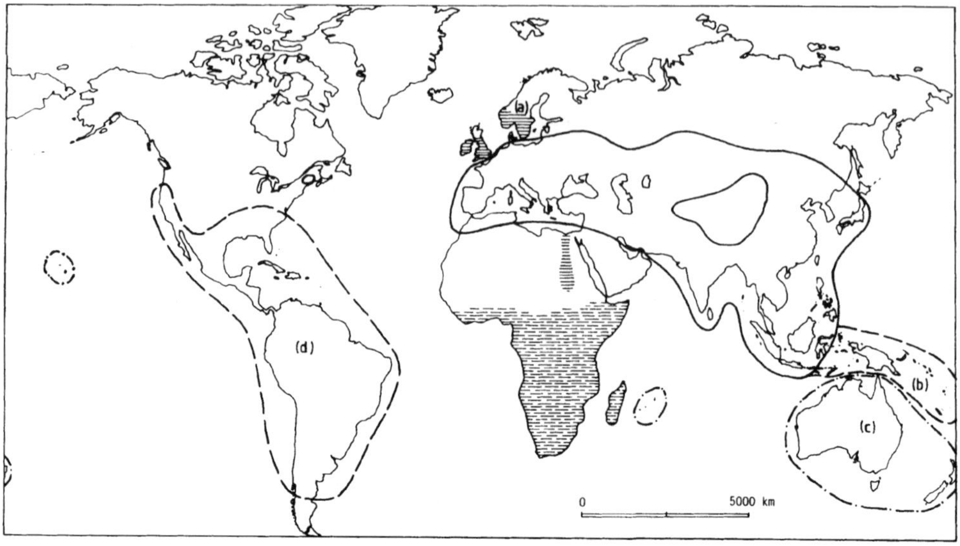

Range of Eurasian wild boar
Feral pigs and wild boar range
Feral pig range
Approximate distribution of bush pig
Areas in which wild pigs have become extinct

NOTES: (a) Small numbers of wild boar have been reintroduced to Sweden.
(b) South-eastern range of wild boar uncertain, since feral mixtures occur. All wild pigs in New Guinea, Melanesia and Polynesia appear to be of feral origin, arriving from 6000 B.P. onwards.
(c) All wild pigs in Australia and New Zealand of feral origin within the last 200 years or so.
(d) Wild pigs in the New World of feral origin and in some cases have interbred with introduced wild boar.

Fig. 1.4 Approximate current world distribution of Sus scrofa *(both the Eurasian wild boar and feral pigs) and* Potamochoerus porcus *the bushpig*

The New World was the next major area to which pigs were introduced. Christopher Columbus is reputed to have introduced pigs to the West Indies in 1493 and in 1539 they were introduced to Florida.[22] By the end of the sixteenth century Spanish colonial settlement of the West Indies, Central America, Mexico, Florida, Peru, Chile and nearby areas was already well advanced and Portugal had established settlements in Brazil. Consequently the domestic pig was introduced into a large part of the New World and spread with continuing settlement. Because the practice of allowing pigs free-range was common in the New World, feral pig populations had undoubtedly formed by the end of the sixteenth century and this population would have grown and been added to with continuing human settlement. European *wild boar* appear to have been successfully introduced to the New World only during this century. As mentioned earlier, they were introduced to North Carolina in the United States of America in 1912.[23] These boar, intended for hunting purposes, escaped and have successfully spread. They have, for instance, formed a substantial population in the Great Smoky National Park and nearby in Tennessee. Wild boar were also imported to Argentina earlier in this century and now reported to have spread to Chile.[24]

The approximate present distribution of the wild pig on the mainland of the United States of America is shown in Figure 1.5.[25] There are two concentrations. One is in the south-east extending from Florida to eastern Texas and covering various wetlands in the south and the lower Mississippi drainage area and extending north to the area of the Great Smoky National Park. The other wild pig population is concentrated in California and appears to date from the introduction of free-roaming domestic pigs in 1769.[26] Wild pigs are also present on Hawaii, where they were introduced by the Polynesians, and in Puerto Rico and the U.S. Virgin Islands.

European expansion into the Indian Ocean and eventually into the Pacific following the journey of Vasco da Gama to India in 1497-8 also contributed to the spread of *Sus scrofa*. It was introduced to islands in the Indian Ocean where Suidae had previously not existed. The Portuguese had brought it, for example, to the Mascarene Islands (Mauritius, Réunion and Rodrigues) by the 1600s. It is reported to have contributed to the extinction of the dodo (*Raphus* sp.) on these islands.[27] The practice by early European navigators of leaving pigs or goats on islands to fend for themselves and to provide a source of future food for passing sailors led to its spread to smaller islands.

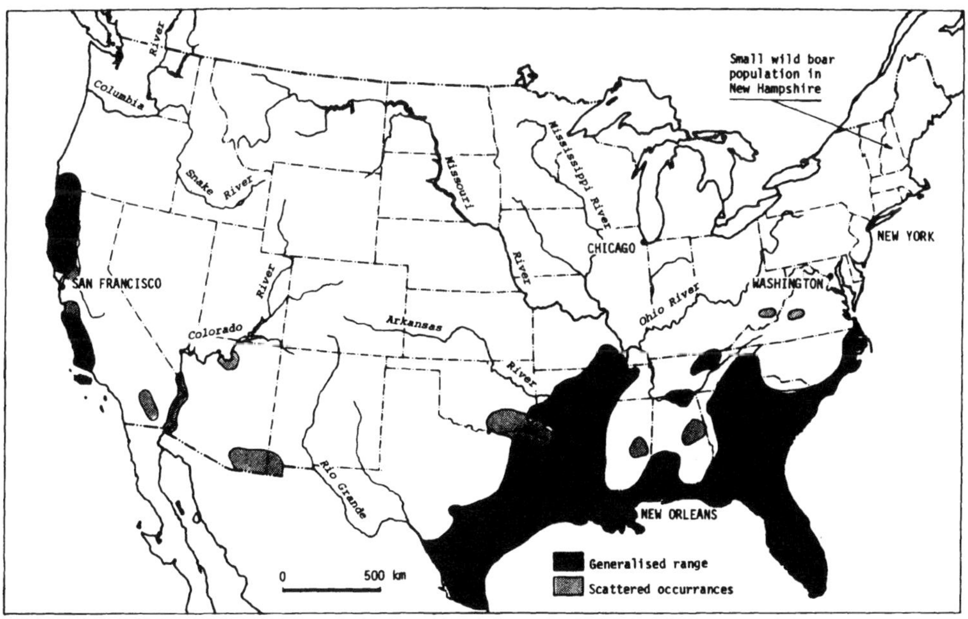

Fig. 1.5 Approximate generalized distribution of wild pigs (feral and wild boar) on the mainland of the United States

However, it is interesting to note that *Sus scrofa* does not appear to have been able to establish itself in the range of the bushpig *Potamochoerus porcus*. Adequate opportunities have existed for its possible spread, for example, from the early European settlements in southern Africa and Madagascar. Although a small feral pig colony may exist in Cape Provence, its presence has not been confirmed.[28] Overall it is clear that *Sus scrofa* has failed to establish itself in a wild state where the bushpig is present. While the reason for this is not known, the bushpig may be a more effective competitor and better adapted to its environment.[29]

Australia and New Zealand were the last relatively large land masses to be colonized by Suidae. Thomson reports that the first pigs were left in New Zealand by De Surville in 1769.[30] They were young ones and were given to the Maoris. What became of them is not known.

Captain Cook was possibly the first to introduce pigs successfully to New Zealand. In 1775 on his second voyage to New Zealand, he left a boar and two breeding sows in the forest of Cannibal Cove on the South Island in the hope that they would multiply in the wild.[31] Also he gave some pigs to the Maoris on this voyage and on his third voyage. By 1839, pigs seem to have become widespread in the South Island. Governor King of New South Wales brought ten pigs to the North Island of New Zealand on his visit of 1793 and gave them to Maoris in the Bay of Plenty.[32] Settlers (commencing in 1840) also brought pigs with them. However, feral pigs seem to have been widespread even before this, because Thomson says that

> 'Even much earlier they [pigs] must have been very abundant, both tame and wild, for nearly every sealing and whaling vessel which visited these islands between 1800 and 1830 took away quantities of pork as part of their cargo to Sydney.'[33]

The tame pigs at that time would have been kept by the Maoris.

Dr Challies points out that 'some of the animals [pigs] from earlier introductions were released directly into the wild in deliberate attempts to naturalize them to provide food for explorers, sealers and whalers, and castaways. Others were either gifted or traded to the Maoris. Pigs soon became an important food item for the Maoris, and were transported around the country by them as articles of barter and plunder. The practice adopted by the Maoris and some of the early European settlers of keeping their "domestic" pigs in a semi-feral state led to frequent escapes into the wild.[34] As a result, feral pigs soon became well established around most Maori and European settlements, from where they readily spread into adjacent areas. By the late 1800s feral pigs had spread throughout all of the districts suitable for them'[35] in New Zealand. Their distribution in 1971 is shown in relative detail in Challies' article.[36]

The 'pedigree' of New Zealand's feral pigs is uncertain. Early introductions of pigs were probably all domestic animals directly or indirectly from Europe. There is no record of true wild pigs ever having been introduced.[37]

While it has been said that Captain Cook introduced pigs to Australia,[38] no evidence has been found to support this contention.[39] It seems probable that pigs were first introduced to Australia with the European settlement in 1788 (at Sydney) and spread with this settlement. The practice of allowing pigs free-range soon resulted, as discussed below, in feral colonies forming, and by the end of the last century feral pigs were widespread in Australia. Their naturalization in Australia will be discussed in more detail in the next section.

Although the Eurasian pig extended its range in the last millenium as a result of Eurasian human migration and increased trading activities, it also disappeared from some parts of its range as a result, it seems, of human activity - closer settlement, economic development of the land and laws allowing greater freedom to commoners to hunt.

In the British Isles, the exact date on which the wild boar disappeared is uncertain but it seems possible from Harting's observations that it could have been around the end of the seventeenth century.[40] However, as Dr Juliet Clutton-Brock of the British Museum (Natural History) has pointed out to me,

> 'It is likely that for hundreds of years the wild boar interbred with domestic swine that lived freely in the woods under the system of pannage [the practice of allowing domestic pigs to forage in the forest]. It would therefore be very difficult to say when the last truly wild population of *Sus scrofa* was replaced by a mixture of wild and feral pigs.'[41]

Britains, Romans, Saxons and Normans all hunted the wild boar in Britain. Its figure appeared on early British coins and it was featured in Celtic works of art. It was incorporated into the coat of arms of some Norman kings of England - Richard III, for example. Following the Norman Conquest wild boar and other game were reserved for the hunting of kings and powerful nobleman. 'William the Conqueror punished with the loss of eyes those convicted of killing a wild boar, stag or roebuck.'[42] Severe restrictions on hunting by other than powerful nobility continued until the reign of King John. At the same time as King John signed Magna Carta he also signed the Charter of the Forests. By this Charter 'all lands which had been converted into woods or parks since the commencement of this reign were disafforested, and the tenants bordering on royal forests secured against spoilation, in a word, the laws made for the protection of the game and wild animals were either partially repealed or considerably mitigated.'[43] Further concessions to allow hunting by other than royalty and powerful nobility were made with the passage of time. Hunting pressures on the wild boar and other game increased, forests were burned to dislodge them and overhunting occurred.

Furthermore, with economic development and population growth, forests were cut down and formerly wild areas cultivated or utilized more intensively for economic production. All of these factors contributed to the extinction of the wild boar in Britain. Dr Juliet Clutton-Brock suggests that '...The reason for the extinction of *Sus scrofa* must be directly related to the destruction of forests and woodland in Britain. In temperate Europe the wild boar is an inhabitant of broad-leaved woodland and in Britain it would be near the northern extreme of its range. As with other large mammals, the elk, aurochs, wolf, etc., the boar could not survive the combination of over-hunting by man and the destruction of its environment.'[44]

Harting cites written British references to wild boar which suggest that they continued to exist in Britain until the end of the seventeenth century.[45] Their original range included not only England and Wales but also parts of Scotland. They occurred in Ireland and existed there down to the seventeenth century. The exact date of their extinction in Ireland, as in Britain, is uncertain.

Attempts were made to reintroduce wild boar to Britain but in the long term these did not succeed because, it seems, the majority of people regarded wild pigs as pests and saw to their destruction. Charles I reintroduced wild boar to the New Forest. It is said that 'they encreased, and became terrible to the travellers. In the civil warres they were destroyed but they had tainted all breed of pigges of the neighbouring partes, which are of their colour; a kind of soot colour.'[46] As late as 1791 there were reported to be pigs in the New Forest with these characteristics. They were also reintroduced to a forest in Essex towards the end of the seventeenth century but it is reported that 'they digged the earth so up, and did such spoyle, that the country would not endure it.'[47] Again when they were introduced later it is said that 'the country rose upon them and destroyed them.'[48] The wild boar has not been successfully reintroduced to the British Isles.

Sweden provides another interesting case of the disappearance of the wild boar in comparatively recent times. In the distant past the wild boar existed there as far north as Uppland[49] though Sweden is situated in the northern limit of the pig's range. In the period 1770-1800 it seems to have disappeared.[50] Possibly (as in the British Isles) overhunting and the increased intensive economic use of the land contributed. At the request of farmers, the wild boar is known to have been hunted to extinction around 1770 on Oland (as a pest) but its exact date of disappearance on the mainland of Sweden is unclear.[51] Indeed, the remnants of its population merged as a result of interbreeding with domestic pigs allowed to forage in the forests.[52] (As observed earlier, a similar thing happened to the wild boar in the New Forest in England.)

From 1940 onwards wild boar were imported to Sweden to be kept *enclosed* in areas for hunting, limited meat production and zoos. Some of these animals escaped from the enclosures and have re-established a scattered wild (unconfined) population of about 200 wild boar.[53] These wild boar have become contentious in Sweden. There was an investigation into them in 1980. Dr Sven-Olaf Lindblom of the National Board of Agriculture, Sweden, informed me that

> 'The question is now whether we should allow these pigs to spread over Sweden or whether we should eradicate them. The question is not yet settled but the Hunters' organization wants the pigs and the National Board of Agriculture together with others do not want them. There are two main reasons why we do not want a widespread population of wild pigs:
>
> (a) Wild pigs do a lot of damage in districts having intensive agriculture, especially to potatoes and similar crops.
>
> (b) An outbreak of a contagious exotic disease (e.g. hog cholera, African swine fever, foot-and-mouth disease) would be very difficult to fight if it became established among wild pigs.'[54]

In Denmark (again close to the limit of the wild pig's northern range) the reasons for the disappearance of the wild boar were possibly similar to those in Sweden.[55]

In the Nile River Valley (in the southern extremity of its range) the reasons for its disappearance are not known but presumably more intensive settlement and agriculture and the hunting of it as a pest played a role.[56]

In Japan, the wild boar seems to have disappeared or virtually disappeared from Hokkaido.[57] The reasons are not clear, because this island is not heavily populated or intensively developed for economic use. Once again, however, it is near the northern limit of the wild pig's range.

In summary, it can be seen that *Sus scrofa* has extended its overall range in the world as a result of the expansion of Eurasian human populations. It was introduced to New Guinea, Melanesia and most of Polynesia by migrating and trading peoples originally from South-east Asia. The Spanish and the Portuguese introduced it to the New World in the sixteenth century and it was brought by the Portuguese to relatively isolated islands in the Indian Ocean also in that century. First transported to New Zealand and to Australia in the eighteenth century, the pig had well-established feral colonies there by the end of the nineteenth century. By the end of the nineteenth century, wild members of the pig family, Suidae, were therefore present in most parts of the world principally as a result of the spread of *Sus scrofa*. *Sus scrofa* had formed colonies in many areas (such as Australia, New Zealand and the Mascarene Islands) that had previously had no pig-like animals. Their presence seriously disrupted pre-existing ecosystems, especially on isolated islands, and hastened the extinction of or continues to endanger a number of previously isolated species.

THE FERAL (WILD) PIG IN AUSTRALIA

Australia is an interesting case in which the pig has been introduced comparatively recently by man. It is the area given greatest coverage in this book.

Confusion exists about the origin of feral (wild) pigs in Australia. The evidence indicates that they stem from domestic stock introduced by European settlers. Before European settlement, no pigs appear to have been present in Australia. This is so despite the fact that pigs were present in New Guinea at least 6000 years ago and New Guinea is close to the Australian mainland. Furthermore, Australian Aboriginals are reputed to have had some contact with peoples to the north. Just why pigs were apparently not introduced is a mystery, but to tend semi-domesticated pigs would have been in conflict with the semi-nomadic life-style of the Australian Aboriginals. The non-introduction may also indicate that there was little incursion into Australia by settlers from New Guinea (or at least not by those attaching much importance to the pig) in the past 6000 years or so.

It is commonly believed that pigs were introduced to the Australian mainland by Captain Cook[58] who is reputed to have liberated them near Cooktown in Northern Queensland when he careened the *Endeavour* for repairs. However, no evidence has been found to support this hypothesis. Pullar[59] after studying the logs of Captain Cook concluded that it is most unlikely that Cook released pigs in Australia. Such a release is not recorded in his log books. The hypothesis may arise from confusion with the New Zealand experience. On later journeys, Cook did release pigs in New Zealand.

Pigs were amongst the domestic livestock brought by the first European settlers of Australia in 1788, and subsequent shipping arrivals brought further stock. As the European settlement of Australia spread, the settlers took domestic swine with them. The common practice of allowing domestic pigs free range to forage for food soon led to a population of feral pigs becoming established. As early as 1795 pigs were a problem at Sydney Cove and 'orders were issued that they could be shot if they trespassed on private property.'[60] Settlers in the interior also allowed their pigs to roam. Cunningham reported in 1827 that

> 'They [pigs] are allowed to run in the bush during the day, just giving each a cob of maize to bring it home in the evening, if not employing a man to look after them. They feed on grasses, herbs, wild roots and native yams, on the margins of rivers or marshy grounds, and also on frogs, lizzards, etc., which come their way.'[61]

While the practice of allowing pigs free range is now uncommon in Australia, it still occurs in areas of sparse human population.

The population of feral pigs has been added to in other ways than by free-roaming pigs becoming wild. Pigs have escaped from enclosures, some settlements have in the past been abandoned leaving the pigs behind, and farmers have occasionally let their pigs 'go bush' in periods when prices for pigs have been so low as not to recoup the cost of taking them to market. In addition, hunters sometimes let pigs loose to upgrade wild pig stocks or establish feral pig populations in new areas.

The composition of Australian feral pigs is not certain. This is because Asian pigs were introduced from Timor to the Northern Territory of Australia in 1827 and it is likely that Chinese brought pigs with them during the Gold Rush of the 1800s.[62] In any case, English domestic breeds were improved by crossing with pigs imported from Thailand (Siam) during the eighteenth century. We can, however, conclude that Australian feral pigs originated from Eurasian domestic swine.

Distribution and Numbers of Feral Pigs in Australia

Feral pigs are now widely spread in Australia and were already widely spread by the beginning of this century. They appear to be most numerous and widespread in Queensland and New South Wales. They occur in significant numbers in the Top End of the Northern Territory and in scattered populations or colonies in Western Australia. Small scattered colonies also occur in Victoria. In South Australia, they are present only on Kangaroo Island. In the state of Tasmania, while a couple of very small colonies have been reported to be present in Tasmania itself, the only substantial colony seems to be on Flinders Island.[63]

Figure 1.6 indicates the general distribution of feral pigs in Australia.[64] The majority of the population is located, roughly, in an arc that commences in the top of the Northern Territory, veers east and extends through Queensland into most of New South Wales, petering out in Victoria. The borders of the distribution alter to some extent with seasonal conditions. For example, in wet years pigs may move down Queensland rivers draining inland to South Australia.

In inland Australia (that is, in drier parts of the mainland) pigs are concentrated near river beds and associated swamps. Possibly the greatest concentration of pigs in Australia occurs in upper portion of the Darling River and its tributaries in north-western New South Wales and southern Queensland. Another large concentration of pigs occurs around the confluence of the Lachlan and Murrumbidgee Rivers in New South Wales. However, in Queensland substantial densities of wild pigs occur over a large part of the state including the northern areas. In the Northern Territory the density of wild pigs is greatest in the north-west. In Western Australia populations of pig seem to be of relatively low density and scattered, with the main populations occurring in or near river valleys. While

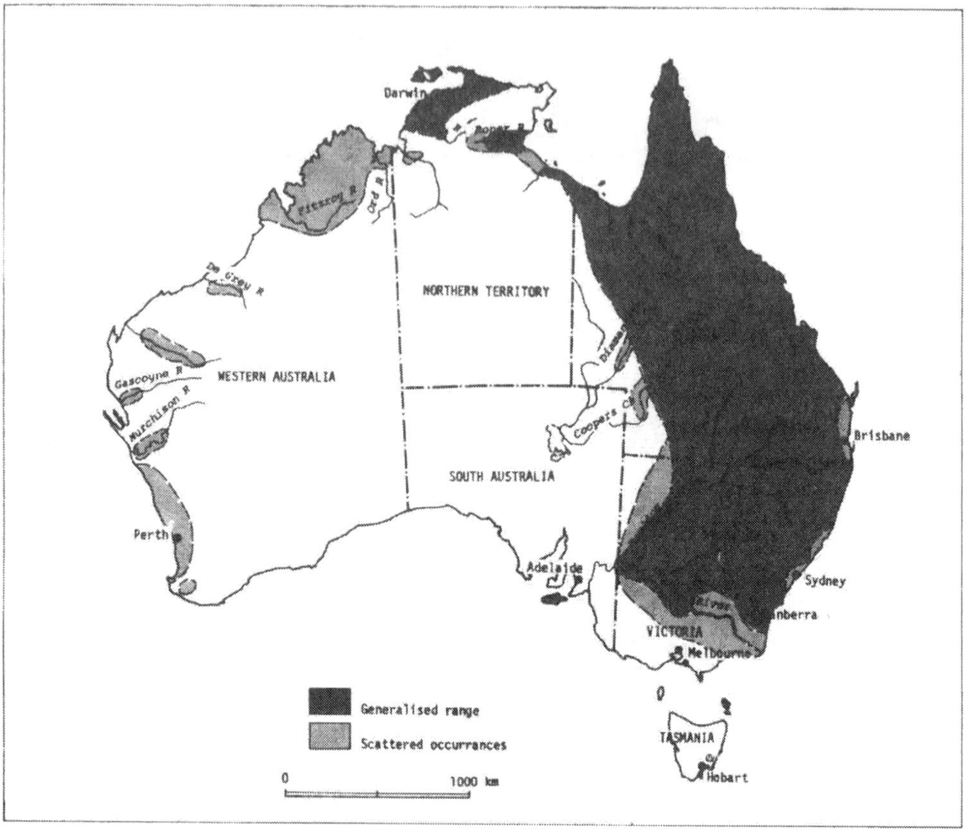

Fig. 1.6 Distribution of the feral pig in Australia

scattered populations occur elsewhere where sufficient water is available, the largest population occurs in the Kimberleys (the north-west of Western Australia) and in the south-west of Western Australia.

The numbers of feral pigs in Australia are uncertain since little scientific work has been done to estimate populations. However, it is clear that numbers fluctuate greatly depending on seasonal conditions. They are much reduced during years of drought which periodically plague large parts of Australia.

McKnight estimated the number of feral pigs in Australia to be 473 000 to 1 470 000.[65] An approximate average of these figures, namely 975 000, has been taken by the CSIRO as its estimate.[66] Later evidence, however, indicates that this figure is a major underestimate. Recently Dr P. M. Flynn (1980) estimated the numbers to be 3-5 million in New South Wales, 5 million in Queensland, up to 200 000 in the Northern Territory, 500 000 in Western Australia and about 2000 elsewhere.[67] This gives an Australia-wide total of approximately 8-11 million feral pigs. This estimate may, in contrast to McKnight's, be on the high side.

Using estimates provided by Pastures Protection Boards in New South Wales for their respective areas and other evidence, I calculated that there were 3-6 million feral pigs in Australia in 1977. The details are given in Chapter 2. However, the population fluctuates with seasonal conditions in Australia and much further work is needed to improve the reliability of population estimates.[68] Nevertheless, there can be no doubt that a sizeable population of feral pigs exists in Queensland and New South Wales and reaches high densities in several parts of these two states.

Consequences of Feral Pigs in Australia

One of my main aims in this book is to explore the consequences for Australia of the occurrence of feral pigs. In many parts of Australia, they are an agricultural pest. They destroy crops, farm property and even kill lambs. The next chapter assesses the extent of such damage. However, at the same time, wild pigs have become an important asset (possibly the most important species) for sporting (amateur) hunters in Australia, and recently commercial harvesting has started in order to export feral pig meat as game meat. This assessment of the feral pigs examines these aspects as well as their impact on forestry and Australian wildlife and natural ecosystems. Particular attention is given to diseases and parasites associated with wild pigs, the risks that these pose in an Australian setting and ways to reduce these risks. Available methods and the economics of controlling feral pigs in Australia are also outlined. Each year considerable amounts are spent in Australia to control feral pigs and government bodies are undertaking research to find improved or more economical methods for their control.

Different groups in Australia have conflicting interests in the wild pig. Most farmers and many conservationists regard the feral pig as a pest. On the other hand, many amateur or sporting shooters find that the feral pig is an important asset and a few landowners can profit by selling hunting rights in it. Furthermore, it is an asset for the developing game-meat industry in Australia. Ambivalent attitudes towards the wild pig, and conflicts about it, add to the interest of this study.

ATTITUDES TO THE WILD PIG WORLDWIDE

Varying attitudes have been expressed about the wild pig in different countries of the world. These are influenced by religion and by the size of the population of wild pigs present in the country. Islam and Judaism portray the pig as unclean, and in countries dominated by these religions the wild pig is likely to be seen as a pest with no redeeming qualities. In countries where wild pig densities are high or have increased (due to, say, environmental factors) the wild pig may also be seen as a pest.

To obtain some general information on the status of wild pigs in different countries and attitudes towards them, I wrote to embassies or consular representatives in Australia with the questions set out in note 69. In most instances, these were referred to an appropriate government body in the home country, and a number of useful replies were received. It is not possible to summarize all the answers here;[70] however, the following samples of replies may be of interest and other replies are drawn on elsewhere in this book.

From Greece, it is reported that wild boar are present in the northern and central part of continental Greece (mostly in dense oak woods) and while feral pigs occur their numbers are not known. Wild pigs cause some agricultural damage especially to maize, potatoes (and other crops) and slight damage to wildlife. No special measures are taken to control wild pigs because amateur hunters provide an effective control. There are believed to be about 300 000 Greek hunters in a population of 9 million! The Head of the Game Management Department, the Ministry of Agriculture, Greece, points out that

> 'Wild (or feral) pigs are extremely important for hunting purposes because they are quite popular game for hunters and their role as an economic resource for this reason is high. In conclusion we may say that wild pigs in Greece, with few exceptions are certainly a resource.'[71]

From nearby Yugoslavia, in Vojvodina Province (in the north of Yugoslavia) wild pigs are said to be neither a significant pest nor a resource because their numbers are so low. However, the population has grown and there has been interbreeding with domestic pigs gone wild. Damage occurs to agricultural crops especially, wheat, potato and beet. The risks of domestic pigs catching swine fever from wild boar seem to have diminished because of the spread of the intensive husbanding of domestic pigs. Some wild pigs are shot for sporting purposes and some are used commercially in this Province.[72] Farther south in Kosovo Province wild boar appear to be more highly regarded. They occur mostly in oak forests but do damage agricultural crops such as wheat and maize which are sometimes guarded at night by guards sitting in front of fires. They are hunted by sporting hunters as big game. Commercial hunting also occurs for meat and skins. It is reported that hunting provides considerable control of the population. The trophies (tusks) of wild boar from this Province have won several medals at international exhibitions.
Eng. Vladimir S. Naumov, Provincial Hunting Inspector, summarizes the status of the wild boar as follows:

'Our wild boar is protected from hunting in the months of February, March and April (time of reproduction), because it is reared in pure line, because it has a big value for trophies and because of its possible sale for meat. According to that we rate it as an economic resource with hunting-economic importance.'[73]

In many other European countries closed seasons apply to the wild boar. This is so, for example, in Germany, Poland, the U.S.S.R., France and Spain, where the wild boar is sought for amateur hunting. Despite the fact that the wild boar causes agricultural damage in these countries, on balance it is regarded as an asset given its existing numbers. If this were not so, there would be no reason to conserve its numbers by means of closed hunting seasons. However, hunting seasons are sometimes extended if wild boar populations build up. In recent years a pest problem appears to have developed in Portugal. Because many peasants have left the land in the past few years, the land has reverted to scrub in some areas and this environmental change has been favourable to an increase in numbers of wild boar.[74]

In Asia, attitudes to the wild pig vary greatly. The wild pig is held in low regard in Islamic countries but this is not so in most non-Islamic countries. Thus on the whole the wild pig is despised in a country such as Pakistan, more favourably regarded in Sri Lanka and Thailand, and considered to be an asset in China and Japan. Consider its status in a selection of Asian countries:

In Japan two subspecies of *Sus scrofa* (wild pigs) occur, namely *Sus scrofa leucomystax* and *Sus scrofa riukivanus*. The former occurs in Honshu (except the northern region), Shikoku and Kyushu and the latter in islands south of these large Japanese islands, namely Amamioshima, Tokunoshima, Okinawa main island, Ishigakijima and Iriomotejima. They sometimes cause considerable damage to crops especially in mountainous areas but are not considered important carriers of disease. Their meat is highly valued and sold at high prices and they are considered to be useful for game hunting, although as Mr Akiyoshi Nawa points out success requires a lot of skill and experience.[75] Hunting of wild pigs is regulated so as to conserve stocks. They can be taken only during a limited hunting period as game. However, licences can be issued for their hunting when they are damaging crops or when required for scientific research.

In Thailand wild pigs occur throughout the country and cause damage to agricultural crops especially near forested areas. They are not a protected animal. Their meat is considered to be a delicacy and is used in the restaurant trade. They are hunted for commercial purposes and little or no amateur or sporting hunting occurs. According to Mr Phairot Suvanakorn, Director, Wildlife Conservation Division, Royal Forest Department of Thailand, they are unimportant in Thailand as an environmental pest and as an economic resource.[76]

Coming now to a sample of countries with sizeable Moslem populations, it becomes clear that a much less favourable view is taken of the wild pig. From Indonesia, it is reported that wild boar are very destructive of agricultural crops and forest seedlings. Hunting, poisoning, trapping, pitfalling and other types of extermination campaigns are carried out and are sometimes organized by local authorities. Ir Harun Alrasjid, Director of the Forest Research Institute, Bogor, points out:

> 'As the majority of Indonesians are Moslems, who consider both domesticated and wild pigs as forbidden, the meat is not so popular for human consumption... Wild boars caught by means of shooting, trapping or pitfalling by farmers or by local hunters are usually sold to their non-Moslem neighbours, but nevertheless wild boar meat is not available in markets. Regulations to protect wild fauna do not apply to wild boars. As far as I know, there is no restriction on the hunting season and the bag is unlimited.'[77]

However, these comments apply to *Sus scrofa* and not to *Babyrousa babyrussa*. Dr Setijatl Sastrapradja indicates that for Java, where numbers of wild pigs are low, the wild pig is more favourably regarded.[78]

In Turkey, wild pigs are widely distributed and occur in scrubland and forests, and in reedy marshy places. Especially when densities are high, they cause damage to crops by trampling them, they eat oak and beech seeds and they damage and destroy seedlings. A Turkish correspondent indicates that

> 'According to our religion, people of our country do not eat pig. No commercial use is made of the meat of the wild pig. However, the hides are used for leather. When wild pigs damage agricultural areas, we allow shooting drives. Local hunters can hunt wild pigs all year round and there is no daily hunt limit. However, foreign hunters must pay $100 for each pig taken.'[79]

While wild boars have little standing, if any, in Turkey, they are firmly regarded as a pest in Pakistan. A communication received from Md Kalimullah Shirazi, Assistant Secretary, National Council for Conservation of Wildlife in Pakistan, says that wild pigs *Sus scrofa cristatus* occur in millions throughout Pakistan.[80] The areas of greatest infestation are in the Punjab, Sind and North West Frontier Province. He points out that

> 'Wild boars destroy large areas of food/fodder and cash crops and plants and damage nurseries, forest plantations and wild birds and their eggs. They also cause occasional injuries/casualties to human beings. Crops damaged by wild pigs include sesame, grain, wheat, peas, mustard, sugar cane, cotton, maize, millet, nuts and vegetables.
>
> Due to heavy damage to agricultural crops and occasional danger of attack of humans from the ever-increasing population of wild boars, the Government of

> Pakistan has launched campaigns from time to time at the national level for the destruction of wild boar in all Pakistan particularly in places of their greatest population.
>
> Hunting and shooting, the catching and destruction of piglets, innoculation of adult pigs with virus [swine fever] and their release have been attempted. Bonuses are given as an incentive for their destruction.'[81]

The communication concludes by pointing out that no commercial use is made of wild boar in Pakistan because the religion of Islam, the state religion, does not permit it. Limited amateur hunting of wild pigs occurs with the goal of destroying them. It is stressed that 'wild boars are a pest and not an economic resource.'

Information about attitudes to the wild pig in the New World are limited. The ambivalent attitude towards them in the United States has already been noted. The position in Latin America is unclear but may be similar to that in the United States. Wild pig populations may not have increased as yet in Latin America to levels to cause concern. Both feral pigs and European wild boar are present in South America. For example, Mr Paulo Nogueira Neto of the Ministry of Interior of Brazil informed me that

> 'There are feral pigs in the Pantanal of Mato Grosso, on the border of Bolivia. There are also small populations of the European wild boar in the States of Santa Catariana and Paraná, we were told by some people. In Paraná they have interbred with feral pigs in the Pon ta Grossa region.'[82]

While *Sus scrofa* does not occur in Africa (except in the Mediterranian area), as mentioned earlier other species of wild pig do. As noted, the habits of the bushpig *Potamachoerus porcus* are similar to those of the wild boar and it is widely distributed in Africa. A reply from Ghana gives some indication of attitudes to wild pigs in parts of Africa. There are no feral pigs in Ghana but warthogs *Phacochoerus aethiopicus*, bushpigs or red river hogs *Potamochoerus porcus* and the giant forest hog *Hylochoerus meinertzhageni* occur. While they do cause some agricultural damage, they are not considered to be an important environmental pest, since numbers are not considered to be high. They are *very important* for subsistence hunting and contribute considerably to supplementation of animal protein supply in Ghana. Mr C. K. Manu, Chief Game and Wildlife Officer of Ghana concludes that '...Wild pigs of Ghana are very important as an economic resource (for their meat) but less important as environmental pests.' Programmes exist for managing this resource throughout Ghana. Management and conservation extends beyond National Parks and Game Production Reserves.[83]

Even this small sample of countries indicates the variety of attitudes towards wild pigs. Within many countries, including Australia and New Zealand, attitudes to the wild pig are often mixed.

BIOLOGICAL ASPECTS AND HABITS OF WILD PIGS[84]

It is useful to bring attention to some biological aspects and habits of *Sus scrofa* (wild boar or feral pigs) even at this early stage.

Wild pigs are omnivorous. While vegetable matter usually accounts for the major portion of their diet, insects, rodents, frogs, lizards, other small animals and carrion can be important. Even crabs and fish are consumed in some localities. Weak animals or the young of animals (such as lambs) may also be killed and eaten. Eggs and young birds on the ground may also be added to the diet. Because wild pigs are omnivorous and eat an extremely wide range of foods, they are a tenacious species even when food is in short supply. Because of their rooting habits, wild pigs find much of their food under the ground when they are present in areas of moist soils or not very hard or compacted soil. Items such as earthworms, larvae of insects, bulbs and edible roots and fungi may be obtained by rooting.

Regular access to water and shade, especially in hot weather, is important. Not only must pigs be able to drink regularly but they have no sweat glands and therefore have difficulty in regulating the heat of their bodies unless they are able to wallow in wet or moist places during hot weather. Shade, for instance from reeds, shrubs or trees also aids their thermal regulation. Hone and O'Grady point out that

> 'Dense vegetation near reliable water sources is the favoured habitat of feral pigs. This is because feral pigs are very susceptible to high temperatures. Where temperatures exceed 30°C water may be required daily. If the temperature exceeds 35°C and there is no protection from direct sunlight, feral pigs will die within several hours from dehydration and exposure.'[85]

It is not uncommon for temperatures in Australia to exceed these levels in summer in areas where feral pigs are present.

In addition thick vegetation helps to protect wild pigs from their predators, especially man. Nevertheless they survive in a wide range of climatic conditions and occur from lowland areas to those of high elevation. They can withstand snow provided that it is not so deep as to prevent their locomotion and they survive in temperatures below freezing. At the same time they thrive in the tropics. They also occur in areas of low rainfall and high temperatures provided that surface water and shade is available. This happens, for example, in inland parts of Australia where temperatures sometimes exceed 40°C but rivers flowing towards the inland or other water sources provide surface water and where shade is provided by low-growing shrubs or reeds.

While pigs have poor vision, they have extremely well-developed senses of smell and hearing. Their sense of smell helps them detect food such as carrion over long distances and also aids them in smelling out food below the surface of the ground. Their well-developed senses of smell and hearing are valuable in helping them avoid predators. Experienced hunters are well aware of this. To increase their chances of survival, pigs may try to walk into the wind or across it. Hunters are usually advised to approach wild pigs against the wind as quietly as possible.[86]

The survival of wild pigs is also aided by the fact that they migrate long distances when under environmental stress. However, normally they are irregular sedentary animals. Unless conditions are unfavourable (that is, food or water inadequate) they remain in a specific but not necessarily exclusive home range. The home range of sows tends to be about 5-30 km^2 and that of males about 50 km^2. Males are most mobile and may move around more in search of breeding opportunities.

It is usual for groups of females and offspring of less than a year old to form groups or sounders. In cold weather up to 15 or so of these pigs may sleep together in beds made of plant material. Males one year and older are not gregarious and are usually solitary except when sows are in heat. At this time they may rejoin a group, chase away males of more than a year old that have not left the group and fight with rivals who also come to service the sow.

Wild pigs *Sus scrofa* are considered to be diurnal by some authorities and to be nocturnal by others. When the weather is *not* hot, it is sometimes possible to observe large groups of these animals from the air in inland Australia. However, their main times of feeding activity especially in hot weather may be in the very early morning and the late afternoon and evening, and they are likely to rest under shade during the hottest part of the day. Nevertheless they can be active at night, especially because vision is not the most important sense for them and they can alter their habits to avoid danger. For example, if hunting of pigs during the day is common, their nocturnal activity and early morning and late afternoon activity is likely to increase at the expense of daytime activity. In Australia, Hone and O'Grady observe that

> 'Feral pigs are active mainly at dawn and dusk. This will vary according to temperature, rainfall and control pressure. Pigs leave their bedding area shortly before dusk and travel along pads to a waterhole or trough to drink and wallow, and then feed.
>
> At a study site north-west of Bourke [in New South Wales] the main period of pig activity observed near waterholes during the summer was from 7.00 to 8.15 p.m. Feeding may be resumed around dawn, after which pigs return to cover.'[87]

Wild pigs are sexually mature at about 8-10 months old and have a gestation period of about four months. While in northern Europe sows usually have only one litter per year, in warmer climates, such as those of Australia and California, they may have two litters per year. The average farrow appears to be about five or six piglets. Thus it is obvious that there is a great potential for expanding the population of wild pigs rapidly. Under *ideal* conditions a wild pig population in Australia could increase by five times in a year and in northern Europe by about two and a half times. However, the mortality rate amongst piglets is usually high. In California, Barrett observed a 70-90 per cent mortality rate in piglets from birth to six months old.[88] However, he points out that at a 90 per cent mortality rate and two litters per year a population can increase in size by a third.[89] Under Australian conditions a doubling (or more than doubling) of the population of feral pigs in a year is quite feasible.[90]

Sows strongly defend their young. When a pregnant sow is about to give birth, she leaves the herd and forms a nest of grass, small shrubs and tree seedlings in which to give birth. The nest helps to keep the piglets warm and dry, which is important because they are small and born with little hair. When the mother leaves the nest to find food or water, she may cover the young with material from the nest. After the first week, the piglets may accompany their mother on short trips but always return to the nest. When they are slightly older they and their mother may join other sows with piglets. After a few weeks the nest disintegrates and eventually the family group may join a herd. Even though piglets begin to eat solid food a fortnight after birth, their main source of food for about three months is the sow's milk. After this the young become increasingly independent in a herd. By the end of the first year, males have generally left the herd.

Pigs may do some slight damage to tree seedlings and other vegetation in building nests for the delivery of their young. Nests may also be built for sleeping in, especially during cold weather when extra warmth is needed. It is also observed that wild pigs may damage trees by repeatedly scraping their sides on them after wallowing. A scraping tree is often close to a wallow. Pigs, by regularly scraping on the trunk of the tree to remove flaked mud and external parasites, may strip the trunk of the tree of bark. In addition, boars occasionally try their tusks on small trees and extremely hungry pigs may eat bark and consequently ringbark or girdle trees. These habits and others mentioned above result in damage to some of the economic resources of man but they have to be offset, at least in some countries, against benefits obtained from the presence of wild pigs.

CONCLUSIONS

Although the wild pig family, Suidae, consists of a number of species, *Sus scrofa*, comprising the Eurasian wild boar, their feral pig descendants and mixtures of the two resulting from interbreeding are geographically the most widely distributed. They occur on all continents but only in the far north of Africa. However, in the remainder of Africa a close relative of similar habits, the bushpig *Potamochoerus porcus* occurs.

Man has possibly been the most important agent in the geographical spread of *Sus scrofa*. First as a result of southward human migration through Indonesia, New Guinea and Melanesia and then into Polynesia, the pig spread throughout a large area in the Pacific but did not reach Australia and New Zealand, as far as it is known. The second important wave of its spread began in the last half of this millenium with increased European exploration and settlement of new countries. Feral pigs soon were present in the New World following the introduction of domestic swine, and subsequently European wild boar were introduced both into North and South America for naturalization. Some of the islands in the Indian Ocean also had pigs introduced for the first time. They were brought to Australia and New Zealand somewhat later, but feral pigs were already well established in both countries before the commencement of this century. It is possible, however, that the Australian population has increased and spread further afield in this century as a result of the availability of a greater amount of agricultural crops appetizing to pigs and man-made schemes to improve the availability of water (for example, inland irrigation schemes and water bores or the tapping of artesian water supplies) so providing permanent sources of water.

Attitudes to wild pigs vary throughout the world. Even in countries where wild pigs are regarded as valuable for hunting and/or for meat, they are often regarded as a pest in agriculture and silviculture, as dangerous carriers of disease and as a threat to other wildlife. Their management or control is usually considered to be important. In many countries, amateur or sporting hunting is an important consideration in their management and control, and populations are managed by altering hunting regulations where necessary, to keep populations within reasonable bounds at the same time as conserving the species. Extra controls may, however, be instituted at times and in places where the species is causing above-average agricultural damage. In Australia this introduced species has not been protected and hunting pressures have not been adequate to exert the same amount of control on the species as has been possible in Europe and Japan. Consequently, feral pigs are considered to be an important pest in Australian agriculture. Their role in this regard and that of wild pigs in other countries is explored in the next chapter.

CHAPTER 2

Pigs as an Agricultural and Pastoral Pest

Wherever wild pigs occur in significant numbers they usually cause agricultural damage. The damage can be very serious, sometimes resulting in the loss of a farmer's complete crop and a noticeable reduction in national agricultural yields. This chapter sketches the agricultural damages caused by wild pigs on a world-wide scale and examines the Australian situation in detail - since considerable damage occurs in Australia from (introduced) feral pigs, which appear to have increased greatly in number in the absence of predators and strong native competitors. The second half of this chapter summarizes a study of agricultural damages in New South Wales.

The difficulties of controlling wild pigs in agriculture and of evaluating the economic damages are discussed in this chapter. Some answers are given to the following questions: What types of agricultural damage do wild pigs do, how extensive is this damage and how does one assess it from an economic standpoint? Is it economically worthwhile controlling the populations of wild pigs and to what extent? What are the difficulties involved in determining and obtaining optimal control of these species?

WILD PIGS AS AN AGRICULTURAL PEST THROUGHOUT THE WORLD

While wild pigs appear to be an agricultural pest wherever they occur, the severity of damage depends upon the type of agriculture and the densities of pigs. Since wild pigs occur both in temperate climates and tropical and sub-tropical climates, a wide range of crops is affected. In some countries, also, pig predation on lambs is a problem. Table 2.1 sets out some of the agricultural commodities damaged by pig activity, mostly rooting and eating; the animal is omnivorous, so a wide range of commodities is affected.

Europe including U.S.S.R.

The amount of agricultural damage caused by pigs in Europe varies from country to country. Damage occurs in all countries where they are present - that is, all areas excluding the British Isles and until recently the Scandinavian Peninsula.

TABLE 2.1 Agricultural Produce Reported to be Damaged by Wild Pigs

Temperate areas:	Grains (wheat, oats, barley, rye, maize, sorghum (milo)) Root crops such as potatoes and turnips Oilseeds such as sunflower, soya beans, linseed but excluding safflower Pastures and fodder crops Lambs (e.g. in Australia and New Zealand)
Tropical and sub-tropical areas:	Rice (paddy), sugar cane, tapioca, yams and sweet potatoes, coconuts, rubber trees, bananas, oil palm trees, maize, millet, pineapples, pastures

For example, the Director of Nature Conservation reported from France that 'Each year, wild pigs cause damage in crops and this leads the hunting authorities to indemnify the agricultural victims for this damage. In 1979, 35 million francs were paid by way of indemnity for the damage caused (to farmers) by large game; 80 per cent of this was attributed to wild boar'.[1] Greatest damage is caused to grain crops. Hunting with an open season, which can be extended for farmers suffering particular damage, is the main means of control. However, in cases of severe agricultural damage the Prefects are empowered to take action to control pigs out of season.

Again, from nearby Belgium it is reported that wild boars are still abundant in L'Ardenne, 'so much so that they are able to cause serious damage to cultivated crops. Their habit of moving constantly at night over very long distances explains how they sometimes suddenly occur in large numbers, ploughing ground on their way. Omnivorous, they not only eat roots and bulbs, acorns and cereals but also grubs (larvae) and young rabbits.'[2] Many of the hunts in Belgium are actually organized to limit the agricultural damage caused by these animals.

Agricultural damage from wild boars in Portugal has increased greatly in the past decade, because wild pig numbers have risen as farms have been abandoned (due to human emigration from agriculture) and have been allowed to revert to shrubland or have been converted to planted forests.[3] The Ministry of Agriculture and Fisheries of Portugal informed me that 'Wild boar hunting has been forbidden in a general way until the present as it was for a long time a rare species due to intense cultivation and close agricultural settlement. Recently, in particular in the past decade, it became increasingly widespread in the areas where emigration has been stronger ... As a consequence, damage on the remaining agriculture grew rapidly, making it necessary to introduce some control measures in the form of specially permitted hunting.'[4] It is intended to introduce an organized hunting scheme in the near future.

In Portugal damage occurs to both crops and livestock but this has not been systematically measured. Maize and other cereals and potatoes are greatly affected. 'Some local damage is caused to domestic pig herds by male wild boar mating with domestic or semi-domestic sows and so producing economically unproductive hybrid litters.'[5]

In Poland, according to Professor Jerzy Krupka 'The wild boar every year do great damage in cultivated fields in the proximity of forests, pasturing on potatoes and maize and uprooting winter-corns and meadow.'[6] Regina Mackin, in her detailed study of the losses to agricultural crops in Poland for the period 1965-67[7], suggested that oats and potatoes are favourite foods of wild pigs. The value of the loss was greatest for potatoes, then (in decreasing value) rye, oats, wheat, mixed grain, meadows, and barley. Amateur hunting is the prime method of controlling wild pig numbers.

Agricultural damage is also caused by wild boar in the U.S.S.R. For instance, Sludskii found that wild boar in the area of Ila River delta in Kazakhstan ate amongst other things millet, wheat, oats, rice, maize, melons and sometimes clover.[8] Sludskii also reported that lambs and weak livestock had been killed by wild boars in the Soviet Union. The extent of the agricultural loss in the U.S.S.R. does not appear to have been estimated.

America

In the United States, feral hogs or wild boar cause pastoral and agricultural damage (as well as damage to tree seedlings) in southern states and California. In Texas wild pigs were found by Springer to have a strong preference for grain. Springer concluded from his study that 'Based on personal communication with farmers and personal observations in grain fields, cultivated grains were a preferred food of wild hogs. Very few wild hogs were collected on the study area during the period when grain was available on adjacent farm fields. Three animals were collected at an average distance of 5 km from fields containing cultivated sorghum and had been feeding on this grain. One animal was collected 7.5 km from the nearest grain field and had consumed a large volume of sorghum grain.'[9] Springer points out, nevertheless, that despite their damage to grain, wild pigs in the United States are a valuable game hunting resource, even though they are not protected. He suggests that 'areas of intensive grain farming where wild hogs are considered a pest, should consider the fact that the wild hog is highly sought after, especially by bow hunters.'[10]

Grain and artichoke losses are also reported from California, and in Tennessee it has been said that wild boar 'damage cornfields, attacked domestic hogs, chased chickens, and broke down fences.'[11] From his studies of wild pigs in California, Barrett concluded that greatest problem in the management of those species was 'to minimize negative impacts on crops, rangelands, native flora and fauna.'[12]

From Argentina (and now Chile) it is reported that introduced European wild boar apart from damaging crops are also killing lambs.[13]

Africa

Sus scrofa is present only in northern Africa though several other species of wild pig are present in the rest of Africa. The bushpig *Potamochoerus porcus*, in particular, causes considerable agricultural damage. The damages are similar in many respects to those reported to be caused in Australasia by feral pigs *(Sus scrofa)*. Fradrich says of the Cape bushpig in South Africa that 'Since leopards have been severely limited in many places, the pigs have so greatly increased that they now endanger the crops ... peanuts are apparently the favourite food of the Cape bush pig. Losses of this crop have been up to 100 per cent. Furthermore they eat corn, pineapple, grapes, pumpkins and watermelons. The Cape bush pig is able to smell carrion over a distance of many kilometres. It also attacks small, ailing domestic animals, and newborn sheep and goats, eating part of them. It also digs its way under "jackal-proof" fences which protect the farms and breaks through barbed wire fencing.'[14] It is regarded as a serious agricultural pest in southern Africa.

Asia and Oceania

In Asia wild pigs (mostly *Sus scrofa*) can also be damaging to agriculture. Diong's observations in Malaysia indicate these agricultural losses occur to tapioca (sago), rice (paddy), sugar cane, yams, oil palm, rubber plantations, coconuts and fruits, the damage varying by locality and with the availability of other foods.[15] A communication from the Department of Wildlife and National Parks, West Malaysia, pointed out that 'Wild pigs mainly *S. scrofa* cause much damage to young oil palm plants.'[16]

Diong noted that wild pigs are wasteful feeders: 'The pig uses its snout to dig for tubers, roots, worms and other materials in the ground. This vigorous nosing behaviour makes it a very wasteful feeder, destroying more than it consumes.

Between March 15-20, over 15 acres of young tapioca plantation near Changkat Chermin were completely uprooted by a herd of about 25 pigs. Examination of the area showed that most of the uprooted tubers were either uneaten or only partially eaten.'[17]

The wild boar has been designated as an agricultural pest in Malaya since 1917. But now only the shooting of boar causing damages is legal because the species may now be threatened. Trapping is illegal.

Agricultural damage caused by wild pigs in countries near Malaysia are likely to be similar to those in Malaysia. I have been informed that 'Wild pigs are very destructive to both agricultural crops and young forest plantations including forest nurseries' in Indonesia.[18] However, the damage on islands that are now being more closely settled is mostly caused by *Sus scrofa*. Specifically, 'Agricultural damage by *S. verrucosus* is insignificant since its population is low. Agricultural damage by *S. scrofa* is not significant in Java, but serious on newly settled agricultural land in such places as Sumatra and Kalimantan. Damage is limited to crops. No damage is caused by *Babyrousa babyrussa*.'[19] Only the latter species is protected. The other species are sometimes hunted by farmers whose crops are being damaged.

In Thailand, 'wild pigs cause damage to agricultural crops near forest areas' and are not protected.[20] It has been observed in Thailand that wild pigs 'eat many kinds of fruit and vegetables, including cultivated pineapples, rice and corn. They typically use their mobile snout for rooting in soft soil, and since their nostrils are at the very end of their snout, they can detect the presence of earthworms, grubs and other invertebrates. Young shoots of *Imperata* (blady) grass, which begin to grow a few weeks after the burning season, are also eaten.'[21]

In Sri Lanka, wild pigs cause damage mostly to crops such as paddy, root vegetables and sugar cane.[22] They are unprotected.

In Japan, wild pigs are mostly in mountainous areas and have caused great damage, mainly to crops in these areas. Ten thousand are caught each year specifically because they are causing agricultural damage and about fifty thousand are taken as game.[23] In South Korea there is some (limited) agricultural damage from *S. scrofa* but the population level is low.[24]

The types of agricultural damage caused by feral pigs in New Zealand are similar to those caused by wild boars in Europe, but in New Zealand lambs have been reported to have been killed by feral pigs.[25] At one time agricultural damage was considered so serious that bounties were paid on the destruction of feral pigs in New Zealand but this practice stopped several years ago.[26] The commercial use of feral pigs (their export as game meat) has reduced numbers of feral pigs considerably in New Zealand.[27]

In Australia, feral pigs cause considerable agricultural damage. They are responsible for losses to a wide range of crops (temperate and sub-tropical crops), to pastures and of new-born lambs. Let us now consider the Australian situation in detail.

AGRICULTURAL AND PASTORAL DAMAGE CAUSED BY WILD PIGS IN AUSTRALIA - AN OVERVIEW

Feral pigs are present in Australia in temperate, sub-tropical and tropical areas and cause considerable agricultural and pastoral damage.[28] Possibly no other vertebrate pest causes so much damage in Australia now that rabbits are not as great a problem as in the past. I have estimated below that the annual economic loss caused by feral pigs in Australia is not likely to be less than $50 million annually. A wide range of agricultural industries are affected: grain production, the sheep industry, the cattle industry, oil seeds, sugar cane, and others.

The main damage is caused in New South Wales and Queensland. Feral pigs are most numerous in these states[29] and in these states are produced high levels of agricultural and pastoral commodities subject to pig damage. As mentioned earlier, feral pigs also occur in the Northern Territory but primary industry there is less subject to damage because the beef industry predominates, sheep are absent and cropping is insignificant. Even beef production in the Northern Territory accounts for a low proportion of total Australian output. In Western Australia, the feral pig population appears to be low and appears to be concentrated in the Kimberleys and other cattle country. In Victoria, feral pig numbers are low and scattered, so that damage is not major at present. Wild pigs are virtually absent from South Australia (with the exception of Kangaroo Island) and Tasmania. The specific assessment of agricultural and pastoral damage below will therefore concentrate on the position in New South Wales and Queensland. At the same time, it is recognized that feral pig populations in Western Australia and Victoria may expand and create larger losses in those states in the future.

Assessing the Damage and Economics of Control of Wild Pigs

To assess the agricultural and pastoral loss caused by feral pigs is not easy. The exact nature of the damage caused by pigs is not always known, and the extent and cost of damage is likely to fluctuate with seasonal conditions and the size of wild pig populations - populations that can vary considerably. From the point of view of farmers and the agricultural sector as a whole, and assuming aversion to risk on the part of farmers, the actual *burden* of damages caused by pigs exceeds the average loss sustained by farmers as a result of wild pig activity.[30] This is a consequence of the following: The amount that one would have to pay a farmer to take on the pig-risk voluntarily would exceed his expected or average loss from wild pig damages. The degree of excess depends upon the variability of the risk and the degree of aversion to risk-taking by farmers.

When a pest sufficiently affects production of commodities to alter market prices, there might be a social loss from pest-damage that would not be taken into account by the loss to farmers from the pest; it is the loss to consumers, as a result of higher prices of the products affected, and is usually measured by the fall in their consumers' surplus.[31] But in the case of pig damage in Australia, rural-product prices are unlikely to be affected significantly at present because pig damage is so widespread across rural industries. While *individual* sheep-growers, grain producers, sugar-cane growers and other agriculturalists suffer severe losses and some regions are badly disadvantaged, the percentage overall effect on the production of each of the rural industries is small. Loss of sugar cane annually, for example, is about one-tenth of one per cent of the whole crop. On the other hand, loss or sorghum might amount in some years to 10 per cent of the whole crop, so a price effect could not be ruled out here. But if demand for sorghum is relatively elastic (because of available substitutes), the price-rise effect may still be small and the loss in consumers surplus small. There are a priori reasons for expecting the rise in rural product prices as a result of feral pig damage in Australia to be small or negligible at present. Therefore in assessing *rural industry* losses from pig damage it seems realistic to concentrate on the extent of losses suffered by farmers and to take into account native wild-life, park and forest losses. The latter are considered in later chapters.

Actual losses or damages stated by farmers as occurring as a result of feral pig activity may understate economic losses from another point of view: In some instances, farmers have altered or adjusted their product specialization so as to cope with the presence of the pig-pest. For example, some graziers have switched from sheep production to cattle production, or from the production of one type of grain to another that is less palatable to feral pigs. As a rule some *profit* or income *is foregone* by farmers as a result of product substitution by them. This loss in profit should be recorded as part of the economic loss from the presence of the pest.

Furthermore, as far as rural industry is concerned, the actual loss due to the presence of the feral pig must be reckoned to be in excess of the actual value of damages inflicted for another reason: The damages are estimated given current control measures against pigs. The current expenses of combatting the pest must be regarded as a further cost of their presence. The total outlay by rural industry on pig control is currently unknown but it is significant and control measures impose environmental costs such as unwanted spillovers on native wildlife from pig-poisoning programs.

The problem of assessing pig damages and the economics of control of feral pig populations will be discussed in detail in Chapter 9 and later in this chapter, but a few general observations on the economics of control may be in order here. The mobility of the feral pig (coupled with its fecundity) reduces the incentive of individual farmers to control the numbers of pigs on their property. On the whole, one would expect each farmer to consider the personal gain from control measures. The quicker any destroyed pig is replaced on a farmer's property (either by inward movement of pigs or by reproduction) the less is the incentive to destroy it, because the smaller is the gain appropriated *by the farmer*. Furthermore, the more likely and more frequently any pig roams to neighbouring properties the less is the gain from an individual farmer's destroying it. Since pigs are mobile, collective action (for example, through a semi-government body such as a Pasture Protection Board) is necessary to ensure the optimal level of control of the pest.

It should also be mentioned that the mere fact that a species is causing damages (is even a 'net' pest) does not mean that unlimited control or eradication of the pest is desirable. The economic level of control of feral pigs is likely to vary from a social point of view in Australia from region to region, even from farm to farm, and with variations in agricultural prices and costs. For example, a rise in the prices of rural products subject to damage will generally increase the optimal degree of control of the pest population. In some regions and on some properties, little or no control *may* be economically optimal from a social point of view. Some of the principles of optimal control are discussed in detail in Chapter 9 but the fallacy needs to be squashed that it is worthwhile controlling or trying to eradicate a pest in all circumstances.

Ways in which Feral Pigs Cause Different Types of Agricultural and Pastoral Damage

Feral pigs cause agricultural and pastoral damage in a variety of ways. They do so by rooting, trampling, pushing away obstructions such as fences, eating crops and pastures, by predation on lambs, wallowing and acting as carriers of weeds, parasites and diseases. These activities and some of the associated damages are summarized in Table 2.2. The nature of the damage and its significance depends on the locality. For example, the rooting of bore drains and bore outlets is a serious problem on properties in dry areas of the Australian continent which depend upon these as the (sole) source of water for livestock. The rooting away of a bore

TABLE 2.2 Cause and Type of Damage done by Wild Pigs in Australian Agriculture

Cause and type of damage	Comment
Rooting and wallowing in bore drains and bore outlets	Source of drinking water for livestock in many places in inland Australia. This water is obtained from artesian basins. Drains may be diverted, water wasted and upkeep increased by pig activity. Bore drains often serve several properties.
Rooting up of pipes and troughs at livestock watering points, pollution of dams and waterholes by digging and wallowing and rooting	Water may be polluted to such an extent that livestock will not drink it. Bogging of sheep becomes more likely.
Rooting of irrigation works	Damage to channels, drains, pipes.
Rooting of farm and public dirt roads and light aircraft strips	Increases road maintenance and road erosion. A light aircraft strip in the Livingston Shire, Qld, was 'entirely rooted up by pigs and rendered completely unserviceable for aircraft'.
Damage to floodgates and levy banks by rooting	
Lifting and holing of fencing by rooting and movement	Most fencing is affected including vermin-proof fencing such as rabbit-proof and dog exclusion fencing.
Rooting of earthen dam walls and overflows and creek banks	Erosion encouraged.
Digging in pastures or crops esp. during dry weather or winter	Results in a reduction in food for domestic livestock or lowered crop production.
Trampling of crops and flattening of these for bedding	Taller crops most affected.
Rooting around trees	May cause bananas to fall and may kill other trees.
Eating of pastures	When green grass is available it forms the bulk of the feral pig's diet. Consequently there is food competition with domestic livestock.
Eating of crops	Pigs are especially fond of grain - for example, wheat.
Predation on livestock	A considerable number of lambs may be taken by pigs.
Onslaughts on temporary grain stores and haystacks on farms for food	

Cont.

Table 2.2 *cont.*

Cause and type of damage	Comment
Damage to harvesting equipment caused by pigs in tall crop (such as maize) failing to move from its path. Incidences seem to be isolated.	
Spread and encourage the growth of noxious weeds	May occur as a result of rooting, eating and spreading the seeds of noxious weeds, by adherence to body of the pig and by trampling.
Induce or accelerate erosion by rooting, wallowing, movement and treading with cloven feet	Water erosion aided by wallowing and rooting especially in water-courses. Wind erosion aided as a result of removal of vegetation cover.
Spread some diseases	Leptospirosis is spread by pigs as well as other animals. Induces abortion in livestock. Anthrax, tuberculosis and other diseases and also parasites spread.
Earthworm and soil life losses	Pigs by feeding on soil invertebrates such as earthworms and larvae (a major food item in some places) inevitably disrupt the cycle of breakdown of organic matter and decomposition. Productivity of the soil lowered.

drain can cut off water supplies to the livestock in neighbouring properties or paddocks with serious consequences. Indeed, any interference with water supplies in these areas is serious for livestock and can have unfortunate consequences for native wildlife.

Complaints are also received from irrigation areas that rooting by pigs interferes with flood irrigation control. Water supply channels can be rooted causing accidental flooding. The even-flow and drainage of water can also be impaired. Control over irrigation is reduced.

Damage by rooting is also reported to flood gates and to levy banks designed to hold back inland rivers from homesteads and agricultural land during flood periods. Not infrequently pipes and troughs at livestock watering points are rooted up, farm dams and waterholes polluted by wallowing and the walls and outflows of earthen dams (for example, turkey dams) undermined.

Rooting and pushing by pigs causes serious damage to fencing. Pigs make holes in the fences and weaken wires and posts. Consequently livestock may stray into crops, into paddocks in which they are unwanted and to neighbouring properties. As a result, crops are damaged by livestock, management of pastures is disrupted and livestock breeding programs can be jeopardized. Pigs sometimes damage vermin-exclusion fences such as rabbit-proof (see Fig. 2.1) and dingo fences, thus further adding to pest damage. All of these costs (damages) can be assigned to the fence damage by pigs, as well as the cost of repairing the fences, the expense of extra mustering of livestock due to increased straying as a result of fence damage, and the costs of more frequent checking of fences.

Digging by pigs can adversely effect rural transport facilities. Farm and state dirt roads are sometimes dug up and road drainage disturbed, causing erosion so the affected road deteriorates rapidly (see Fig. 2.2). Farm airstrips have also been rooted up, on occasions to such an extent as to be unuseable.

Pastures and crops are frequently dug out and are eaten. Taller crops and young crops are often trampled by pigs walking in these and by their bedding in tall crops such as maize. Grain crops appear to be especially attractive to pigs but pigs appear to eat a high proportion of green matter when it is available. Crops appear to be vulnerable at all stages but pigs will travel long daily distances for grain when it is available. Giles reported diurnal forays by feral pigs of up to 10 km from bedding areas in the grain belt of New South Wales in order to feed on grain. Apart from the crops actually eaten or destroyed by pigs, much of the trampled crop may be lost because of the difficulty of harvesting it with mechanical harvesters and because it becomes more readily accessible to rodents. Although this would be an infrequent occurrence, according to one farmer the feral pigs have damaged his harvester in tall crops by not getting out of its path and so getting caught in it.

It has been reported for inland Queensland (Charleville area) that 'Pigs are a problem mainly in the frontage country (to creeks and rivers). This is also the best lambing and calving country in Mulga areas. Pigs appear to be attracted mainly by water, *green feed* and more productive and easier rooting (digging).'[32] Another report from Queensland claims that as far as damage to the natural environment is concerned,

> 'The greatest damage has been in rooting of substantial areas of pasture and consequent destruction of feed and herbage. Much of the country has been in the better class lands along river and creek flats. On one station of 200 sq. miles, the owner estimates that 20 sq. miles was rooted (10 per cent of property). Landholders in melon hole country say that their best feed grows in the depression where water accumulates and lies for some time. Landholders shut up the paddocks but find when they are wanted, that feral pigs have used the melon holes as wallows, destroying what is often the only feed available.'[33]

Fig. 2.1 Rabbit-proof fence rooted up under lower straight wire by pigs. The landowner has tried to block the hole with logs but these have been pushed aside by pigs. The size of the hole is likely to increase with time, letting not only rabbits through but also livestock.

Fig. 2.2 Public dirt road dug up by feral pigs in south-west Queensland.

These reports suggest that feral pigs may cause significant pasture loss. This loss may be especially noticeable during periods when pastures are under stress (for example, during dry periods or winter) and in demand for livestock. Growth and regeneration of the pasture may be retarded when the season improves.

There has been little systematic study of pasture loss in Australia from the activity of feral pigs, although an isolated short-term study has recently been completed by J. Hone at Tenterfield on the northern Tablelands of New South Wales.[34]

About 200 hectares were studied, comprising 58 ha of introduced pasture, 80 ha of native pasture and the balance forest. He found extensive rooting by feral pigs. Seventeen per cent of his randomly selected quadrants were rooted, with the proportion ranging from 14 per cent in introduced pasture to 21 per cent in native pasture. Furrowing was severe in some cases: up to 50 m in length and 20 cm deep.

On introduced pasture, Hone observed in areas rooted by pigs a 74 per cent reduction in green *grass* matter and a 36 per cent fall in standing dry matter. However, there was a tenfold increase in the dry weight of non-grass matter, mostly *broad-leafed weeds*. On native pasture green grass matter was reduced by 98 per cent and standing dry matter by 74 per cent. But in this case, there appears to have been a net reduction in weed species, since pigs disturbed bracken fern *Pteridium esculentum* and blady grass *Imperata cylindrica*.

Hone concluded from his study that 'These results indicate that feral pigs can seriously reduce pasture production by rooting in the soil, as found by Barrett (1971). Growth of introduced and native pastures in the Tenterfield environment is minimal in winter (Lazenby and Swain 1973). Consequently, the adverse effect of pig rooting is likely to be very important at this time and may reduce the potential peak pasture production in the spring.'

The effect of pig rooting and eating of pastures on the capacity of properties to carry domestic livestock does not appear to have been studied in depth. However, in the light of the above, reductions of up to 10 per cent in the livestock carrying capacity of some properties do not seem improbable.

The long-term impact of pig activity on pastures is not clear, although it seems that it can alter the dominant species, and of course continual rooting reduces ground cover. Hunters have also suggested to me that regeneration of ground rooted by pigs is slow (see Fig. 2.3). Evidence on the encouragement of weeds by pig activity is conflicting. Hone's study suggested an increase in weeds in introduced pasture and a possible reduction of weeds in native pasture. However, pigs are widely believed by farmers and others to spread and encourage weed growth on the whole. This they are reputed to do by rooting so creating favourable seed beds for various types of burr, for example, by eating seed and passing it in their faeces such as blackberries and by the adhesion of seeds to their bodies.

On some properties feral pigs have caused damage to haystacks and temporary grain stores and fodder stores in foraging for food. Again, some farm trees may be lost as a result of pig-rooting and fruit trees are affected in some areas. Losses of bananas can be significant. Pigs undermine trees in some areas and eat the fallen bananas.

Erosion of paddocks and watercourses is liable to be accelerated by pig activity. Their rooting, the impact of their cloven feet and wallowing (see Fig. 2.4) all loosen the soil increasing the likelihood of water erosion and (to a limited extent) increased wind erosion. Reduction of ground cover also contributes. Again there has been little systematic study of this. It could, however, reduce agricultural and pastoral productivity and add to silting over a long period. From Queensland, for example, it has been reported that

> 'In the far south west along the Bulloo and Wilson Rivers, landholders report extensive areas rooted up by pigs and resembling ploughed paddocks. It is feared that with the next flood, great quantities of soil will be washed away causing erosion and silting problems.'[35]

Pigs may also interfere with the decomposition cycle in soils. They are especially fond of earthworms and may affect microbiota in the soil through their eating and rooting habits. This could have some adverse effect on soil fertility as suggested by Barrett, Smith and Diong.[36]

Pig predation on livestock also occurs in Australia. This appears to be confined to lambs but may include enclosed sheep in special cases. Claims that calves may be taken by pigs do not appear to have been substantiated despite the extravagant suggestion in the novel *Pig* that wild pigs may kill fully grown cattle.[37] It has been claimed that in some areas the lambing percentages have been reduced by as much as 60 per cent due to pig predation. This matter will be discussed below when considering damages in the pastoral industries.

Wild pigs are suitable vectors for spreading livestock diseases and parasites. Calving percentages in Australia could be reduced by pigs (not the only carriers) spreading *Leptospira*. These result in leptospirosis which can cause abortions in cattle. Anthrax can also be spread by pigs in Australia and tuberculosis has been observed in pigs in the Northern Territory. A major worry is that if an *exotic* disease such as foot-and-mouth disease were accidentally introduced to Australia wild pigs would help spread the disease and make it more difficult to contain or eradicate. Problems associated with the role of feral pigs in the spread of disease and parasites are discussed in detail in Chapter 8.

Agriculturalists attribute very few benefits to feral pigs. Apart from providing a source of some extra income from hunters (see Chapter 4) the only benefits from feral pigs that I have heard Australian farmers claim is their

Fig. 2.3 Regeneration of pastures after pig rooting can be slow. This pasture shows the scars from feral pig rooting 12 months previously on D. & J. McDonald's Newstead property at Woodburn, northern coastal N.S.W.

Fig. 2.4 A pig wallow on the side of a track in Australian Paper Mill's eucalyptus plantation near Coff's Harbour, N.S.W. Water erosion occurs as a result of such wallowing.

destruction of rats and mice *Rodentia*, removal of carrion, and reduction of some weed pests such as Paterson's curse *Echium lycopsis*, bracken fern *Pteridium esculentum* and cutty grass *Imperata* sp. In addition, pigs may make a minor contribution by devouring the larvae and adults of some insect pests. For example, pigs at times include grasshoppers in their diet. Reports from Belgium indicate that wild pigs sometimes dig young rabbits out of their burrows and eat them.[38] Whether they do this in Australia to any significant extent where rabbits are a pest is not known. Agricultural benefits from feral pigs in Australia seem relatively minor, so on balance it is a major agricultural pest.

LOSSES CAUSED BY FERAL PIGS TO PARTICULAR AUSTRALIAN RURAL INDUSTRIES

Only limited data are available on the losses caused to particular Australian rural industries by feral pigs. However, in terms of the magnitude of the value of the loss it appears to be greatest for the wheat and grain industries (as in Europe), followed by sheep industry losses, and losses to the beef, oilseeds, sugar cane and fruit industries. Some damage is also caused to a wide variety of other crops including potatoes and other vegetable crops, and cotton in the Wee Waa area, N.S.W. Table 2.3 gives rough estimates of the value of production loss by Australian rural industries on the value of 1979-80 gross production. It gives some idea of the relative order of magnitude of damage. Wheat losses account for almost half the value of lost agricultural production due to feral pig activity, and total grain losses account for about 56 per cent. About a quarter of the loss is borne by the sheep industry as a result of pig predation on lambs and as a result of loss in sheep carrying capacity on pastures eaten or damaged by pigs. Just under a fifth of the total loss may be inflicted on the cattle industry, again as a result of reduced cattle carrying capacity because of competition for feed and because of management problems arising from the presence of pigs. These relative losses are indicated in Fig. 2.5. The annual loss in agricultural production appears to be of the order of $70-80 million.

However, this figure takes no account of the cost of pig control programmes, fencing losses, increased erosion, and so on. On the other hand it might be argued that the value of the loss is overestimated - for example, running fewer sheep or cattle can actually reduce costs, and a lower grain harvest may reduce costs. I shall take this point up again presently. But it does seem that even my updated estimate, mentioned earlier, of a annual loss of $30-40 million (based on Pastures Protection Board returns) understates annual agricultural losses due to feral pigs.

TABLE 2.3 Value of Estimated Loss to Australian Agricultural Industries as a Result of Feral Pig Activity, 1979-80

Industry	Australian production gross value $m *	Assumed loss in output**		Loss $m	Loss as % of total loss
Wheat	2 275	3% of crop N.S.W. & Qld (1.5% Aust.)		34.0	46.38
Sorghum	106	5% of crop		5.0	6.82
Barley	447	1% of crop N.S.W. & Qld		1.5	2.04
Oats	129	1% of crop N.S.W. & Qld		0.5	0.68
Maize	14	3% of crop		0.4	0.54
Rice (paddy)	86	Some loss likely		?	?
TOTAL GRAIN				41.4	56.46
Oilseeds (incl. soya beans)	70	2% of crop		1.4	1.9
Sugar cane	497	0.1% of crop		0.5	0.68
Sheep Wool	1 665	Lamb predation based on Benson's estimates	6.0	17.0	23.19
Sheep Meat	605	1% of wool and meat N.S.W. & Qld (0.5% Aust.)	11.0		
Cattle Beef	2 580	1% of beef N.S.W. & Qld (0.5% Aust.)		13.0	17.73
Miscellaneous (potatoes, vegetables, cotton, fodder, etc.)	-	Some loss		?	?
TOTAL				73.3	100.00***

*As stated in Bureau of Agricultural Economics, *Quarterly Review of the Rural Economy*, Canberra, May 1980, pp. 228, 229.

**It is assumed that practically all of the loss from feral pigs is accounted for by lost production in New South Wales and Queensland where pigs are most numerous. The basis for this particular loss percentages chosen are discussed in the text.

***Does not add exactly due to rounding.

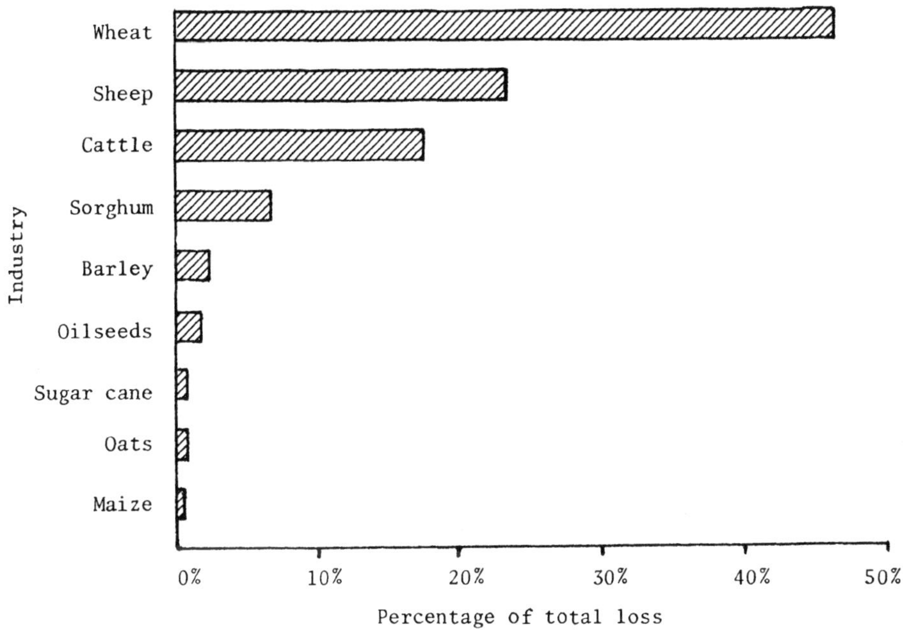

Fig. 2.5 The estimated loss in value of production of Australian agricultural industries as a proportion of the total loss in value of Australian agricultural products caused by wild pig activity

An annual agricultural loss of at least $50-60 million does not appear improbable. Let us consider each of the industries in detail.

Grain (wheat, barley, sorghum, oats, maize) and Oilseed Losses

Wheat seems to be subject to the greatest overall loss from pig activity. On the basis of a sample survey, Benson[39] suggested that in the 11 Pastures Protection Board districts in north-western New South Wales wheat losses because of feral pig activity amounted to 5.6 per cent of the crop. Given that the density of wild pigs is lower in *some* of the other wheat-growing areas of New South Wales, it seems reasonable to reduce this figure on a statewide-basis. I have selected 3 per cent as plausible and have assumed the position to be similar in Queensland. But even if the loss is as low as 2 per cent for these states (1 per cent Australia-wide), the value of wheat lost as a result of pig activity exceeds that for any other agricultural industry.

Non-bearded varieties of wheat are more susceptible to damage by pigs than bearded varieties[40] and the loss of non-bearded crops tends to be much higher than

Fig. 2.6 A large feral boar shot in a wheat crop near Moree, N.S.W. Pigs appear to cause greater aggregate losses to the wheat industry than any other.

Fig. 2.7 Some of this maize has been flattened and eaten by feral pigs. Large losses of maize occur in some areas but overall Australian production of maize is relatively low.

for bearded varieties. Nevertheless, bearded varieties are susceptible. Peter Pavlov in a study of crops in the Girilambone area has estimated losses to the bearded wheat varieties Condor and Timgalen at 7 per cent in 1977 and 4 per cent in 1978.

The percentage losses for barley in Table 2.3 are based on Pavlov's finding that in the same area barley damage from pigs in 1977 was 1 per cent and in 1978 2 per cent.

In 1978 Pavlov also conducted experiments to determine the preference of pigs for different types of grain. He compared the bearded wheats Condor and Timgalen with non-bearded Kite and also barley and oats. He found that pigs preferred non-bearded Kite and oats to the bearded varieties and these in turn were preferred to barley.

There is little information on sorghum losses. However, it has been reported from Queensland that 'sorghum suffers the severest damage from pigs. Approximately 20 per cent of the crop in each Shire suffers to some degree.'[41] Nearly all Australian sorghum production is in Queensland and New South Wales. In some areas (for example, Hay district, N.S.W.) large proportionate losses of maize have been caused by feral pigs.

No reliable data seem to be available on oilseed losses. However, it is reported from Queensland that 'safflower is the only oil crop which pigs appear to avoid.' Most of Australia's oil seed production is in New South Wales and Queensland.

Sugar Cane Losses

Sugar cane in Australia is grown on the eastern coast, from around Grafton in northern New South Wales to Mossman in northern Queensland. Approximately 95 per cent of total Australian production is from Queensland and the remaining 5 per cent from New South Wales. Pigs cause damage to cane in all areas but the greatest losses occur in the northern Queensland areas.

The Director of the Bureau of Sugar Experiment Stations, Queensland, reports that, in relation to cane lands, 'Water is usually available freely, or within reasonable distance, for most of the year. Food in forested areas seems readily available also. Pigs which develop a taste for cane come back to it regularly - some bite chunks out, chew it well and spit out a lot of the fibre; others may just chew the stalk enough to extract the juice, and leave the stalk mangled but in places but entire.'[42] They may also knock cane down.

In Queensland, the Bureau of Sugar Experiment Stations estimates that about 20 000 tonnes of cane have been lost annually in recent years. This amounts to a

Fig. 2.8 Pigs damage sugar cane by chewing it and trampling it. Feral pigs have knocked a part of this crop down and have chewed some of it.

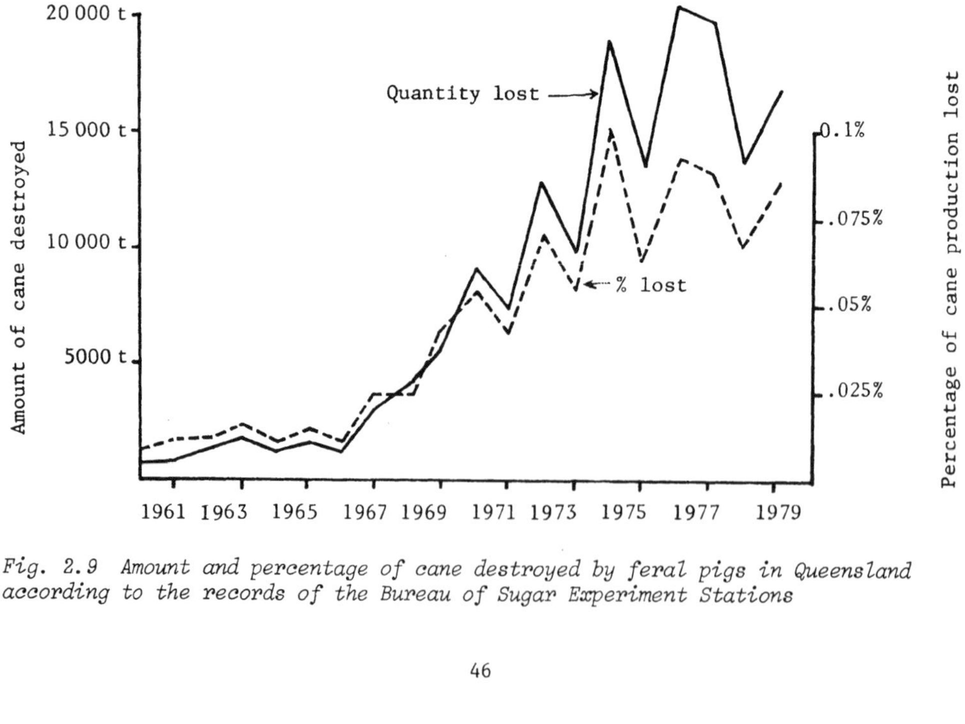

Fig. 2.9 Amount and percentage of cane destroyed by feral pigs in Queensland according to the records of the Bureau of Sugar Experiment Stations

little less than 0.1 per cent of the total Queensland crop, and this is the percentage given in Table 2.3. The actual percentage average loss in the six years 1974-79 was 0.082 per cent. As Figure 2.9 indicates, the quantity of cane and the percentage of production lost to pigs increased until 1974.[43] Since 1974 there is no clear trend, but on average the amount of cane damaged and the percentage loss remains higher than in earlier periods. In particular, percentage losses do not appear to be declining.

The loss of cane is less in the southern area of Queensland than in areas farther north, where up to 0.5 per cent has been lost in some years. The actual percentage losses in different years has varied considerably, though the percentage (and total) loss of production in all districts was significantly higher in the 1970s than the 1960s. Average percentage losses of cane in the six years 1974-79 inclusive were:

0.165 per cent in Mossman/Babinda district in the far north;

0.186 per cent in Innisfail/Tully district;

0.186 per cent in Herbert River district;

0.065 per cent in Burdekin district;

0.036 per cent in the Central district;

and the loss in the Southern district was negligible.

A very useful assessment of feral pig control in canefields has been communicated to me by the Director of the Bureau of Sugar Experiment Stations. He pointed out that

> 'Feral pigs are declared vermin throughout Queensland. Bounties of $2 to $5 per head have been paid by Cane Pest and Disease Control Boards for many years to encourage shooting. This may be supplemented by the farmer ($10 to $50), depending on how much damage the individual pig is causing. Bounties help the shooters to keep and doctor his pig dogs - virtually no one shoots pigs for the money, but mainly for sport.
>
> Shooting has little effect on populations, but it is a remedy for those pigs which take a liking to sugar cane. Baiting with 1080 has been tried to a limited extent near cane land and it has been found that the pigs will take the poison if it is inserted into animal fat. However, the hazards to non-target species is great and the benefit of limited duration. The best control is to exclude the pigs from the crop with fences. Netting fences have been used for many years, but are too expensive for extensive use. More recently, relatively cheap and effective energizers for electric fences have become available and electric fences for the protection of cane crops are proving very successful. Limited trapping is also practised, but this would have little impact on the overall population.'[44]

Fig. 2.10 This banana tree at Bucca Creek near Coff's Harbour was undermined by feral pig rooting and fell.

Fruit (bananas, pineapples), Potatoes, Vegetables and Miscellaneous Crops

As far as I am aware there is no quantitative data on losses to the above agricultural crops. However, it is known that feral pigs do cause some damage to bananas in the Coff's Harbour area of New South Wales. Trees are undermined by pig rooting, and pigs eat bananas from the fallen trees. Banana stools are also rooted out. Pineapples (nearly all the Australian crop is grown in Queensland) are also reported to be damaged: 'Pineapple damage has included eating of fruit, damage to suckers and the eating of the growth centre out of plants which is their ruination and general rooting and trampling damage. Much of this damage is occurring in urban areas along the east coast.'[45] Reports have also been made of some damage to fruit trees such as apple trees by rooting in the Stanthorpe area of Queensland.

In some localities potato crops are affected. One farmer in the Nundle area near Tamworth, N.S.W., told me that he was forced to give up potato production because the potatoes were eaten by pigs coming from a forest adjoining his property. Lettuce crops have been damaged in the Hay district, N.S.W., and other vegetable crops may suffer. Japanese millet and other fodder crops are also a target for pigs. Even cotton crops have been damaged by pigs in the Wee Waa area of New South Wales.

Fig. 2.11 Mr Rup Hanna with a lamb and sheep-killer which he has shot on his property. The size of the boar can be appreciated when it is realized that Rup Hanna is 6 ft tall. This boar had one tusk, had killed a pig dog, and is reputed to have killed not only lambs but penned wethers in good health.

Sheep (meat and wool) Losses

Sheep numbers may be reduced by pig predation, by pigs lowering the amount of pasture available to sheep and by limiting the carrying capacity in other ways - for example, by fouling the water. Thus meat and wool production may suffer.

Pig predation appears to be mainly confined to lambs but one farmer in the Hay area has had fully grown sheep attacked in an enclosed area by a boar.[46] Occasional attacks on larger sheep cannot be ruled out.

Two field trials have been made by different researchers of lamb predation by pigs. A reduction in lambing percentages of 31.6 per cent (from 117 per cent to 80 per cent) from the presence of feral pigs was found by Plant and colleagues, comparing two mobs of 500 pregnant ewes.[47] 'Using a similar design, but having both pregnant and non-pregnant ewes in the paddocks, Pavlov, Hone, Kilgour and Pedersen (unpublished data) found significant effects of feral pigs on lamb survival in two of four lambings. Eight kills of healthy lambs by feral pigs were observed during the study.'[48] Pavlov and his colleagues were lucky to observe pig predation in this study in western New South Wales, for predation generally takes place at night; few if any signs of the predation remain the next morning since the pigs devour all of the carcass. The fact that pigs were seen killing healthy lambs is significant, since some farmers and a few scientists believe that pigs only kill weak or unhealthy lambs.

A survey of shires in sheep breeding areas in Queensland found that total lamb predation from foxes and pigs ranged from an average by shires of 13 per cent to 42 per cent of lambs born. From a survey of 133 landholders in 1978, in western New South Wales Benson estimated a net decline of 7 per cent in lambing percentage due to pigs.[49] Benson used this percentage to calculate the total annual loss in the 11 Pastures Protection Board districts in north-western New South Wales and, by valuing each lamb lost at $15, estimated the loss to be $3.2 million. For Table 2.3, I increased this figure by approximately doubling it for New South Wales and making a small addition for Queensland. (Queensland sheep numbers are low in comparison to those in New South Wales.)

Nevertheless, it is pertinent to draw attention to Benson's caution about his estimate. He says:

> 'These estimates of a net decline in lambing percentage due to pigs must be viewed with a fair degree of caution, as landholders have a minimum of information with which to work. Pigs kill at dusk and early morning and a few hours later little evidence may exist of what was once a live lamb. Other predators such as foxes kill lambs and there are those that die from lambing difficulties and ill thrift. The difficulty in landholders estimating lambs killed is seen in the 23 landholders who knew pigs were taking lambs but could not estimate how many.
>
> Whilst there was a considerable number of lambs lost to pigs, one of the biggest costs is the time spent at lambing patrolling ewes. Some landholders spend six weeks at lambing in the saddle, shooting pigs and deterring them by their presence. Some landholders ran cattle or wethers to avoid the losses but incurred some income loss in doing so. Nineteen landholders reported changing all or part of their ewe flock over to wethers or cattle although the presence of pigs was only partly the reason for this in many cases. Seven landholders lambed their ewes in open country where there was no pig cover.'[50]

In Table 2.3 I have also allowed for a reduction in the capacity of properties to carry sheep as a result of pasture being eaten by pigs and lost as a result of their rooting. About a 1 per cent loss of wool and sheepmeat is assumed for New South Wales and Queensland, or about 0.5 per cent loss for Australia as a whole, due to this element. A loss of this order would suggest that food competition and food damage by pigs could be a greater source of loss to the sheep industry than lamb predation. But clearly scientific evidence is needed on this matter.

Fig. 2.12 Pasture dug up by feral pigs in sheep and cattle country in south-west Queensland, thus reducing pasture available to livestock and facilitating wind erosion

Cattle Industry Losses

The main current loss in production of the cattle industry from feral pigs probably stems from food competition. The diet of pigs principally consists of green grass when it is available. One report indicates that 'Cattle and pigs eat basically the same food when pastures are good. The real effect, however, becomes apparent when pastures dry off and pigs root over the country destroying feed in the process.'[51] Hone's Tenterfield study, mentioned earlier, confirms that feral pigs can substantially reduce the amount of pasture available to domestic livestock. Since there is no information for Australia as a whole, the 0.5 per cent loss in value of beef production for Australia as a whole assumed in Table 2.3 must be regarded as a guess. Nevertheless, even if the percentage loss of beef due to food competition is low, the magnitude of the loss is not.

Interestingly a recent study, *Feral Animals in the Northern Territory*, expressed the economic losses due to the presence of feral animals in terms of cattle equivalents.[52] (Practically all agricultural production in the Northern Territory is of cattle). It was assumed that the trade-off between 10 pigs was 1 head of cattle. Since 100 000 feral pigs were estimated to be present, it was calculated that 10 000 cattle equivalents were displaced. In turn it was claimed that an *annual* potential turn-off of 2 000 head of cattle valued at $200 000 annually is foregone because of the presence of feral pigs. This was taken as a measure of the economic cost of feral pigs in the Northern Territory.

Fig. 2.13 Pasture rooted out by pigs coming from the pine plantation in the background and searching for grubs. Rooting and eating of pastures by feral pigs reduces the available food for cattle and other domestic livestock so may reduce the capacity of properties to carry such stock.

In relation to cattle equivalents, the N.T. Board of Inquiry into feral animals said:

> 'There is a limit to how many free loaders a business can carry. The most obvious and direct effect of many feral animals on the pastoral industry is that they occupy a place which could be used by a more productive unit. Table 1 (Feral Animals Expressed in Cattle Equivalents) is adapted from figures provided by the Primary Industry Division and McKnight. The Board designed this table and recognises its limitations. The technique could perhaps be refined by other authorities. Conservatively the total for feral animals represents 500 000 cattle equivalents, or nearly 30 per cent of the total Territory herd. On the basis of the assumptions and approximations we have made, it appears that the Northern Territory beef industry is losing more than $12 million per annum through the reduced turnoff which stems from the presence of a host of unproductive feral animals.'[53]

The exact assumptions are not spelt out in the report but seem to be influenced by weight comparisons between feral animals and cattle. Clearly straight weight comparisons as a test of possible trade-off between pigs and cattle would be crude because the diet of these animals overlaps only to a certain extent. While the cattle alternative may be a realistic basis in the Northern Territory for assessing economic loss, this will not be so elsewhere where cattle are not the main economic alternative. The figures of the Board of Inquiry might overstate losses if the market value of cattle were taken as the measure of loss. The profit lost (market value less costs) rather than revenue lost by not being able to run the extra cattle would appear to be a more appropriate measure of the economic cost or loss imposed by the presence of feral animals. It is worthwhile discussing this matter more generally, even though the non-specialist may wish to skip this discussion in the next subsection.

Problems of Valuing Cropping and Livestock Harvesting (turn-off) Losses

In order to determine the economic loss due to the presence of a pest, it is necessary to subtract maximum profits in the presence of the pest from maximum profits in its absence (ignoring general spillovers). If Π_A represents maximum profit in the absence of the pest (for example, pigs), and R_A and C_A the corresponding levels of total revenue and total costs

$$\Pi_A = R_A - C_A.$$

Indicating the corresponding variables by subscript B in the presence of the pest, maximum profit if the pest is present is

$$\Pi_B = R_B - C_B.$$

Hence, the economic loss L, due to the presence of the pest is

$$\begin{aligned} L &= \Pi_A - \Pi_B \\ &= (R_A - C_A) - (R_B - C_B) \\ &= (R_A - R_B) - (C_A - C_B). \end{aligned}$$

From this it is clear that the economic loss from the presence of the pest is only equal to the revenue lost if total costs are the same whether the pest is present or not.

It has been suggested that costs might be lower if the pest is present because there is a smaller crop to harvest and less livestock to rear and a possible cost saving.[54] This is a *possibility*. On the other hand, the presence of the pest may impose additional costs that ought to be taken into account such as extra fencing costs, control costs, etc., so it is *feasible* for total costs in presence of the pest to exceed those in its absence. The cost change however needs to be specifically estimated.

It might also be noted that the economic loss is proportional to the quantity of (optimal) output lost due to pests *if* total costs of production are the same whether pests are present or absent, if quality of the product is unaffected by the pests, and if sellers are unable to influence the price at which they sell their product (are price-takers). In this case, where P represents the price of the product, X_A indicates the quantity of its supply in the absence of the pest and X_B its supply when the pest is present, the economic loss (reduction in profit) caused by the presence of the pest can be expressed as

$$L = (PX_A - PX_B) - (C_A - C_B)$$
$$= P(X_A - X_B) \text{ since } C_A = C_B \text{ by assumption.}$$

The annual economic loss is proportional to the reduction in *profit-maximizing* output caused by the occurrence of the pest. This loss equals or exceeds the loss in revenue (or percentage loss in output) caused by actual pest damage when the pest is present.

Concluding Comments on Agricultural Loss due to the Pig Pest

While one can identify theoretically the economic loss due to a pest, evaluating it in practice, especially on a significant nation-wide scale, is no easy task. In the case of feral pigs in Australia, data are inadequate on the damages actually caused and of course are not costless to collect and compile. Furthermore, damages and economic losses are liable to fluctuate from year to year as agricultural economic conditions (prices of produce and weather conditions) and the population level of feral pigs fluctuate. Nevertheless, it is useful to try to establish orders of magnitude of economic loss as long as their imprecise nature is recognized. Despite imprecision, it does seem that at present in Australia greatest total economic loss is caused to the wheat industry by feral pigs. Damages in proportion to the crop are also significant for other grains such as sorghum and oilseeds except safflower. The next greatest loss is suffered by the sheep industry and then apparently by the cattle industry. It is difficult to imagine that the current economic loss to the whole Australian agricultural industry as a result of feral pig activity could be less than $50 million per year and may range as high as $80 million per year. This loss is approximately equivalent to the total value of output of some of the smaller agricultural industries (for example, rice (paddy), apples and sorghum).

THE PIG PROBLEM AND WAYS OF CONTROLLING IT AS SEEN BY PASTURES PROTECTION BOARDS IN N.S.W.

To obtain a broad picture of the wild pig problem as it affects farmers and graziers in New South Wales (and could similarly affect landholders in Queensland and other relevant states) I surveyed all Pastures Protection Boards in New South Wales in August 1977 using a mail questionnaire.[55] Excellent co-operation was received from

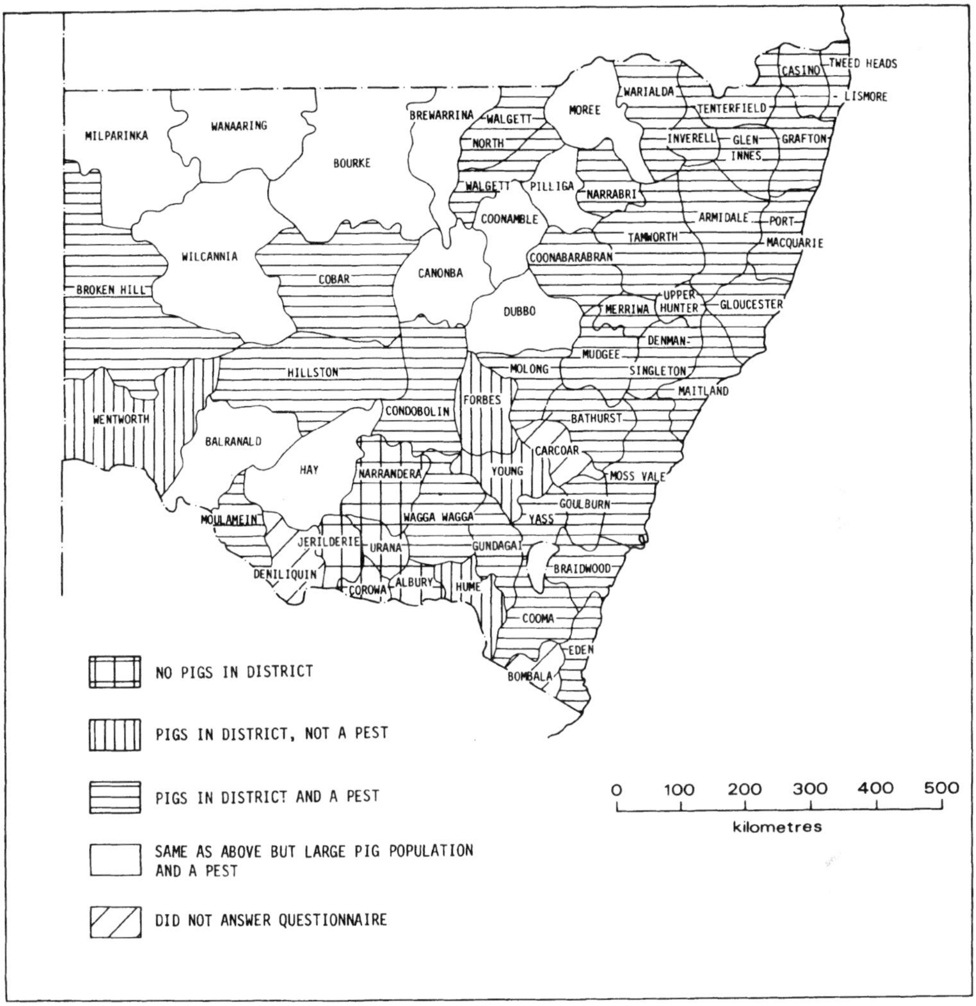

Fig. 2.14 Pastures Protection Board districts in New South Wales, and the presence of wild pigs and their occurrence as pests, September 1977

the Boards. The Boards consist of landowners elected by rate-payers (owners of livestock paying rates to the Board) and are amongst other things responsible for the enforcement of regulations concerning control of noxious animals in their districts. There are fifty-eight such districts and Boards in New South Wales and together they cover the whole state. Each democratically elected and representative Board has a small group of permanent staff working for it, such as rabbit inspectors and staff able to assist landholders with pest control measures such as the use of 1080 (sodium monofluoroacetate) poison. The activities of the Boards are financed

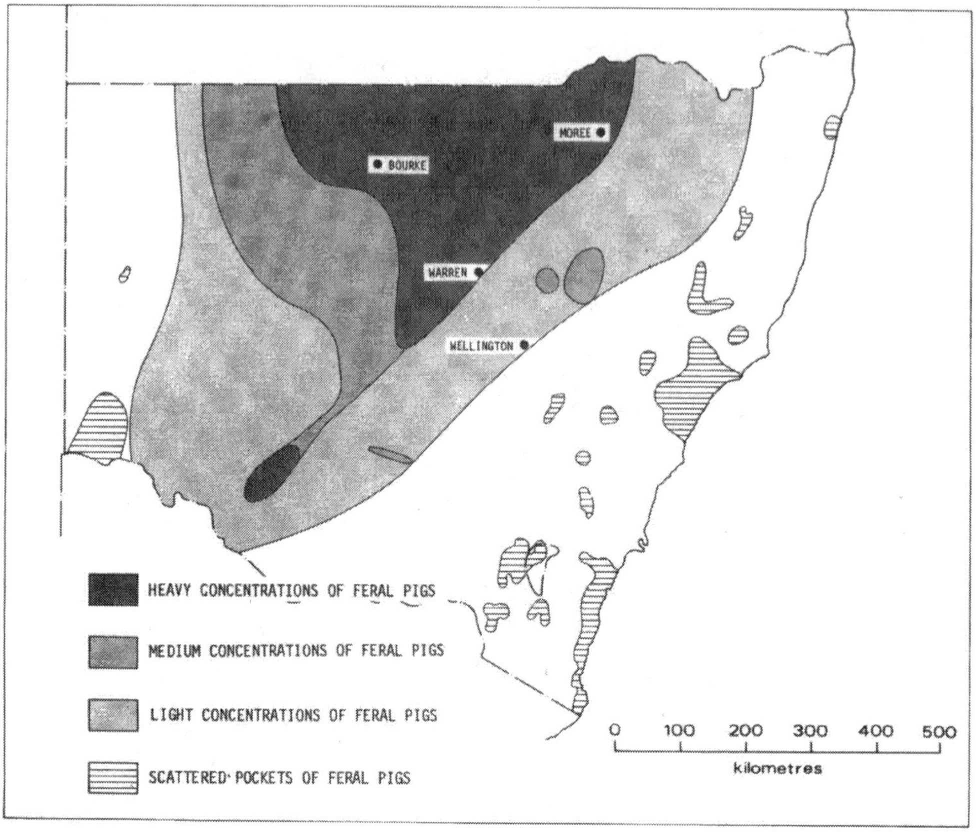

Fig. 2.15 The distribution and density of wild pig population in New South Wales (after Giles[57], 1977)

mainly from the levies (taxes) that they impose upon owners of livestock in their districts.

Of the 58 Boards in New South Wales, 54 completed and returned the questionnaire and one responded to some of the questions by telephone, so the response rate was 93 per cent. More than 90 per cent (49 out of 54) of responding Boards reported the presence of feral pigs.

Feral pigs appear to occur everywhere in New South Wales except for a small part of the state near the Victorian border (see Fig. 2.14). In all except four of the districts where they are present, wild pigs were reported to be an agricultural pest. Thus allowing for non-responses, it can be concluded that wild pigs are present in at least 84 per cent of the Pastures Protection Board (P.P.B.) districts in New South Wales and are an agricultural pest in more than 75 per cent. It may be that pigs have spread in New South Wales, as suggested by the Department of Agriculture, so that the problem is now even more widespread.[56] Fig. 2.14 maps the P.P.B. districts of New South Wales and indicates:

(a) where feral pigs are present and are regarded as a pest;

(b) where wild pigs are present but not considered to be a pest;

(c) the five P.P.B. districts reporting the absence of feral pigs.

The latter districts are situated on the northern flank of the upper Murray River in New South Wales. The map also indicates the districts in which pigs are present but are not considered to be a pest. Except for Wentworth in the south-west corner of New South Wales, these districts are near those without feral pigs and have low populations of wild pigs.

It is interesting to compare the map in Figure 2.14 with Giles' suggested distribution of feral pigs in New South Wales in 1977, which is like that shown in Fig. 2.15.[57] The broad responses from the Boards support this pattern, but reported populations of feral pigs are even more widespread than suggested by Giles.

In 82 per cent of the P.P.B. districts where pigs are considered to be a pest, Boards were taking measures of some kind to help control the number of feral pigs. The Boards taking control measures and those not doing so are shown in Fig. 2.16. Boards were reported to be taking no control measures in the Western Division of the state and in the P.P.B. districts of Brewarrina, Condoblin and Wagga Wagga where pigs are reported to be pests. It appears that in marginal agricultural areas of the state, such as the far west where holdings are larger, least effort is being made to control pigs because of economic factors.

Pastures Protection Boards were asked about the desirability of different types of control of pigs. Most were opposed to the payment of bounties on scalps as a method of control. Of the 53 respondents to the questionnaire, 43 (or 81 per cent) opposed the payment of bounties, six (or 11 per cent) favoured such a payment, and four did not answer this question. Again most of the Boards having wild pigs in their district were opposed to the idea of amateur shooters playing a bigger role in the control of wild pigs. Thirty-two (66 per cent) of the Boards (of the 48 reporting pigs) were opposed to the increased use of amateur shooters, whereas 16 (33 per cent) favoured the idea. Most Boards reported that farmers and graziers in their area were not taking adequate measures to control pigs, that effort to control pigs in their district varied over time, and that least effort was likely to be made to control pigs in economically bad times. (The last result is consistent with the economic model of pest control that I discussed in *Simple Economic Models of Pest Control - Models with Possible Application to the Control of Feral Pigs and other Wild Animals.*[58]) But let us consider the results in detail. These provide information about the level of the population of feral pigs, damages caused

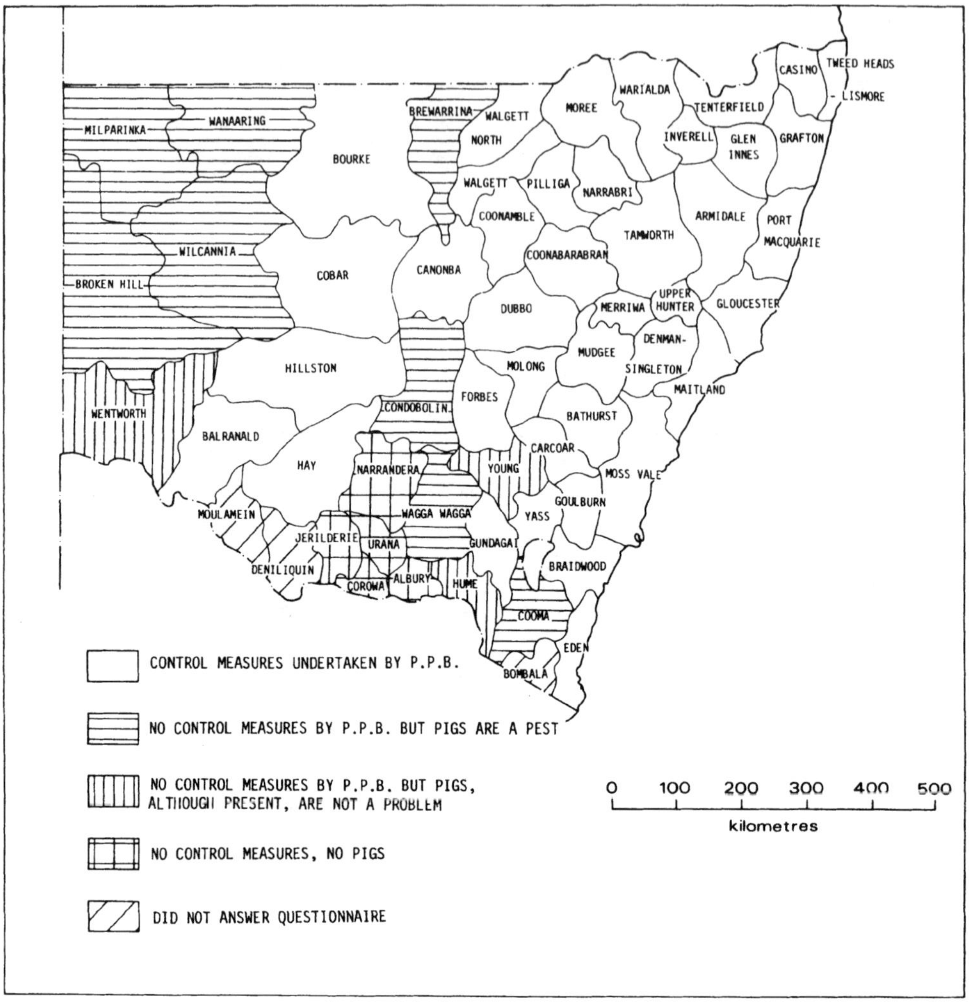

Fig. 2.16 N.S.W. Pastures Protection Board districts classified according to whether Boards are involved in the control of pigs and whether pigs are pests, September 1977.

by pigs, the costs of these damages, control measures taken by the Boards, control measures adopted by farmers and graziers, the role of amateur shooters in pig control, bounties on pigs destroyed as a control measure and government policies desired by Boards for control of feral pigs.

Number of Feral Pigs, Damages Caused and Cost of Damages

Boards were asked to give a rough estimate of wild pig numbers in their district. From the estimates received (shown in Figure 2.17) it appears that there are *at least* 1.55 million wild pigs in New South Wales. However, given that estimates were not received from all Boards, some of which are in areas of heavy wild pig populations, the actual number of wild pigs in New South Wales could exceed 2 million.

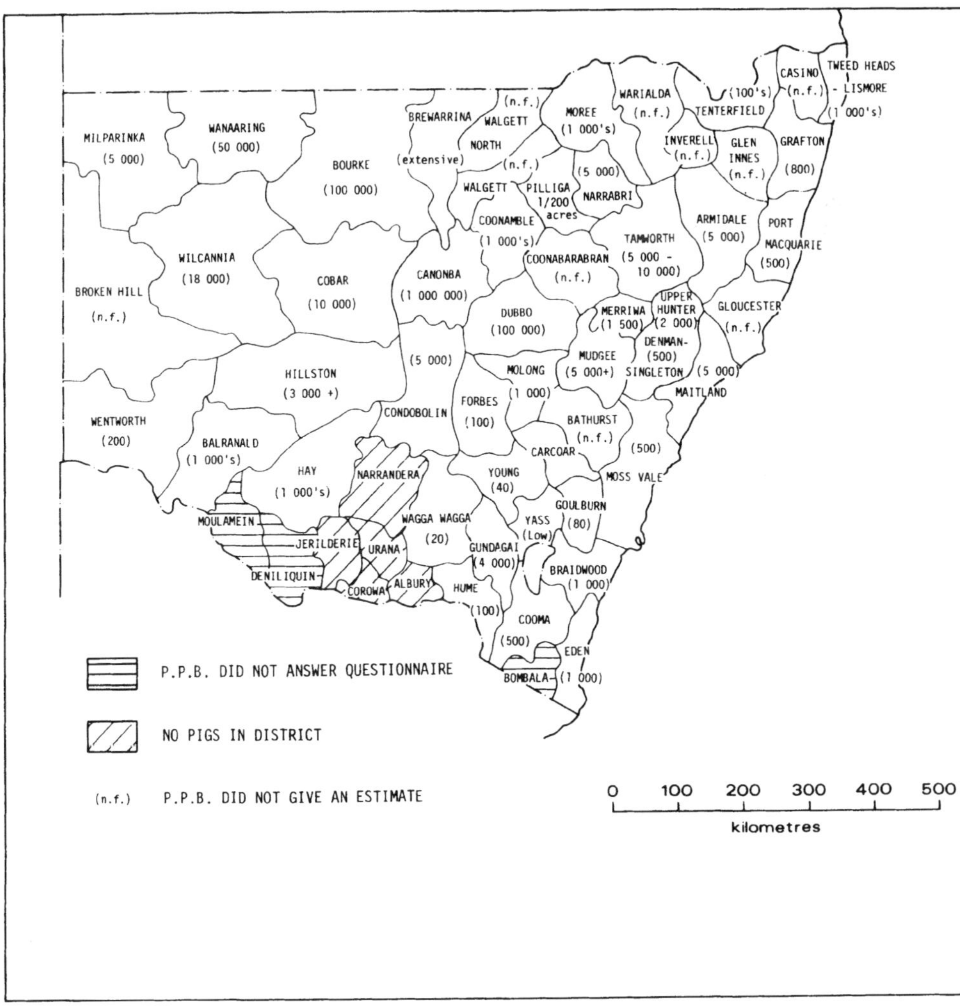

Fig. 2.17 Estimates given by Pastures Protection Boards of the number of wild pigs in their districts, September 1977.

In 1976, the number of domestic pigs in New South Wales was 709 000. Hence the number of wild pigs in New South Wales was more than twice the number of domestic pigs if the estimates by the Boards of numbers of wild pigs are roughly reliable. The reliability of the estimates is difficult to determine - some Boards may exaggerate to stress their problem and others under-estimate because of lack of data. The above estimate is heavily influenced by the rough estimate from Canonba P.P.B. district, where 1 million pigs are claimed to be present. Nevertheless one can be reasonably confident that the wild pig population in New South Wales is likely to be between 1-3 million and that the numbers are therefore considerable. There is, however, a need for a population census to be undertaken by an independent authority. But it seems clear to me that the estimate given by McKnight[59] of 975 000 wild pigs for the whole of Australia is well below present figures.

Boards in districts where pigs are a pest were asked to indicate the type of damage the pigs were causing. The responses are summarized in Table 2.4. The most frequently reported damage was destruction of crops by rooting and eating. This was often associated with trampling and flattening of crops, such as wheat, thus making harvesting more difficult. The crops damaged covered a wide spectrum and ranged from sweet potatoes, sugar cane, potatoes, through to wheat and milo. Uprooting of pastures was also mentioned with great frequency and can also be considered to be a major damage.

The next most frequently mentioned damage was the killing of new-born lambs, lambing ewes and weak sheep. It accounts for 20 per cent of the damages mentioned by all Boards. There is little doubt that the belief is widespread that pigs are killing sheep, although Fenessy believes that reports may be exaggerated.[60] Several Boards indicated lamb losses of 30-40 per cent because of the presence of wild pigs.

Other types of damage mentioned in decreasing frequency are: damage to fencing, fouling of water supplies, soil erosion, damage to gardens and lawns in urban areas, disease-carrying risks, the destruction of habitat in State forests, damage to irrigation layouts and destruction of pines in State forests.

The disease-carrying potential of wild pigs has obtained considerable publicity. They are capable of transmitting a number of exotic diseases including African swine fever, foot-and-mouth disease, rinderpest, Aujezsky's disease, Teschen disease, and will increase the difficulty of controlling these diseases should they be introduced to this country.[61] At present in Australia they help to transmit leptospirosis, tuberculosis[62] and spargonosis.

Another important source of damage is the destruction of native habitat by pigs as a result of rooting and wallowing. This is being studied in the Northern Territory, for instance, by the CSIRO[63] and is taken up in a later chapter.

An examination of detailed comments about damages indicates that loss to

TABLE 2.4 Types of Damage Caused by Wild Pigs as Reported by Pastures Protection Boards in New South Wales

Type of damage	Frequency mentioned		Relative frequency %	
Destruction of crops by rooting & eating	36)	) 65	24.7)	) 44.5
Uprooting and eating of pastures	29)		19.8)	
Killing new-born lambs	18)			
Killing lambing ewes	6)	) 29	29.8	
Killing weak sheep	5)			
Damage to fencing	23		16.0	
Fouling water supplies	12		8.2	
Soil erosion	6		4.1	
Damage to gardens & lawns in urban areas	2		1.4	
Disease-carrying potential	2		1.4	
Destruction of habitat in state forests	2		1.4	
Damage to irrigation layouts	2		1.4	
Orchard damage by rooting	2		1.4	
Destruction of pines in state forest	1		0.7	
TOTAL	146		100.0%*	

*May not add exactly because of rounding

grain crops are especially prevalent. Some of the more interesting detailed P.P.B. comments about damage caused by feral pigs are:

(Balranald) 'Excessive damage to all irrigated crops - sorghum, barley, wheat, etc. - particularly where adjacent to main watercourses. Killing of lambs - at times grown sheep particularly, weak sheep.'

(Walgett and Walgett North) 'They damage bore drains and water holes where stock water. Costly damage to fences. They kill young lambs and cause damage during the lambing season. They cause serious damage to crops. Extreme depredation to lambings and grain crops plus carring diseases such as leptospirosis which can be transmitted to both livestock and humans.'

(Upper Hunter) 'Damage to crops, e.g. potatoes. Damage to fences. Soil erosion. Pasture devastation particularly on the plateau plains where intense cold prevents sufficient growth in a season to replace groundcover. Some predation has been reported.'

(Maitland) 'They are causing damage to crops by eating and destroying what they do not eat. They also destroy pasture and grazing lands by rooting up large areas thus killing the grass. Thousands of acres are ruined annually in this way.'

(Braidwood) 'Destruction of corn (maize) and sorghum crops, grazing lucerne, oats and established improved pastures. Depleting pastures by uprooting. Where sub-clover roots eaten off the seed potential is lost. Creating soil erosion in gully areas. Damaging gardens and lawns in urban areas. Damaging netting fences.'

(Coonamble) 'Damage to crops, pastures, water supplies (bore drains and ground tanks). Numerous lamb losses attributed to feral pigs.'

(Coonabarabran) 'Mainly crops damaged (wheat, oats, barley, lupins in the winter and in the summer sunflower and sorghum). Also damage to pasture by rooting.'

(Denman-Singleton) 'In one area damage is mainly to grain crops in another to grape crops and citrus. Some damage occurs in other areas to pasture by pigs digging up roots and insect larvae.'

(Gundagai) '(1) Destruction of habitat in Kosciusko National Park, (2) Rooting up pasture, crops, e.g. oats. Damaging netting fences in conjunction with kangaroos.'

(Milparinka) 'Digging up natural pastures and bogging water holes. Something we could well do without in this area.'

(Grafton) 'Destruction (1) of pasture by rooting, (2) maize and sugar cane by rooting and eating, (3) bananas by rooting.'

(Armidale) 'Pasture damage. Damage to dingo and rabbit-proof fences. Soil erosion in steep country. Damage to young pines in state forest.'

(Condoblin) 'Water supplies, pastures, fences and some stock losses particularly lambing ewes when sheep are poor and weak.'

Boards were asked if possible to give some estimates of the annual cost of damages to crops, sheep, etc., caused in their district by wild pigs. Most were unable to give an estimate. Only 19 Boards were in a position to give us a rough estimate of annual losses in their districts and, as most emphasized, these estimates are subject to considerable uncertainty. A total annual cost of damages of \$5.5 million was reported in the districts reporting estimates. Many said the loss was considerable but could not put a figure on it. It is reasonable to conclude from these results and answers that losses due to wild pigs in New South Wales probably exceeded \$10 million in 1976-77. But in the light of the earlier discussion, even this may be an underestimate. Allowing for inflation and the rise in the prices of agricultural produce, the annual current economic loss in agriculture from wild pigs in New South Wales is about \$30 million on the basis of estimates given earlier in this chapter. Given the earlier estimated number of wild pigs in New South Wales (1.55 to 2 million), this works out to an *average* annual economic cost to agriculture on the basis of 1976-77 figures of $5 -$6.50 per pig. However, currently the cost of damages appears to be around $15 per pig. It could therefore be worthwhile spending up to this amount annually and on *average* to reduce the pig population for the year by one pig. However, the optimal level of control expenditure is likely to vary from district to district, area to area and from industry to industry, and the average damage per pig will also vary with such factors.

The greatest losses *reported* from feral pig activity were from the Walgett and Walgett North districts where, in response to our request for an estimate of the costs of damages, the Board advised:

> 'This would be impossible to estimate as over 50 per cent of the owners claim that pigs are responsible for large losses. Estimates range from two to three million dollars. The most serious losses occur in lambings and crop damage. Serious damage in lambings would cost approximately $750 000, crop damage would cost $750 000, damage to bore drains, ground tanks and pollution of water would cost approximately $500 000, and damage to fences would cost approximately $250 000 per year.'

The total cost of damages caused by pigs in this area was $2.25 million in 1976-77 and similar losses were probably experienced by districts bordering this area. Losses because of pigs along the coastal strip of New South Wales seem to be small, but large losses are occurring in the area of the North-west Plains and Northern Slopes - that is, in the areas where wild pigs are prevalent (as indicated in Fig. 2.14).

Some of the comments received on the magnitude of the losses help to illuminate the type of problems faced by farmers, as a result of the presence of feral pigs. These comments include:

> (Narrabri) 'Main loss is from cereals (wheat and milo). Main damage to fences would be where pigs break into crops.'
>
> (Cobar) 'Over the past four years lambing figures have generally been down by about 30 per cent. There is fair evidence that pigs are responsible for the poor results.'
>
> (Balranald) 'Cannot estimate economic loss. Two property owners have been forced to pig-mesh fence irrigated areas and install offset electric fencing; other farmers forced to night patrol crops. Chief damage to crops comes from pigs flattening large areas.'
>
> (Wanaaring) 'Approximately 40 per cent of lambings are believed to be lost to wild pigs.'
>
> (Tamworth) 'Destruction in some cases includes the whole of grain crops and

in many cases the whole of potato and turnip crops.'

(Gundagai) 'It appears that a group of pigs caused severe damage to substantial areas of pasture in the area of Kosciusko National Park. I have seen areas of approximately 100 yards diameter (100 square metres) almost ploughed within two weeks of snow melting.'

(Gloucester) 'They are causing a very serious soil erosion problem in the higher country of the Board in non-stable soils.'

(Milparinka) 'Being where we are, there are no crops or sown pastures and with the average paddock being from 15 000 to 20 000 acres in size (6000-8000 hectares), it is very hard to estimate the number of lambs that are eaten by pigs, but we have no doubt that it is considerable.'

(Wilcannia) 'Lamb losses of about 2000 annually and about 800 acres (320 hectares) of grain sorghum and sunflower lost annually.'

(Bourke) 'Probably 40 per cent loss of new born lambs to pigs.'

Control of Wild Pigs by Pastures Protection Boards

In just over 80 per cent of the P.P.B. districts where pigs are reported to be a pest, Boards were taking control measures of some type. But in almost 20 per cent of the P.P.B. districts where pigs are a pest, Boards report that they were not involved in control measures. These districts are mainly in the Western Division where properties are large and in many cases only marginally commercial. (The Boards taking pig control action in 1977 are shown in Fig. 2.16).

Table 2.5 sets out the reported types of assistance given by the Boards in the control of wild pigs. In more than 80 per cent of the P.P.B. districts where pigs are a pest, Boards advise and assist landholders with poisoning and trapping. They may provide some labour and poison baits for landholders. Some Boards merely provide advice on control measures. In many cases Boards also assist by helping the formation of groups of landholders to co-ordinate their action against pigs. This can clearly be useful, given that wild pigs move from one property to another. Other forms of assistance in decreased order of frequency mentioned are: property inspections for pests, sending field officers for training at pig schools, arranging the supply of ammunition exempt of sales tax or at cost, the free supply of poisons to landholders to enable them to treat their own grain pellets, survey and monitoring of pigs, assistance to landholders with shooting and the issuance of destruction orders. In many districts control measures are indirectly subsidized by Boards by providing services for landholders destroying wild pigs. In a few cases poisons are supplied free and ammunition is supplied at cost.

All Boards assisting in the control of wild pigs finance their involvement

TABLE 2.5 Assistance Provided by Pastures Protection Boards for the Control of Wild Pigs

Type of involvement	Frequency mentioned		Relative frequency %
Advice and assistance (e.g. provision of labour) with poisoning and trapping	20)	37	41.0
Poisoning baits for landholders	17)		
Advice (only) on control measures	17		19.0
Formation of groups of landholders to control pigs collectively	14		15.4
Property inspections for pests	4		4.4
Sending field officers to schools dealing with pig control*	4		4.4
Arranging the supply of ammunition *exempt* of sales tax	3)	6	6.6
Supply of ammunition at cost	3)		
Supply of poisons to landholders free of charge to enable them to treat their own grain pellets	3		3.3
Survey and monitoring of pigs	3		3.3
Shooting assistance	2		2.2
Issuing of destruction orders	1		1.1
TOTAL	91		100.0%**

* Frequency is greater than reported

**May not add exactly because of rounding

from their general funds. These are principally derived from rates or levies on livestock. This income may be supplemented by small amounts from rabbit destruction, registration of earmarks and brands, stock permits and travelling stock charges. One Board reported that the C.S.I.R.O., the Forestry Department and the National Parks and Wildlife Service had given them some financial assistance. Livestock levies vary considerably between P.P.B. districts and seems to be no clearly defined relationship between the amounts the individual landholders pay in levies to Boards and the benefits they receive from the activity of their Board. Payments do not accurately reflect ability to pay nor benefits received from Board activities, only some of which involve the control of pests. In particular, there is no clearly defined relationship between the assistance individual landholders receive with wild pig control and their contribution to the funds of their Board. It could be, however, that costs and benefits to individual landholders do balance out when the whole range of a Board's activities is taken into account.

Efforts by Farmers and Graziers to Control Feral Pigs

In more than three-quarters of the P.P.B. districts where pigs exist, Boards reported that farmers and graziers were not taking adequate control measures in 1977. In elaborating on this, several Boards claimed that the financial situation of many farmers and graziers was such that they could afford to pay for control, that no effective economic method of control was available and that use of 1080 poison was subject to too much restriction. Some Boards claimed that control was inadequate because of the presence of pig refuges on crown land, state forests, national parks and in swamps and other areas with limited accessibility. In some districts wild pig numbers had built up only recently and farmers therefore had little experience with the problem. In some areas where the population of pigs was not dense, farmers did not realise the potential problem and wished to retain pigs for their sporting value.

In most districts (over three-quarters), efforts to control pigs vary over time. Effort appears to be greatest (as a rule) when most immediate damage is likely to occur - that is, at lambing time and when crops are growing. It is also influenced by seasonal and weather patterns in some districts. In some districts, pigs move out of rough country towards the end of winter to obtain better feed, but the time of year of most control activity varies between P.P.B. districts. Control effort seems to be closely geared to the possibility or occurrence of immediate damage. Given the degree of mobility of pigs, this is likely to be a rational response from the point of view of the individual landholder. A farmer may be able to control pig numbers for a short time but numbers will tend to be replenished over a longer period and will build up unless control effort is maintained. In times when damage by pigs is of little economic consequence, it is unlikely to pay a landholder to spend heavily on the control of pigs; but it may pay the landholder to do so when the opposite is the case.

Most P.P. Boards where pigs are pests reported that local effort to control pigs eases off when economic conditions are bad for farmers. This is consistent with the behaviour of any profit-maximizing firm that has to deal with pests. One model of pest control which I have considered[64] suggested that where pigs are pests there is likely to be less profit incentive (other things equal) for landholders to destroy pigs on the margin of large pig refuges or harbours, since the numbers destroyed by landholders are likely to be rapidly replaced. Boards were asked in the survey whether greatest effort to control pigs in their district is made in areas of greatest pig infestation, and they were asked to give some particulars about this. The majority answering this question said that greatest effort was made in areas of heaviest infestation but a substantial minority said that this was not so. The results were not conclusive from the point of view of the above hypothesis. For instance, factors varied. In some instances, the most profitable commercial opportunities for using the land were in areas of heaviest pig infestation; in other instances, the least profitable commercial opportunities were in these areas. In the former case, one would expect a greater incentive for control than in the latter. Furthermore, theory suggests that control will be absent if pig numbers are small and do not exceed some threshold.[65] In most areas where pig numbers are small, there is no attempt by most landholders to control pigs. This is rational from their private point of view.

Some of the Boards indicating that most effort to control pigs was not made in the most heavily infested areas said that this was so because heaviest infestations are located in forests, crown land, national parks and relatively inaccessible land such as swamps. Several Boards were critical of the lack of control of pigs by the government on lands owned by it.

Some of the answers to the question 'Are greatest efforts to control pigs in your District made in areas of greatest pig infestation?' are worth considering. The following help to illustrate the diversity of influences on pig control:

> (Armidale) 'No. Heaviest areas of infestation are in the southern areas of the district where properties are large and terrain is difficult. Other pockets of pigs are in areas of low land value.'
>
> (Braidwood) 'Yes. The areas of greatest infestation are generally those of favourable habitat and breeding grounds adjacent to most attractive crops and/or pastures. Light infestations are generally nomadic and in closer settled areas roam over many properties, hence damage of lesser significance and no landholder effort.'
>
> (Cooma) 'No. The greatest pig infestations are found in the Kosciusko National Park where virtually no control is carried out.'
>
> (Coonabarabran) 'Yes. This Board goes into areas showing greatest infestation and organizes poisoning campaigns on a group basis covering 50 000 to 80 000 acres (20 000-32 000 hectares) at a time - that is, from 20 to 40 properties depending on their size.'
>
> (Coonamble) 'Both. Continuous eradication is carried out by landholders adjoining the Macquarie Marshes and state forests (areas of heavy infestation), and in other areas where intensive farming is being carried out.'
>
> (Gundagai) 'Both. Effort usually proportional to damage on individual properties, no control programme in forests but some shooting and trapping

by rangers in the National Park.'

(Hay) 'No. Where numbers are thickest landholders often use shooters to kill them as a means of income.'

(Inverell) 'Yes. This answer only stands to reason. Very few landowners work on noxious animals unless they become heavily infested or are forced to to so.'

(Maitland) 'Yes. Problem areas are always handled (by Board) first.'

(Molong) 'Yes. Pigs occur mainly in four areas of the Board where they move out of crown and forest lands.'

(Mudgee) 'No. Mountain country mainly affected.'

(Tamworth) 'No. The greatest effort is where pigs are doing the greatest damage and there are some areas where very little damage is done but there is a heavy pig population.'

(Tenterfield) 'Yes. The greatest pig infestation is where pastures and crops are grown.'

(Walgett) 'No. The largest pig infestations are in watercourses and low-lying country which is subject to flooding, and crops are not grown in these areas; but cattle are usually run on these areas and if sheep are run they are moved out before lambing commences. So the landholder mostly sees poisoning (of pigs) as an extra expense as he considers that pigs are not costing him much in damage.'

(Warialda) 'Yes. Pigs are in greatest number when fodder (such as cereal in head) is readily available.'

(Wilcannia) 'Yes. Along the heavily infested areas such as the Paroo and Darling River systems.'

(Yass) 'No. Most control work takes place in areas adjoining the national parks.'

On the whole these responses are consistent with control being heavily influenced by the economic gain from it, as far as the landholder is concerned. This is particularly so for the responses from Armidale, Braidwood, Coonamble, Gundagai, Tamworth, Tenterfield, Walgett and Warialda, for example. Efforts to control wild pigs seem to be related to the economic benefits and cost of controls experienced by landowners. Just because the number or density of pigs in an area is high, it does not follow that it is economic to control them. The gains from control (value of produce saved, etc.) may be low and the costs high, for example, if the area of infestation is remote and rugged. The economics of control will be looked at in detail in Chapter 9.

Role of Amateur Shooters in Controlling Pigs

Boards were asked: 'Do you think that amateur shooters, say through shooting clubs, should play a bigger role in controlling pigs?' Just over two-thirds of the Boards said no and the remainder said yes. Most Boards felt that amateur shooters should *not* play a bigger role. Those who opposed an expanded role for amateur shooters mainly felt that such shooting dispersed pigs and made control more difficult - for example, it may make pigs shy of bait. Others said that shooters were a nuisance *only* (implying amongst other things that they accounted for the removal of few pigs), that they accounted for few pigs, or were only of use when pig numbers were small. It was also pointed out, but not very often, that indiscriminate shooters caused stock losses. One Board noted that shooters were ineffective in crown or forest land because of the nature of the terrain and its size. The responses are summarized in Table 2.6. Those favouring greater use of amateur shooters qualified their answers. Most emphasized that amateur shooters should be controlled and organized, and some suggested that they could be usefully employed in mopping-up operations or when pig numbers were small.

TABLE 2.6 Summary of the Comments by Pastures Protection Boards About the Role of Amateur Shooters in Controlling Pigs

Value of shooters	Frequency mentioned		Relative frequency %
Boards Opposed to Greater Role for Shooters			
Disperse pigs making control difficult	11		42.0
Nuisance only	7)		
Account for few pigs	2)	11	42.0
Only useful if pigs are few	2)		
Cause stock losses	3		12.0
Ineffective in forest and crown land	1		4.0
TOTAL	26		100.0%
Boards Favouring a Greater Role for Shooters			
Of assistance if organized	13		81.25
Useful in mopping-up after poisoning	2		12.5
Useful when pig numbers are small	1		6.25
TOTAL	16		100.00%

The general view of the Boards appears to be that amateur shooting has little control value unless pig numbers are small and that it frequently hampers control. If shooting does occur it should be organized and controlled. One Board suggested that amateur shooting clubs should make it known that, subject to the approval of landholders, their members are available for pig shooting.

Payment of Bounties on Wild Pigs Destroyed

Only one Board (Pilliga) indicated that it was paying a bounty on wild pig scalps or snouts in 1977, though a number of Boards have paid bounties in the past. Most Boards (more than 80 per cent) did not favour the payment of bounties on scalps as a means of controlling the number of wild pigs. They opposed the payment of bounties on the grounds that it would be too costly, not effective in reducing the number of pigs and open to abuse (for example, a bounty might be collected upon the death of a domestic pig). Some also said that bounties have not worked in the past. They failed to provide an incentive to reduce feral pig populations. On the other hand, the Boards favouring bounties felt that these would provide an incentive for a reduction in the number of wild pigs. The reasons given for favouring or opposing the bounty are set out in Table 2.7. Incidentally the respondent from Pilliga was opposed to continued payment of the bounty by his Board. One problem apart from misrepresentation of items as wild pig scalps or snouts is that these may be brought from outside the district so that pig control elsewhere is subsidized.[66] Moreover, a subsidy of this nature favours shooting and trapping in comparison to poisoning, since poisoned pigs are rarely recovered so making it difficult to obtain their snouts or scalps.

Special Problems and Views of Pastures Protection Boards on the Economics of Controlling Wild Pigs

Most Boards mentioned that there were special problems in controlling wild pigs in their district. The most frequently mentioned special difficulty was the presence of pigs in inaccessible areas such as swamp and in lignum or dense undergrowth or heavily timbered areas. The next most frequently mentioned special problem was the existence of state forests, national parks and crown lands in P.P.B. districts. Control is absent or inadequate in these areas, and they are a source of infestation of local properties. This was mentioned in 12 cases. Other special difficulties mentioned, in decreasing frequency, are: the widespread existence of suitable habitats for pigs in the district, lack of suitable poison or bait for pigs, the existence of large properties with a sparse human population,

TABLE 2.7 Reasons Given by Pastures Protection Boards for Favouring or Opposing Payment of Bounties on Wild Pig Scalps

Reasons	Frequency mentioned	Relative frequency %
Boards in Favour of Bounties		
Way of keeping pig numbers down	2	33.3
Incentive to landholders to eradicate feral pigs from their land and crown land	3	50.0
Encourage stock owners to destroy pigs where these are in large numbers	1	16.6
TOTAL	6	100.0%*
Boards Opposed to Bounties		
Too costly (e.g. means higher rates)	10	23.8
System open to abuse	9	21.4
Bounties have not worked in past	6	14.3
Bounty would encourage landholders to harvest and manage wild pigs	6	14.3
Landholders *should* be responsible for destruction on their property and no bounties should be paid	4	9.5
Invites shooting which is an ineffective control measure	3	7.1
Shooters cause nearly as much damage as pigs	1	2.4
Entice shooters who frighten pigs and so make poisoning more difficult	1	2.4
Control should be the responsibility of groups and bodies, not individuals	1	2.4
Ineffective because pigs too scattered and inaccessible in the district	1	2.4
TOTAL	42	100.0%*

*May not add exactly because of rounding

shortages of manpower or resources for control purposes, the presence of a small number of pigs widely scattered, and the poor state of fencing (as a result of the economic circumstances of farmers) which restricts the use of poison because it is hard to keep sheep away from it. The comments received are summarized in Table 2.8.

Some of the specific answers to the question 'Do you believe that there are special problems in controlling wild pigs in your District?' are worthwhile considering to provide a deeper perspective. The following were some of the answers

TABLE 2.8 Special Problems mentioned by Pastures Protection Boards in Controlling Wild Pigs in P.P.B. Districts

Problem	Frequency mentioned		Relative frequency %
Inaccessible swamp and other areas of infestation	15)	27	50.9
Dense undergrowth, lignum, heavily timbered	12)		
Infestation from state forest, crown land or national parks where pig control inadequate	15		21.4
Widespread existence of suitable habitat for pigs	5		8.9
Lack of suitable poison or bait	4		7.5
Presence of large properties and sparse human population	3		5.7
Shortage of labour or resources to control pigs	2		3.6
Small number of pigs but scattered	2		3.6
Fences unrepaired due to economic circumstances and inadequate to keep sheep from poison	1		1.8
TOTAL	56		100.0%*

*May not add exactly due to rounding

from different Boards:

'Yes. To the degree that the average area of properties in this district would be about 50 thousand acres and the population quite scarce over an area of 10 million acres plus.'

'Yes. Lack of suitable poison. When SAP (poison) was available control was far better than it is today.'

'Yes. Conditions for the use of "1080" are necessarily restrictive. Paddocks have to be emptied of stock. This is not always convenient.'

'Yes. The main problem is the type of country that pigs have got into which comprises large swamps surrounded by thick scrub.'

'Yes. Many landholders in poor economic circumstances may have problems because of the poor state of fencing - difficulty in keeping sheep away from poison.'

'Yes. The terrain is not easily crossed (except the plains which are free of pigs). The areas with pig populations are heavily timbered. The Kosciusko National Park is a special problem.'

'Yes. Mainly because of the large areas of forest and national parks within

Fig. 2.18 An amateur shooter with a feral pig shot near Warren, N.S.W. Most Pastures Protection Boards feel that amateur shooting has little control value unless pig numbers are low.

this Board.'

'Yes. Because of protected areas such as Macquarie Marshes and state forests which form natural habitat for pigs. Only a small percentage of pigs are killed when they leave these areas; the majority use these areas for breeding and protection.'

'Yes. The numbers are small but scattered.'

'Yes. Much of the pig country is flood country and can only be gotten to on foot.'

'Yes. The apparent reluctance of the authority in question in not taking action to destroy noxious animals on crown land, forestry reserves, and national parks and wildlife areas.'

'Yes. The vast areas (average holding 130 000 acres approx.) and so little manpower available would make the task practically impossible unless a very effective and reasonably cheap poisoning programme can be devised.'

'Yes. The main infestation areas are rough and heavily timbered to swampy with little access and these are the normal pig habitats.'

Given the inaccessibility of areas of pig infestation in some districts, it will be very difficult to exterminate wild pigs in New South Wales. Even the dropping of bait from the air does not seem a possible solution since some pigs are shy of bait and baits are not specific to pigs (they can affect native wildlife, for example). Any policy of extermination would also need to be simultaneously carried out at least in Queensland.

Boards were encouraged to make any general comments or observations about the *economics of the control of wild pigs*. Only a few Boards did comment and it is difficult to tabulate their answers. The following comments were made by different Boards:

> 'All funds of the Board come from ratepayers. There is no subsidy for Boards from the Government. Most Boards need additional inspectors for wild pig control. The economic situation of Boards cannot withstand additional appointments. There is an upper limit to the amount of rates that can be charged against landholders - particularly in the present economic climate in the pastoral industry.'

> 'In view of the fact that some Boards supply 1080 poison for pig eradication free of cost to landholders and it may be presumed that all Boards could do so, it is considered reasonable to expect that the full cost of free feed and bait material should be borne by the Department of Agriculture...'

> 'A firm definition of a wild pig is necessary in the first instance and then a proclamation declaring it a noxious as such. Property occupiers of Western Land Leases who have little equity over the land would require investigation for additional Government assistance if the wild pig is to be brought under control in a reasonable time. Many of these leases carry a high infestation which would require heavy costs for eradication.'

> 'Pigs are in small numbers in this district but increasing. Therefore, the sooner that something is done in areas like this district the cheaper it will be.'

> 'Wild pigs are far more difficult to control than rabbits as they have no visible point of abode. An order can be issued on a landholder to remove rabbits as rabbit burrows are concrete evidence that an infestation exists. If wild pigs are seen on a landowner's property he most nearly always answers truthfully that their real home is on the adjoining crown land and they are not his responsibility to eradicate. Hence a prosecution against a landowner on whose property wild pigs are seen would seldom lead to a conviction.'

> 'If the district ever had an outbreak of foot-and-mouth disease pigs could readily spread it to (adjoining districts) ... Restrictions on the use of 1080 hamper economic control. Much more work on satisfactory poisons must be

done to allow most economic control.'

'I am sure that the pig problem would be considerably greater only for the destruction of wild pigs by shooters.'

'Feral pigs could play a role in the spread of exotic diseases ... The pigs living in close proximity to Newcastle harbour are of particular concern as they have been observed foraging on the very harbour shores, possibly making Newcastle harbour the first place where some of the diseases could be introduced.'

'The greatest threat is that wild pigs will carry foot-and-mouth disease if it is introduced into the country.'

'The Board is convinced that the most economical way of controlling wild pigs is by landholders forming themselves into destruction groups and with the support, help and advice of P.P.B. officers to carry out a "1080" poisoning campaign. This is quite within the economic resources of many areas of infestation. The Board considers a number of steps to be essential: (a) Restriction should be placed on the transport of wild pigs (by individuals) to stop them being brought into free areas. (b) There should be a definition of wild or feral pig. While it may be impossible to do this biologically it should be possible to do it by defining conditions under which domestic pigs should be kept and it would follow that any pig not kept under those conditions would be considered to be wild or feral. (c) All piggeries should be registered.'

'A dog problem is associated with feral pigs. Amateur shooters abandon pig dogs which kill sheep *in order to survive*. Consequently the recognised sheep area in this district is contracting westward.'

'Landholders and also governments must be made more aware of the economic losses now caused by pigs and the disaster that awaits the livestock industry when foot-and-mouth disease finally enters Australia. Unfortunately not all landholders will operate or organize on their own initiative and there is a need for direction and guidance from governments through Departments of Agriculture and local control through Pastures Protection Boards. There also needs to be co-operation with other states, e.g. Queensland, as pigs are no respecters of border fences.'

Government Policies Desired by Pastures Protection Boards

In commenting on the economics of control of feral pigs, some Boards made suggestions for government policy, as indicated above. But Boards were also specifically asked 'How in your view should the government assist with the control of wild pigs?'

Most Boards gave their views about ways in which the government should assist with the control of wild pigs and these are summarized and tabulated in Table 2.9. The most frequently mentioned means of assistance were: the issue of permits for the transport of pigs, provide more research into control methods, provide funding to enable eradication of pigs to be achieved, establish a Wild Pig Destruction Authority or Board, register of all domestic pigs and piggeries, supply bait or poison free of charge, spot poison with '1080' or anti-coagulant baits when feedings are localized, encourage awareness of control measures by holding more field days for landholders, use armed forces to control, ensure that adequate control measures are taken in national parks, state forests and other crown lands, give more publicity about the danger of wild pigs to public health, legislate against open-ranging of pigs, and allow the re-introduction of Sayer-Allport phosphorus, a proprietary poison that has been used to kill pigs.

Here are some of the comments from Boards about ways in which the government should assist in the control of wild pigs. They highlight a number of the difficulties in controlling this pest:

'It would be a costly exercise, but the only way I can see would be for the Government to finance the registration of piggeries and so control pig numbers and movement through the permit system for pig movement. A large factor in the numbers of wild pigs in this area has been caused by domestic pigs being let loose when pig prices drop, i.e. by domestic pigs being allowed to breed with true feral pigs.'

'The Department of Agriculture has done considerable research into control of pigs and I feel it is now up to the landholder under the guidance of P.P. Boards to put the research information into effect.'

'By enforcing the Act more rigidly and causing such amendments to the Act to make the legislation work.'

'Pigs should be in the same class as rabbits, and the government could appoint pig officers to contact owners, arrange pig poisoning programmes through the state ... Governments should allow, under adequate supervision the greater use of meat baits using 1080; encourage the development of a bait specifically attractive to pigs and not other livestock and native fauna.'

'Governments could possibly assist by making grants to Pastures Protection Boards instead of taking money from them at present.'

'Eradicate pigs from all crown land including forests and national parks. Full approval by government for use of 1080 on all land including forests and national parks and wildlife land.'

'Organize intensive control measures in national parks, state forests and other crown lands.'

TABLE 2.9 Pastures Protection Boards' Views about the Ways in which Government Should Assist with the Control of Wild Pigs

Form of government assistance	Frequency mentioned	Relative frequency %
Permit should be required for transport of pigs	11	15.5
Provide more research into control methods	6	8.4
Provide funding for *eradication* not simply control	5	7.0
Establish a Wild Pig Destruction Authority or Board	5	7.0
Registration of all domestic pigs and piggeries	5	7.0
Supply bait or poison free of charge	5	7.0
Spot poison with "1080" or anticoagulant baits where pigs are localized	4	5.6
Hold field days to inform landholders of control measures	3	4.2
Ensure adequate control in national parks, state forests and crown land	3	4.2
Use armed forces to eradicate or control	3	4.2
Publicize the health danger from wild pigs	2	2.8
Pass proposed amendments to P.P. Act	2	2.8
Legislate against open-ranging of pigs	2	2.8
Allow the re-introduction of SAP	2	2.8
Provide more positive training for P.P.B. officers	1	1.4
Give P.P.B.s more power and backing	1	1.4
Supply helicopters to drop bait during flood time	1	1.4
Rebate on rates to graziers undertaking pig control or eradication	1	1.4
Income tax deduction at a greater rate than normal for landholders undertaking control	1	1.4
Subsidize local and quasi-government instrumentalities involved in pest control on a $ for $ basis	1	1.4
Undertake a national eradication campaign, costs being shared 25% by Boards, 25% by State government and 50% by Federal government	1	1.4
Supply materials and ammunition to landholders at cost or less	1	1.4
Legislate against shooting	1	1.4
Define the difference between wild and domestic pigs	1	1.4
Organize hunting and trapping	1	1.4
Employ specialist gangs for eradication	1	1.4
Provide professional full-time trappers	1	1.4
Tattoo pig's ears	1	1.4
TOTAL	72	100.0%*

*May not add exactly because of rounding

'Clean up areas under government control by regular poisoning.'

'More money should be spent on a determined research programme to cover eastern coastal areas in addition to the inland ... Open ranging of pigs should be made illegal. Many pigs in western areas are allowed to range at will and fed once or twice daily. Some of these pigs eventually go wild and build up the feral numbers.'

'By the introduction of sound legislation defining a feral pig, regulations on movements of pigs and prohibition in the carrying, selling and butchering of feral pigs.'

'Financial support such supplying of bait material. Also restrictions on the movement of wild pigs.'

'More research into control methods. Particularly, in coastal and tableland areas. Most of the research already undertaken is in western areas of the state in watercourse country. These do not necessarily apply to the tablelands.'

'Possibly grant some sort of rebate of rates or portion of rates to those graziers who carry out annual pig control or eradication programmes.'

'By providing professional full-time trappers such as used by the Dingo Protection Board.'

Conclusions from the Survey

The survey indicates that wild pigs are causing substantial economic damage in New South Wales, even though the cost of these damages needs to be assessed more accurately. The nature of the problem is different in different sections of the state. In a small portion of the state wild pigs are not present or are not pests. Population densities of pigs vary greatly in P.P.B. districts where they are present, heaviest infestation being in the north west of the state. Also the extent to which wild pig populations within P.P.B. districts are spread or clustered in groups varies and this has implications for the difficulty of control or eradication of wild pigs.

Although numbers of wild pigs on the coast are generally less than in the west of the state, numbers may be rising, and these areas have a high probability of being first points of contact for the introduction of exotic diseases which can be transmitted by pigs. This adds to the need to consider collective control measures in the coastal area, even though numbers of wild pigs may be below the threshold of economic damage as far as most landholders are concerned. A number of Boards along the coast pointed out that comparatively little research has been done on the pig problem (studies of ecology, etc.) as it affects eastern areas of the state.

Given the widespread nature of damages caused and the potential of wild pigs for spreading disease, there is the difficult question to be decided of whom should pay for control or eradication measures. This is complicated by the presence of externalities and the fact that most landholders on properties where pigs are pests were not responsible for the release of domestic pigs. Collective or group action against wild pigs appears to be a useful means of control given their mobility. It may also be that a government subsidy for the control of wild pigs or their eradication is justified. The question of whether it is more economical to control or to eradicate wild pigs needs attention. However, while domestic pigs continue to be kept, it is most likely that some will escape and form feral colonies even if registration of piggeries is necessary and other controls are introduced. Furthermore, it remains a moot point as to whether wild pigs can be eliminated without extraordinary cost given their inaccessible habitat in some areas and the limited methods of control available. Biological control would appear to be ruled out because of the presence of domestic pigs and pigs will not always take bait. Whether or not pigs should be eliminated in selected areas and these areas be protected by barriers (for example, electric fencing) may be worth investigation.

A number of the suggestions for control put forward by the Boards are worthy of consideration. The damage caused by wild pigs roaming from state forests, crown lands and national parks because of inadequate control of pigs in these areas is a particular bone of contention amongst Boards. The suggestion that a Wild Pig Destruction Board be established to work in conjunction with Pastures Protection Boards has something to recommend it. But the wild pig problem in New South Wales is not likely to be overcome quickly and easily, even though the size of the damages and the risks associated with the pest provides an incentive.

Development in Government Assistance since the Survey

The main development in government assistance since the survey has been completed has been the implementation by the New South Wales State government of a pilot scheme to control feral pigs. It allocated approximately $100 000 per year for a three-year period from July 1978 to assist 11 Pastures Protection Boards in the north-west of the state. The aim of the scheme is to stimulate greater control of feral pigs and obtain more data about the economics of control. The scheme operates through Pasture Protection Boards and funds are allocated 'primarily for the employment of the officers, whose duty is to stimulate interest and co-ordinate recommended control measures with groups of landholders. The Pastures Protection Boards have to provide cars and provide for any other on-the-job expenses.'[67]

Indeed the grant is hard stretched to cover even salaries.

In some areas, also, Wild Pig Advisory Councils have been voluntarily formed on a regional basis. These consist of representatives of adjoining Pastures Protection Boards, forest and national parks and wildlife representatives and meet about every two months to exchange information.

DIFFICULTIES OF CONTROLLING WILD PIGS IN AGRICULTURE

The nature and extent of agricultural and pastoral damage caused by wild pigs has been outlined above. It is worthwhile now summarizing the main difficulties of controlling them, even though this is discussed in depth in Chapter 9.

Economics of Control Generally

From a social economic point of view, the optimal level of control of a pest is that which maximizes net social gain from control - that is, the social benefits from control less the costs of control. While this rule sometimes implies that it is socially optimal to eradicate a pest species, more often it implies that the level of population of the species should be reduced to the level where the marginal social cost of extra control equals the marginal social gain from it. Although it is unlikely to be economic to eradicate feral pigs in Australia, control of their population is likely to be economic. The most economic level of control of wild pigs from each landholder's own point of view is that level maximizing each landholder's own private gains. But individuals' private gains from their controls of wild pigs usually differ from the *social* gains from their controls. Consequently if individual landholders are left to their own devices, control of wild pigs is not likely to be socially optimal. For example, spillovers as a result of the mobility of pigs may cause a landholder's private gain from his or her control to be less than the social gain, because social gain consists of a benefit to the landholder and to the landholder's *neighbours* of control. This mobility spillover (and others to be discussed below) cause private and social gains from control to diverge, and private control is not sufficient to ensure optimal control of wild pigs from a social point of view.

Generally it is very difficult to specify social gains when these diverge from private gains, and specifying the social gains from the control of wild pigs is not easy. This is especially so in Australia where feral pigs are an introduced species and threaten some native wildlife and natural environments. Assessment is further complicated by conflicting claims upon the wild pig pest/resource. For example, the demands of hunters may not parallel those of farmers and conservationists.

Sometimes policy-makers ignore the economics of controlling a pest species and press for control, greater control or even eradication of a species merely because the species is causing damage. But it is irrational from society's point of view to ignore the economics of control because resources used in pest control have valuable alternative uses. One needs to be sure that the value received from using resources in pest control is not less than the value of these resources when used in the next best alternative way from society's point of view. The rule also applies to marginal or incremental control of a pest. Control should only proceed to the point where the value of the best alternative use of the *extra* resources employed in control is just equal to the value obtained from controlling the pest.

Properly considered economics takes into account what human beings *want* from the use of their resources and is just as much concerned about desires for marketed commodities as for wanted non-marketed (or non-commercial) commodities. Thus the desire to preserve wildlife and natural environments (largely not marketed) or a hunting resource are just as much a part of the economic problem as the desire of the community for marketed commodities such as wheat, wool and meat. Economics can *assist* society to choose:

(a) between alternative pest control methods and procedures;

(b) how and when to institute controls; and

(c) the optimal intensity of control.

For example, economics can help in deciding whether it is worthwhile trying to eradicate feral pigs, exclude them from some areas or reduce their population and to what extent. Or, again, should poisoning of wild pigs be encouraged, what available poison should be used, how economic are shooting or habitat removal (for instance, reduction in tree or shrub cover) as measures to control wild pigs? But if economists are to assist, basic technical information must be provided by ecologists, biologists and other scientists. Decisions about pest control require a co-operative multi-disciplinary approach.

Spillovers from Wild Pig Control

Control of wild pigs results in at least four types of spillovers:

(a) Spillovers in benefits because of the mobility of pigs.

(b) Externalities or side-effects arising from actual control techniques. For instance, poisons designed to kill pigs may incidentally destroy wildlife.

(c) The ability of pigs to carry domestic livestock diseases may mean that even farmers with few or no pigs on their property may indirectly benefit from control of wild pigs elsewhere.

(d) Transport and release of wild pigs by humans, for instance, to build up hunting stock in a new locality, and the release or loss of domestic pigs from piggeries have spillover effects on farmers and other landholders damaged by the presence of wild pigs.

Wild or feral pigs are very mobile. They have been observed to travel up to 12 kilometres daily to suitable food and to visit several properties.[68] Longer migratory movements occur in some areas with the seasons, and pigs may move or spread permanently to entirely new areas, even though female groups appear to have temporary home territories. These movements imply that no landholder is able to appropriate all the benefits from controlling pigs that happen to visit or be on his or her property. Other landholders will benefit to some extent if pig numbers are reduced on one property but no landholder will take this external benefit into account. Hence, collective effort at least by adjoining landholders is needed. Farmers' complaints about pigs roaming from national parks, forests and crown lands and causing damage on farms illustrate the problem. It is further reinforced by the fact that foresters and park rangers complain that farmers permit feral pigs to stray into forests and parks.

Different techniques for controlling wild pigs have various side-effects or spillovers. These need to be taken into account from a social point of view and the technique maximizing net *social* gain needs to be selected. Such a technique might, it should be pointed out, still have undesirable side-effects. For example, poisons (such as 1080, sodium monofluoroacetate) when used for feral pig control can kill native animals (and, if one is not careful, domestic stock too), the impact varying with the type of poison and the bait used as a carrier of the poison (for example, grain or meat), measures to exclude wild pigs from water may destroy native animals, removal of tree or shrub cover to reduce pig numbers may cause environmental damage and erosion, and even some pig traps and exclusion techniques can have an adverse impact on other animals. These side-effects cannot be ignored from a social point of view.

There is an even more widespread spillover to consider. Feral pigs, like some other animals, are carriers of domestic livestock diseases. In Australia, feral pigs could become a major carrier of exotic diseases if they happened to find their way here. The outbreak and spread of such diseases - for instance, foot-and-mouth disease - has consequences for farmers and graziers not only in areas where feral pigs are carriers but also in other areas where domestic livestock can be indirectly affected. Furthermore, in Australia's case an outbreak and continued presence of a disease such as foot-and-mouth disease would reduce all meat exports affecting *all* meat producers because, for example, the United States will not import meat from countries having foot-and-mouth disease and similar diseases.

Pig releases by design or human neglect are not uncommon. Hunters sometimes deliberately introduce feral (or wild) pigs to new areas to establish hunting stocks without heeding the side effects on nearby farmers, wildlife and so on. New pig stock is also sometimes released to improve or upgrade existing feral stock. Again, some keepers of domestic pigs allow their pigs to stray and have in the past uneconomic times allowed their remaining domestic pigs to go wild. As a consequence new feral pig colonies are established and existing stock replenished and damage to other farmers is perpetuated.

These spillovers increase the difficulty from a social point of view of controlling wild pigs and cause social conflict. But there are also other difficulties and conflicts involved in control.

Other Barriers to Wild Pig Control and Sources of Social Conflict

The economics of controlling wild pigs is affected by their high fecundity, adaptability and tenacity as a species. The economic level of control is lowered by these characteristics because a pig population is liable to recover quickly from control measures. Generally, the more quickly a pest population recovers from control measures the less economic is it to control the species. A rate of increase of 60 per cent (per year) for feral pigs in Australia is quite realistic.[69] This rate is somewhat in excess of that for the European wild boar since feral pigs breed continuously in Australia, whereas they are usually limited to a single season in Europe.[70] Because the wild pig is omnivorous in its habits, because it appears to occupy a position in Australia duplicated by no native animal and has no predators of consequence here and is present in remote and inaccessible places, it is a very resilient species in Australia. The fact that it is suspicious of man and confines much of its activity to the night and early morning also restricts the hunting of it. The chances of eliminating feral pigs from Australia using currently known techniques appear to be negligible. While the pig needs an environment with shade and water, there is no practical way of denying this everywhere in the vast area of Australia.

Social conflict is a common problem in wild pig management. Conflicts arise between different landholders, for example, because some find that wild pigs do not damage their industry whereas others do find that their industry is damaged. Lack of control of wild pigs by the former causes bitterness on the part of the latter. Conservationists and landholders may also come into conflict over control methods - for example, pig control methods that poison wildlife. Similarly, pig-hunters (amateur and commercial) can have conflicting interests to landholders because wild pig stocks are of value to hunters as game. Hunters may oppose the use of poisons for wild pig destruction and any proposed elimination of the species But up to a point, the interests of landholders, pighunters and conservationists in feral pig control run parallel. The next chapter examines the claims of amateur hunters on wild pig stocks and their relationship to the interests of landholders.

CHAPTER 3

Recreational Hunting of Wild Pigs

Feral pigs cause damage to the property of agriculturalists and pastoralists, and from their point of view can be regarded as pests. But from the point of view of hunters interested in recreational hunting, wild pigs are an important asset. This is a source of potential and sometimes actual conflict between hunters and landholders. Yet, as pointed out in this chapter, the interests of hunters and landholders are not always in conflict. Their interests sometimes run parallel even if hunters do not pay for hunting rights, because hunters make a significant contribution to controlling the population of a pest species such as pigs.

In evaluating the social desirability of control or amount of control of a species such as wild pigs, the costs of damages to landholders caused by the species must be compared to their value for recreational hunting. From a general economic point of view, recreational activity is no different to any other commodity such as the output of wool or wheat. One purpose of this chapter is to outline methods of assessing the hunting value of a species for recreational purposes and in particular to assess the value of wild pigs in Australia for this purpose. The chapter also provides information on the apparent relative importance of different species for hunting purposes in Australia and reports on the results of a survey of pighunters. The survey provides data on expenditure on pighunting, methods used by pighunters, the costs of taking pigs by the methods used, the influence of hunters on the pig population, attitudes of pighunters to various social issues, and other matters such as the extent to which hunters eat wild pig and the factors they take into account in deciding to eat it.

HUNTING AND PIGHUNTING AS RECREATIONAL ACTIVITIES

Hunting is an important recreational activity in most of the developed world. Some idea of its relative significance as a recreational activity can be seen from the amount of time allocated to it in the United States. The total amount of time spent by Americans annually in hunting is nearly twice as great as the total amount of time spent by them as spectators and participants in organized sport such as baseball, football and basketball but excluding tennis or golf. Or to make another comparison, it is twice as great as the total amount of time spent by them in boating of all kinds.[1] Although similar leisure-time budgets are not available for Australia, hunting is an important recreational activity in Australia.

This is indicated by the fact that more than half a million Australians possess guns for sporting purposes[2] and in New South Wales alone more than 200 000 shooter's licenses are in force.[3] In Australia as a whole, about half a million residents appear to engage in recreational hunting on a regular basis. Assuming that most of the hunters are males, this amounts to one in ten males 15 years of age and over. My sample survey of hunters indicates that 50 per cent of hunters go pighunting in New South Wales, Victoria and Queensland on average.[4] If this is accurate, it suggests that there are about 200 000 pighunters in Australia or about one in 25 of the male population 15 years and over. Even if one reduces these estimates by a half, one still ends up with a sizeable population of pighunters - namely about 100 000, or one in 50 males 15 years and over.[5]

In most countries where wild pigs occur, they are popular game for amateur and recreational hunters. For instance, this is true in continental Europe, including Russia, and countries settled by Europeans. So great is the demand for amateur hunting in Europe relative to the wild boar stock that hunting is regulated and the wild boar population is managed. This is true, for example, in Poland, Russia, Hungary, Spain, Belgium, Austria, and France.[6] It is reported that in Belgium, for example, sows (of the wild pig) in pig are husbanded and let go free with their young once the young have reached a reasonable size. In many countries, game reserves are set aside and are used for pighunting. For instance, a communication to me through the Austrian Embassy in Australia reports that in Austria 'wild pigs are on the one hand kept in great numbers in enclosed hunting reserves (for example, the Lainz animal park near Vienna), and on the other hand many of them still occur in an absolutely wild state'. The communication also points out that while wild pigs are significant for hunting purposes in some areas in Austria, they are of about fourth rank in hunting significance if this is judged by the number of animals killed annually; their numbers are exceeded by those of two species of deer and by chamois.

The President of the Polish Hunting Association, Professor Dr Jerzy Krupka, informed me (in June 1980) that the breeding of wild boars in Poland is conducted by hunters organized in hunting clubs and the hunting of boars is managed to control the population. Every year a game inventory is done, and the aim is to shoot off at the end of the summer the increment to the wild pig population from the previous spring - which is usually a number of pigs about equal to the initial spring population. The population is selectively culled if possible by age.

Hunting of wild pigs (organized through clubs) is very popular in Poland. 'The summer fatting in the forests (killing of boars in forests towards the end of summer) is popular with hunting clubs. This method of control decidely limits the damages caused by wild pigs in the fields.' The wild boar is of great value to Polish hunters and yields valuable trophies, amongst the best in the world.

In most European countries there are closed seasons on the hunting of wild boar and regulations covering this hunting.[7] In West Germany, for example, the open season is from 16 June to 31 January. On the whole, the wild pig is regarded as an asset to be managed. The Spanish Ministry of Information and Tourism in a recent brochure feature the wild boar as one attraction for hunter-tourists to Spain. It is considered to be a part of the big-game and is described as follows:

> 'Very abundant in Spain, the wild boar sometimes causes crop damage. The boar is a highly prized trophy because of the danger in killing him; his powerful muzzle and sharp tusks make him a formidable foe. He is hunted with beaters or the hunter may lie in wait for him.'[8]

On the whole despite the damage to crops, the wild boar appears to be regarded as an asset in Spain and other European countries because closed seasons and other regulations apply to the hunting of the species.

But recreational pighunting is not confined to Europe and Russia. Wild or feral pigs *Sus scrofa* have become established in parts of North and South America and in Australia and New Zealand where their population is used for recreational hunting. In the United States, for instance, *feral* pigs were introduced as early as 1539 into Florida,[9] and European wild boars were introduced to North Carolina as late as 1912 specifically for recreational hunting[10] and have since spread and have interbred with feral pigs.[11] Barrett reports from the U.S.A. that recreational hunters now pay considerable sums to be able to hunt wild pigs. He reports:

> 'Regarding the situation here in California, I can briefly say that (1) the pig population is increasing; (2) it causes minor problems to the cattle, sheep and small grain industries; (3) it is the *second most important game in the state*; (4) most large landowners with pigs on their land permit paid recreational hunting for pigs to control them, obtaining up to $600 U.S. per pig in the process.'[12]

The European wild boar was also introduced into Argentina for hunting purposes.[13] It has thrived and has crossed the continent and entered Chile.[14]

Recreational pighunting in Australia and New Zealand is based on the exploitation of feral pig stocks. Many of these pigs have reverted in their characteristics to become more like their ancestors, the European wild boar, than domestic pigs (see Fig. 3.1). Because there is no other dangerous game in Australia, pigs are

Fig. 3.1 An Australian amateur hunter with a feral boar shot near Burren Junction in Western New South Wales. Recreational pighunting is popular in Australia.

of special importance to Australian amateur hunters. They are the most widely available introduced game animal in Australia and as Frith suggests the most important for amateur-hunting purposes. He claims:

> 'Pighunting offers more chance of personal danger than any other hunting in Australia. Wounded or cornered pigs do sometimes turn and attack sportsmen. There are few if any fatalities, but many groups of hunters are scattered, occasionally some are slashed.'[15]

The relative value of pigs and other species for recreational hunting is discussed in detail below, but it is clear that wild pigs in Australia and most countries where they occur are an important asset for amateur hunters. However, wild pigs are so abundant in Australia relative to recreational hunting pressure that they are not protected from hunting in any way unlike in Europe.

SOCIAL CONFLICT AND SOCIAL GAINS FROM HUNTING

Because wild pigs are an asset for recreational hunting, this can be a source of

Fig. 3.2 Fifteen pigs shot by amateur hunters in a recently harvested barley paddock at 'Curringindi', 50 kilometres north-west of Moree, N.S.W. The hunters were especially invited by the landholder to deal with the pigs.

conflict between hunters and landholders and other members of the community or, depending upon the circumstances, a basis for mutual benefit. If a species is regarded by landholders as a significant pest and the species is sought after by hunters, both parties may benefit from (recreational) hunting. On many agricultural properties in Australia, hunters and landowners mutually gain from recreational pighunting. Many landowners encourage or request amateur shooters to come and shoot on their properties to control their pig numbers. Control is at little or no cost to the landowner (see Fig. 3.2). On the other hand, where a species is considered to be a minor pest by landholders or neither a pest nor an asset from their point of view, conflict is likely to occur between landholders and hunters about hunting, because hunters (by accidentally shooting domestic stock, trampling crops, interfering with fences and installations, and so on) may do more damage or impose greater costs upon landholders than the species concerned. The conflict is liable to be intense if hunters are unwilling to pay for hunting rights in these circumstances. On those Australian cattle properties where pigs are not regarded as a significant pest, attempts are sometimes made to exclude hunters because possible damage and costs imposed on cattle-owners by hunters may exceed that caused by pigs. Similar reasoning leads to exclusion of recreational hunters from some Australian state forests (for example, Queensland state forests) and from national parks and nature reserves. Naturally, if a species is an asset to landholders and is in demand for hunting, there is likely to be conflict between landholders and hunters over the exploitation of the species. The three crude possibilities for co-operation and conflict are summarized in Table 3.1 and illustrated in Chapter 4 by actual

TABLE 3.1 Possible Social Conflict and Mutual Benefit Between Hunters and Landholders over Utilization of a Species

Status of species for hunting	Status of species for landholders (or community)	Social relationship between landholder and hunter
Asset (+)	Significant pest (-)	Co-operation
Asset (+)	Minor pest; Neither pest nor an asset (0)	Conflict
Asset (+)	Asset (+)	Conflict

examples including methods used to resolve some of the conflicts.

The possibilities that exist for conflict and co-operation or mutual benefit between landholders are well recognized. For instance, Clawson and Knetsch touch on many of these matters when they point out that

> 'The hunter wants freedom to move where he thinks his chances of kill are greatest without having to bother about property lines. The landowner often wants a reasonably adequate harvest of the annual crop of game animals so that their numbers do not build up to a destructive point. But he is also often rightly concerned over the possibility of damage to livestock, to fences, to machinery, to roads. Damages, even if caused by a small minority of hunters, can be serious to the landowner. Large forest and other landowners are often sensitive to general public relations and dislike the adverse publicity that often accompanies the closing of lands to hunting. Landowners are thus caught between opposing or conflicting forces, and sometimes they are not sure how to treat hunting on their land. In any case, some problems of equity arise in pricing. Hunting privileges represent a valuable service which the hunter hardly has a right to expect free. Costs may be imposed on the landowner, but there are often offsetting advantages.'[16]

While Clawson and Knetsch raise important issues they do not follow through with a discussion or analysis of these. Their normative contention about the equity of charging for hunting rights appears to be a 'red herring'. It seems more pertinent to ask positive questions such as the following:

(a) Under what circumstances can landholders be expected to charge (or not to charge) for hunting rights, and what factors determine the level of their fee?

(b) If the target species is considered to be a pest by landholders, how does this influence the fee for hunting rights?

(c) If hunters impose costs on landowners, what effect does this have on the hunting fee?

(d) Are there circumstances in which landowners or the community can gain by subsidizing the hunting of species?

These questions are covered by the theory outlined in the Appendix to this chapter. This theory indicates that a hunting fee is likely to be charged (i) if hunters impose a net cost on landholders after allowance is made for their contribution to pest control, (ii) if demand for hunting rights is high relative to available hunting opportunities, *even* if hunters confer a net benefit on landholders. On the other hand, if demand for hunting rights is low relative to available hunting opportunities, no hunting fee emerges if hunters confer a net benefit on *individual* landholders. However, because many animal species are mobile a landholder cannot appropriate all the benefits from pest control on his property. Some of the pests taken by hunters might have moved to other properties if they had not been killed or they might have spent only part of their time on an individual's property. Removal of pests therefore partially benefits other landholders, and these *spillover benefits* are not taken into account by an individual landholder. Again, if pests taken on a property are quickly replaced by immigrants from other properties, the net benefit of pest control to the individual landholder is lowered even though there could be collective benefit to all landholders. The private benefits of individual landholders from pest control is liable to diverge from the collective or social benefit of their pest control to all landholders or the community. Thus where recreational hunters play a role in pest control, individual landholders considering only their own individual gains may *excessively* restrict recreational hunting from the collective point of view of all landholders and the community. One possible way to compensate for this would be to subsidize the supply of hunting rights - that is, pay a subsidy to landowners giving hunting rights for a pest species. But this is on the assumption that the level of hunting is below capacity for hunting on properties. If hunting is at capacity, the subsidy has no impact on the level of hunting and is uncalled for from a collective point of view.

The above questions are of particular importance in the control of feral pigs in a country such as Australia, where feral pigs are regarded as a pest by most landowners (indeed, a serious pest by many) and recreational hunting is important.

METHODS OF ASSESSING THE VALUE PLACED ON RECREATIONAL HUNTING BY HUNTERS

Recreational hunting has economic value, and in this respect it is no different to any other scarce commodity and needs to be taken into account in deciding on the use and allocation of a nation's resources. Hunting is a form of outdoor recreation for many citizens but, as Knetsch points out,

> 'Discussions of values of outdoor recreation have been beset by many misunderstandings. The first is that the only values that are worth considering are those accounted for commercially. A second and related source of error is a belief that outdoor recreation experience is outside the framework of economics, because the relevant values have an aesthetic, deeply personal and even mystical nature. Economically, the use of resources for recreation is fully equivalent to other uses, and the values which are relevant do not necessarily need to be determined in the market place. This last condition does indicate that indirect means of supplying relevant measures of the values produced may be necessary. But this is an empirical problem albeit one of some considerable dimension.'[17]

From the point of view of pighunters, wild pigs are an asset for recreational purposes, and a cost/benefit study of wild pigs requires the value of the wild pig population to recreational hunters to be taken into account. Various methods have been proposed for measuring the value of a species or its importance for hunting. The more theoretically inclined may wish to have a brief outline of these methods, while the less theoretically inclined may wish to skip this discussion and turn to the next section.

The following are some of the proposed methods[18] for measuring the hunting value of a species:

(1) Determine the *willingness* of hunters *to pay* for the privilege of hunting species. For example, determine the maximum amount that hunters would be prepared to pay rather than forgo the opportunity of being able to hunt the species. The aim of this approach is to determine the consumers (hunters) surplus; the maximum amount that hunters would be willing to pay to hunt a species over and above what they do pay.

(2) Determine the *gross or total expenditure* of hunters incurred in hunting a species. These expenses include the travel costs, equipment costs and ammunition outlays by hunters.

(3) Take the *market value of the amount of the species taken*. For example, take the market value of fish caught by amateur hunters as an indicator of the value of fishing to them. This method can be dismissed immediately since most hunters are not hunting for subsistence. Their enjoyment is not based just on amount or commercial value of their catch.

(4) Use the number of individuals hunting the species or the total number of

Fig. 3.3 A group of amateur hunters make a stop on their travel to a pighunt in western Queensland. On average, Australian pighunters travel 800-1000 km per hunting trip. The Clawson method of estimating the demand for outdoor recreation relies on travel costs to estimate demand.

hunting trips made for the species as a rough indicator of its value. Willingness-to-pay approaches of estimating the value placed on hunting require that the demand curve for hunting a species be estimated. But estimation of demand curves, especially if hunting rights are unpriced is difficult. Some of the possible ways of estimating the demand curve for hunting of a species[19] are:

(a) the travel cost method, originally introduced by Clawson for estimating the recreational value of parks; this is an indirect method in which demand is estimated from the willingness of recreationists to incur travelling costs to engage in their outdoor recreation (see Fig. 3.3);

(b) the direct interview method, in which one directly asks hunters how much hunting they would undertake at alternative levels of fees for hunting rights;

(c) the estimation of demand curves from time-series or cross-sectional variations in hunting fees;

(d) or licences.

The Clawson method has a number of limitations. For example, a part of the willingness to pay travel costs may reflect the utility obtained from driving and not just the value of hunting at the destination of the trip. As for the direct interview method, hunters may not be truthful or they may not genuinely know their reactions to different fees for hunting. The econometric approach of using time-series and cross-sectional data faces the difficulty that such data are not always available and econometric problems such as those of identification have to be

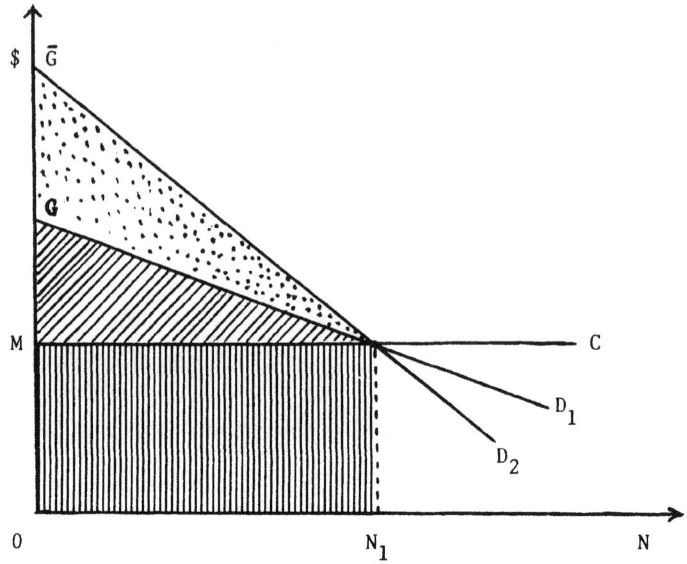

Fig. 3.4 The amount of hunters (consumers) surplus from hunting depends upon how inelastic demand is and how great the level of demand is.

solved. Pearse[20] attempts to by-pass some of these difficulties by using differences between greatest expenditure by a hunter in an income class and the expenditure of other hunters in the *same* income class to calculate the consumers surplus obtained from hunting by members in an income class. Total consumers surplus is found by summing the surpluses for all the income classes. But Pearse's method has a number of shortcomings. To take one, the size of the surplus is influenced by the chosen size of income classes, and this choice tends to be arbitrary.

If we suppose no difficulty in determining the demand curve for hunting, the consumers surplus (hunters surplus in this case) is found in the crudest approach by taking the difference between what hunters have to pay for hunting and the maximum amount they could be forced to pay for this hunting. In Fig. 3.4, where GD_1 is the demand for hunting and MC represents the per unit cost of travel, ammunition and so on, the consumers or hunters surplus is as represented by the hatched area if hunting rights are free. The steeper the demand curve (that is, the more inelastic the demand) for hunting, other things equal, the greater is the hunters surplus. For instance, if $\bar{G}D_2$ is the demand, the hunters surplus is equal to the hatched area plus the dotted area. Also, other things equal, the greater the demand (the further to the right the demand curve) for hunting, the greater is the hunters surplus. The size of the surplus is a measure of the value placed on hunting *in comparison* to other opportunities that hunters have for spending their income.

Whereas the consumers surplus approach measures the economic value of a species for hunting by the area of the triangles indicated in Fig. 3.4, the gross expenditure method measures this by the area of the shaded rectangle ($OM \times ON_1$). This measure takes no account of the surplus received and fails to take into account that hunters may be indifferent (or show some indifference) about spending this money on hunting of the species or in some other way. In relation to this method, Knetsch has said:

> 'These values have some usefulness in indicating the amount of money that is spent on a particular type of outdoor recreation, but as justification for public expenditure on recreation, or for determining the worth or benefit of the recreation opportunity appended, they are of little consequence.
>
> The values we seek are those which show not some gross value, but the net increase in value over and above what would occur in the absence of a particular recreation opportunity. Gross expenditure do not indicate the value of losses sustained if the particular recreation opportunity were to disappear, nor do they show the net gain in value from an increase in a particular recreation opportunity.'[21]

On these grounds the consumers (or hunters) surplus method of measuring the value of hunting is superior.

Of course, the hunters surplus measure of value may exceed the gross expenditure on hunting. The more inelastic is the demand for hunting, and the greater is the demand for hunting, the more likely is hunters surplus to exceed gross expenditure on hunting. The crude indicator of number of hunting trips or total amount of activity in hunting thus has some value in this context.

The hunters surplus from pighunting may well be in excess of annual gross expenditure on pighunting in Australia. Assuming 100 000 pighunters in Australia (as suggested earlier, as a lower limit) undertaking on average one trip per year at the average cost determined in my survey[22] (around $50), annual gross Australian expenditure on pighunting is $5 million annually. But possibly the average expenditure is now slightly higher than in 1979 and possibly the average number of trips is in excess of one per year. It seems likely that annual expenditure on pighunting by hunters is in the range of $5-15 million annually and is most likely towards the upper part of the range. But it is quite likely that annual hunters surplus is in excess of this sum, because demand for pighunting in Australia can be expected to be relatively inelastic since there is a lack of other game posing hunting dangers, native species are regarded as inferior for recreational hunting, deer are in short supply and from now on are not likely to be available to recreational hunters except at high prices. There are no satisfactory available substitutes for pighunting in Australia, and the annual hunters surplus from pighunting could run to at least $20 million. If Arizona data are any indication of the demand for game-hunting, it is likely to be inelastic in Australia.[23]

I should point out, however, that there are dangers in blindly using consumers or hunters surplus to determine policies. The surpluses, for example, reflect to some extent the way income is distributed in the community and may not capture all values - for example, do not take account so-called option values of using a resource. In our problem, for example, a *non-hunter* may be willing to pay something to keep the option open of being able to hunt pigs in the future. Thus the existence of this opportunity has a value that will not show up by using the demand curve for pighunting to compute consumers surplus, because the potential hunter has not expressed his preference by going hunting. These criticisms and other specific ones have been levelled against the consumers surplus and similar evaluation methods by Seckler. He says,

> 'Conservationists, sportsman, welfare workers and others outside the economic profession have argued that such intangibles as recreation benefits cannot be accounted for quantitatively and that an attempt to do so misses the qualitative essence of such experience. We confess a strong sympathy for those who argue the qualitative aspects of recreation experience. Empirical measures of recreation benefits do seem at times to stretch a valid argument beyond the breaking point. But this is not our major concern. Our major concern is with the use of economic science in this area and the danger that economics may seem to imply policy prescriptions which are not an integral part of that theory nor necessarily in the public interest.'[24]

Nevertheless there is scope for rational evaluation of wants in the recreational field, including pighunting.

HUNTING VALUE OF WILD PIGS IN AUSTRALIA

As already indicated, pighunters in Australia probably spend $5-15 million annually on their sport and probably obtain an added value from their sport in excess of this sum. These sums can be expected to be larger than for the recreational hunting of any other species in Australia. From a hunting point of view, the wild pig in Australia seems more important than any other species. Some evidence for this comes from the relative coverage given to different species in Australian hunting magazines. Australian hunting magazines were examined to obtain some

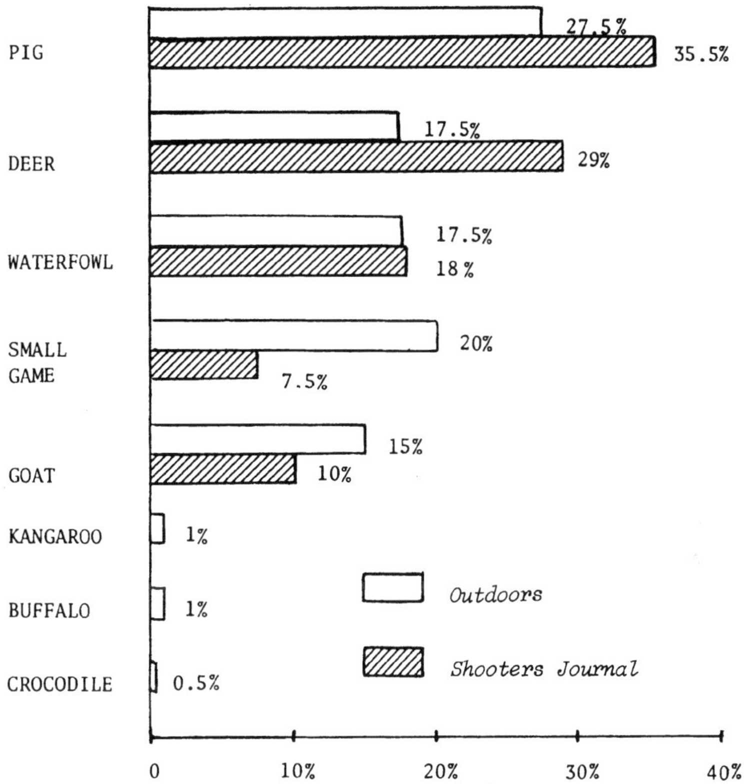

Fig. 3.5 Proportion of all Australian wildlife hunting articles in Outdoors *and in* Australian Shooters Journal, *January 1969 to December 1978, dealing with animals specified above.*

evidence about the apparent relative importance of different species of wildlife in Australia for hunting purposes. The main magazines, the *Australian Outdoors and Fishing* magazine and the *Australian Shooters Journal*, were considered. It was difficult to assemble complete sets of these but with co-operation from various hunters we eventually obtained a complete set of both publications.[25] The set of the *Australian Outdoors and Fishing* magazine covered the period January 1951 to December 1978 and the set of the *Australian Shooters Journal* covered the shorter period, January 1969 to December 1978.

The following counts were made:

(a) the total number of *articles* dealing with the hunting of Australian wildlife and their breakdown between species;

(b) the total number of *pages* used for the above purposes; and

(c) the total number of *photographs* dealing with the hunting of Australian wildlife and their breakdown between species; photographs appearing in advertisements were also included in the count.

TABLE 3.2 Proportion of Pages of All Australian Wildlife Hunting Articles in *Outdoors* and *Shooters Journal*, 1969-78, Dealing with Game Specified Below

Game	*Outdoors* %	*Shooters Journal* %
Pig	26.5	40.5
Small Game	23	6.5
Deer	17	28.3
Waterfowl	17	15.5
Goat	14.5	9
Kangaroo	1	0
Crocodile	1	0
Buffalo	0	0
	100%	100%

Using this data, it was possible to determine the proportionate coverage given to each species or group of species. The overall results for the nine-year period 1969-78 in which both *Outdoors* and the *Shooters Journal* were available are set out above. Fig. 3.5 sets out the proportion of articles appearing during this period dealing with pig, deer, waterfowl, small game, goat, kangaroo, buffalo and crocodile. Table 3.2 gives the relative coverage in terms of pages and Table 3.3 gives this

TABLE 3.3 Proportion of All Australian Wildlife Hunting Photographs, 1969-78 in *Outdoors* and *Shooters Journal*, 1969-78, Dealing with Game Specified Below

Game	*Outdoors* %	*Shooters Journal* %
Pig	24.4	36
Small Game	23	20
Deer	18.5	21
Waterfowl	14.5	13
Goat	13.5	10
Kangaroo	3.5	-
Crocodile	2.0	-
Buffalo	0.5	-
	100%	100%

Fig. 3.6 A large trophy boar shot in a wheat field near Moree, N.S.W. Wild pigs are given more coverage in Australian Outdoors *and the* Australian Shooters Journal *than any other species.*

in terms of photographs.

By all rankings (by proportion of articles, proportion of pages and proportion of photographs) wild pigs are the most important game animal in Australia. More than a quarter of each of the articles, pages of articles on wildlife hunting and photographs in *Outdoors* dealt with the hunting of wild pigs, as did more than one-third of those in the *Shooters Journal* in the nine-year period 1969-78. While the proportions for wild pigs are slightly lower in the *Australian Outdoors and Fishing* over the longer period 1951-78, wild pigs still rank overall as number one by all characteristics. This finding supports Frith's view about the importance of wild pigs for amateur hunting in Australia.

The type of game of second apparent importance differs between *Outdoors* and the *Shooters Journal*. While deer occupies second place in the *Shooters Journal*, it competes with waterfowl for third place in *Outdoors*. Small game is ranked second in *Outdoors*. The hunting of goats, as indicated by coverage, ranks fifth in the list. Even at this level, coverage is significant, taking around 10 per cent of space given to Australian wildlife hunting in the magazines.

Fig. 3.7 Fallow deer at Ross in Tasmania. Deer are given the second greatest coverage of species in Australian hunting magazines.

Fig. 3.8 Goat and kangaroo shot on outskirts of Warrumbungle National Park, Coonabraraban, N.S.W. While goats rank about fifth in coverage in Australian hunting magazines, they obtain much more coverage than kangaroos.

Kangaroo, crocodile and buffalo obtained very little coverage in the period 1969-78. This may reflect the restrictions on hunting of these animals. But in relation to kangaroo, it may also indicate that, as Frith suggests, Australian amateur hunters prefer to hunt European-type game in comparison to native species.[26] Coverage of European-type game (deer, pig, some of the small game such as foxes and rabbits, goat and some of the waterfowl) in these hunting magazines is in excess of 75 per cent of all the space given to Australian wildlife hunting.

One must be careful in drawing conclusions from the amount of coverage given to the hunting of different species. The coverage given to different species might not correlate perfectly with the amount of time spent by amateur hunters in hunting the various species or the amount of their spending on hunting these. For instance, the amount of coverage given to deer may overstate the number of individuals hunting deer, the amount of spending on deer hunting and the time given to deer hunting in Australia. However, this is not to deny that deer hunting seems to be an important recreation for amateur hunters in Australia.

Again, hunting magazines are not purchased by all amateur hunters. Those not purchasing these magazines might put rather different weights on species to those purchasing the magazines.

Furthermore, it might be claimed that coverage is likely to be influenced by individual editorial preferences and that these may change.[27] On the other hand, sales of magazines are likely to depend on the extent to which the wants of purchasers are satisfied. Editorial policy cannot therefore be completely independent of the market-place unless editors are prepared to forgo income. Consumers' sovereignty can be expected to operate to a certain extent on the choice of coverage.

Apart from taking account of the quantity of sales of their magazine, editors may also consider their sales of advertising space. The nature of coverage may be modified to some extent in order to sell more advertising space. But on the whole, one might expect journals to reflect the interests of the amateur hunters who are likely to purchase them

Trends in the coverage of species y these hunting magazines were also assessed.[28] The data support the importance of wild pigs for amateur hunting in Australia: in the past decade, wild pigs have been given more coverage than any other species in the main Australian hunting magazines. The relative attention given to wild pigs has tended to increase; but since 1951 at least, if one can judge from *Outdoors*, they have been very important to amateur hunters. In 1951-55 in terms of proportionate number of articles in *Outdoors*, they ranked third just behind kangaroos and waterfowl, they rose to the first rank in 1955-60, then fell to second place in 1961-65 being slightly exceeded by waterfowl. But since 1965, pigs have retained first rank.

The data also indicate the lack of attention given by amateur hunters to native species for hunting purposes and the major role played by introduced species in amateur hunting. In the past decade, introduced species have obtained more than 70 per cent of the coverage allocated to Australian wildlife species in the leading hunting magazines. While kangaroos were given considerable attention in the early 1950s (they were ranked number one by proportionate number of articles in *Outdoors* in 1951-55), the amount of consideration given to them declined rapidly. At present, practically no attention is given to the hunting of kangaroos in Australian amateur hunting magazines. This may reflect changing attitudes to the conservation of Australian native species and legislation restricting kangaroo hunting.

There is a need to do much more work aimed at estimating the value of various types of Australian wildlife for amateur hunting purposes. An estimate of these values is needed to plan rationally the conservation of, control and management of different species. With this in mind, I undertood a survey of pighunters in Australia.

FINDINGS (ECONOMIC AND OTHERWISE) FROM A SURVEY OF PIGHUNTERS

This survey of amateur hunters of wild pigs was conducted by me in Australia between October 1978 and the end of March 1979. Its main purpose was to obtain data about the economic involvement of hunters in pig shooting, their hunting methods and their costs of taking pigs. The survey also yielded data on the role of amateur hunters in controlling numbers of wild pigs, on whether or not they could be more effectively organized, on their attitude to the extermination of wild pig populations, on whether or not they were having difficulty in obtaining access to properties for hunting purposes, and on whether or not hunting reserves should be established. The survey also yielded information on the possibility of wild pigs playing a part in commercial tourism, the extent to which wild pig hunting is widespread amongst hunters and on whether or not hunters eat wild pig and are selective, bearing in mind health risks of eating feral pigs.

Survey forms[29] were sent to secretaries of clubs of the Sporting Shooters Association for possible distribution to club members, some were distributed through the Australian Bowhunters Association, some were distributed by private contacts and by stores selling hunting equipment in Newcastle, and others were distributed in the September 1978 issue of the *Deerstalker*, the journal of the New South Wales Deerstalkers Association. The response rate was impossible to calculate, except for the N.S.W.D.A. where it was approximately 12 per cent.

Since the overall response rate was low and the main outlet for survey forms was through clubs, one must be wary about generalizing from the results. On the whole, those answering the questionnaire are likely to be individuals with most interest in wild pighunting. Nevertheless, the survey does give the experience of 101 individuals, 95 of whom hunt wild pigs for sporting purposes, and provides results worthy of further investigation.

I shall first give a broad overview of the results and then provide more detailed analysis taking into account differences between the groups of respondents.

Location of and Characteristics of Respondents

Respondents were geographically concentrated in the eastern states, with most respondents being in New South Wales, Queensland and Victoria. The distribution is shown in Table 3.4. The absence of responses from South Australia and Tasmania is indicative of the virtual absence of wild pigs in these states. While there are few wild pigs in Victoria, most Victorian hunters travel into New South Wales to hunt where many more wild pigs are available.

Almost three-quarters of the hunters in the sample are located outside the capital cities of the states. This may indicate that there is greater interest in wild pighunting outside these cities than within them, since the major proportion of the Australian population lives in capital cities. The location of respondents by degree of urbanization of their place of residence is shown in Table 3.5. On the other hand the response rate may indicate my easier access to potential respondents in Newcastle and a more co-operative attitude on the part of country people. About a quarter of the respondents were from Newcastle. The geographical distribution of respondents by towns or cities is indicated on the map in Fig. 3.9.

Respondents were asked whether they were members of hunting clubs. Seventy-four individuals (73.3 per cent) said they were a member of a hunting club and 27 (26.7 per cent) said they were not. Consequently the sample is weighted heavily in favour of club members. Seventy-two of the respondents named the hunting clubs or clubs of which they were members. The frequency of different clubs mentioned is set out in Table 3.6.

Purpose for Hunting Wild Pigs and Methods Used

Ninety-five of the respondents (94.1 per cent) said that they go hunting wild pigs for sporting purposes. The proportion of respondents hunting wild pigs for sporting purposes is high in this sample. Some of the 95 respondents hunting wild pigs for

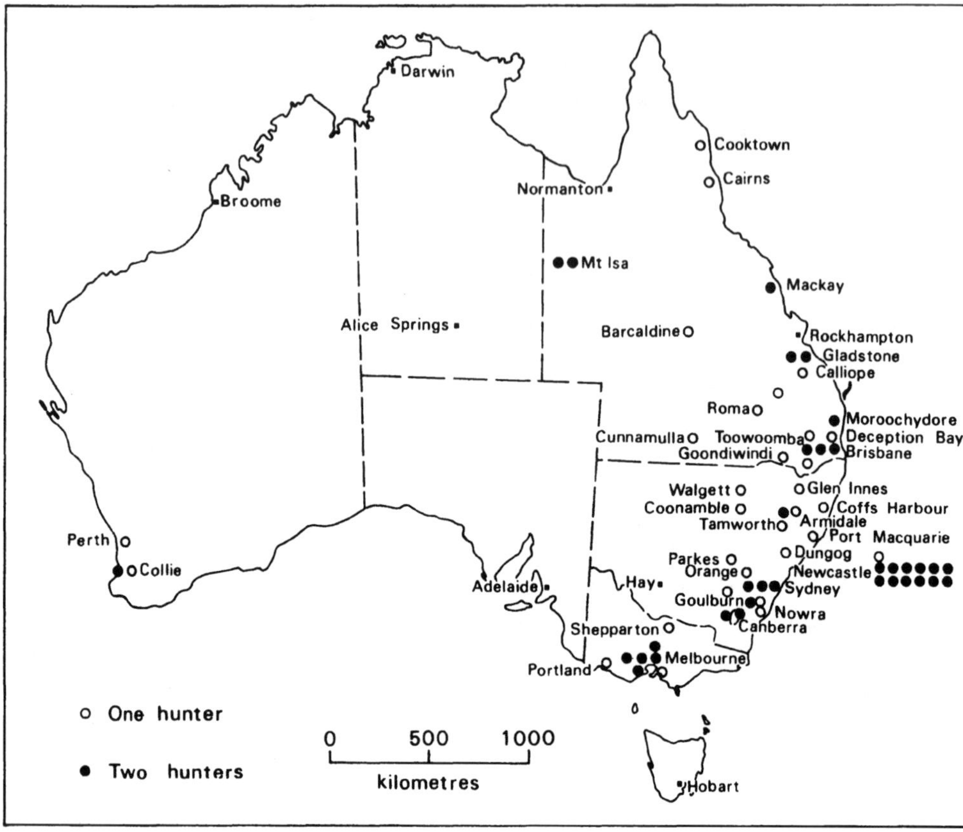

Fig. 3.9 Home location of pighunters responding to the survey

sporting purposes also incidentally mentioned subsidiary purposes for hunting them: for meat (7), to preserve native wildlife (3), for trophies (2) (see Fig. 3.10) and to sell (1).

The most frequently mentioned means of taking wild pigs was shooting, followed by the use of the bow, dogs and trapping. The frequency with which these were mentioned are set out in Table 3.7. Some hunters used more than one method or a combination of methods - for example, shooting and dogs (see Fig. 3.11). The proportion of bowhunters represented in the sample is likely to be higher than amongst Australian pighunters generally since we were fortunate in having responses from a co-operative group of bowhunters in Queensland.

TABLE 3.4 Geographical Location of Respondents by State

State or territory	Number	Percentage of total
New South Wales	50	49.5
Queensland	29	28.7
Victoria	14	13.9
Western Australia	4	4.0
Australian Capital Territory	4	4.0
Other	0	0
	101	100.0*

*May not add exactly to 100 because of rounding

TABLE 3.5 Location of Respondents by Degree of Urbanization of their Place of Residence

Degree of urbanization	Number	Per cent
State capital city	25	24.8
Non-capital city (population < 50 000)	27	26.7
Town to city (≥ 20 000; < 50 000)	14	13.9
Country (population < 20 000)	35	34.7
	101	100.0

TABLE 3.6 Frequency of Membership of Hunting Clubs Mentioned by Respondents

Club or Association	Frequency	Relative frequency %
Sporting Shooters Association of Australia	31	26.3
N.S.W. Deerstalkers Association	27	22.9
Australian Bowhunters Association	10	8.5
Australian Deer Association	8	6.8
Other	42	35.6
Total times mentioned	118	100.0*

*May not add exactly due to rounding.

Fig. 3.10 A couple of feral boars shot in a cotton crop at Wee Waa, N.S.W., by an amateur hunter. Shooting is the most popular method of hunting pigs in Australia, but bowhunting and dogging have their followers..

Fig. 3.11 A dog holding a feral pig by the ear for an amateur hunter on the Warambool River between Collarenebri and New Angeldool, western N.S.W.

TABLE 3.7 Methods Used by Amateur Hunters of Wild Pigs in Sample

Method	Frequency with which mentioned	Relative frequency %
Shooting	71	59.7
Bow	24	20.2
Dogs	15	12.6
Trapping	7	5.9
Poison	1	0.8
Other	1	0.8
	119	100.0

Annual Number of Pig Trips per Hunter, Kill and Costs

All respondents hunting wild pigs (95) stated the number of pighunting trips they had taken in the last year. The average number of trips was nine but this was biassed upwards by a few hunters (mostly living close to pig-infested areas) with a large number of trips. The most frequent number of trips (the mode) was two per year and the median number was around five per year. The average number of pigs taken per trip was 8.1. The most common annual level of expenses incurred per hunter (these expenses include travel costs, ammunition, etc.) was $500 and the mean expenditure was $446.73. *The average cost per pig taken by this group was $6.10.* These statistics are summarized in Table 3.8.

While the expenditure on wild pighunting on average per wild pig hunter per year is considerable, the cost per pig taken might be regarded as moderate in the light of my earlier estimate from Pastures Protection Boards that each pig on average per year caused $6.00 of damage to primary industry[30] and the more recent estimate (given in Chapter 2) of at least $15.00 annual agricultural damage per pig. It is also moderate in view of the fact that amateur hunters obtain enjoyment from hunting wild pigs.

Figs 3.12 and 3.13 respectively show the relationship between the expenditure per pig taken and number of pigs taken by amateur hunters from New South Wales/ Australian Capital Territory, and Queensland. A decreasing per unit cost relationship is indicated in each case (that is, the cost per pig taken tends on average to fall with the number taken) despite the degree of variation in the observations.[31] The Victorian and Western Australian situation is similar in this respect. The two main methods of hunting used by amateur hunters in this sample were shooting and

TABLE 3.8 Data on Pig Trips per year, Kill and Costs

	Item	Average or mean	Mode	Median
1.	Trips per pighunter per year	9	2	5
2.	Pigs taken annually per pighunter	73.2	10	-
3.	Annual expenditure on pighunting per pighunter	\$446.73	\$500	
4.	Cost per pig taken	\$6.10	-	-

the bow. There was little overlap between the use of these methods; hunters tended to use one or the other. Examining the cost per pig taken as a function of the number of pigs taken, the hypothetical average cost function based on Queensland data is decreasing in both cases.[32] Queensland, in my sample is the most useful state to take for comparative purposes, since it had most bowhunters, and only two hunters used both the bow and shooting. The Queensland results indicate that bow-hunting tends to be more costly per pig taken than shooting. Average expenditure per pig taken was \$6.44 for bowhunters from Queensland and \$3.66 for amateur shooters from Queensland.

Initiative for Pighunting Trip and Co-operativeness of Landowners

Respondents were asked where the initiative for a wild pig trip usually came from. Some mentioned more than one source, but the majority indicated that a trip was generally on their own initiative. However, in almost 30 per cent of the cases requests from landowners were important. Presumably landholders were hoping to benefit from pest control by the hunters. The responses are summarized in Table 3.9.

Ninety-three of the wild pig hunters responded to the question of whether landholders were co-operative in allowing them to hunt pigs. Eighty-seven (93.5 per cent) said they were and six (6.5 per cent) indicated that they were not co-operative. However, 40 of the 87 who said that landholders were co-operative qualified this by saying that the landholders were co-operative if the hunter was known to the landholder. Eight respondents specifically mentioned that landholders were co-operative because pigs were a pest and this form of control cost the landholder nothing. Because of this, some landholders provided free accommodation to hunters such as that shown in Fig. 3.14.

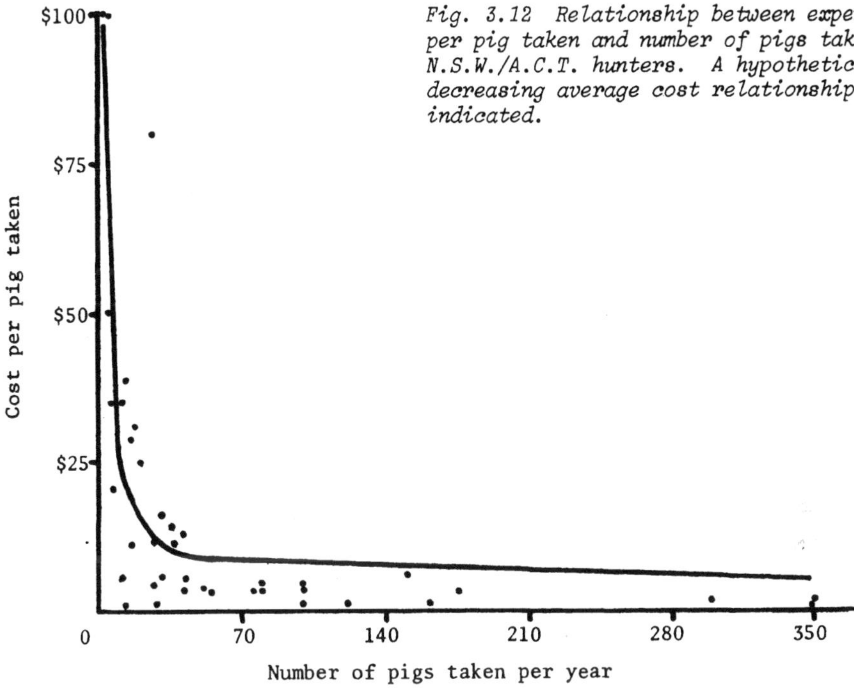

Fig. 3.12 Relationship between expenditure per pig taken and number of pigs taken by N.S.W./A.C.T. hunters. A hypothetical decreasing average cost relationship is indicated.

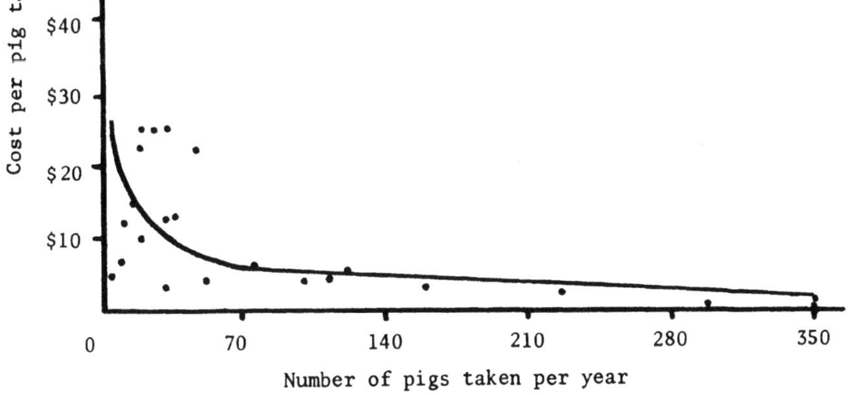

Fig. 3.13 Relationship between expenditure per pig taken and number of pigs taken by Queensland hunters. A hypothetical decreasing average cost function is indicated.

TABLE 3.9 Source of Initiative for a Wild Pig Hunting Trip

Source	Times mentioned	Relative frequency %
Own initiative	82	70
Landowners	33	28.2
Public bodies	1	0.85
Other	1	0.85
	117	100.0

Fig. 3.14 Some Australian landholders provide free accommodation for pighunters, whom they encourage in order to benefit from control of wild pigs as pests on their property. In this case, amateur pighunters are being provided with free overnight accommodation in the shearers' quarters of a Queensland outback station.

Elimination of Feral Pigs and Control by Amateur Hunters

All except two respondents answered the question on whether or not they would like to see wild pigs wiped out. Eighty-six per cent of the hunters (85) individuals said 'no' they would not like to see feral pigs wiped out and 14 per cent (14) said 'yes'. On the whole, this group of hunters was opposed to the complete eradication of wild pigs.

Fig. 3.15 Results of night shooting by amateur hunters in north-western New South Wales. Most amateur hunters believe they play a useful role in helping to control the numbers of wild pigs but believe they could be more effectively organized for this purpose.

Ninety-two of the respondents gave their reasons for answering as they did. The main reason given by those against eradication was the sporting value of pigs for hunting purposes. This was mentioned by 64 individuals but was qualified in 17 instances by the rider that although they did not want to see wild pigs eradicated they would like to see them controlled more effectively. Thirteen of the 14 hunters in favour of eradicating feral pigs said they favoured this because feral pigs were a pest. Six hunters also added as a further reason for favouring eradication that wild pigs were carriers of disease.

Most of the respondents felt that amateur hunters could play a useful role in controlling wild pig numbers. The distribution of their answers is as summarized in Table 3.10. About half (42) of the hunters saying that amateur hunters had a useful role to play in controlling pigs said that they stopped populations of pigs from exploding. They acted as a useful brake on numbers. Eleven said that amateur hunting was preferable to poisoning pigs because it did not have so many side-effects on other species. Five said that amateur hunting was useful because there was no cost to landowners for this form of control. Ten hunters (those saying 'no' plus one other) said that amateur shooters were ineffective or that their hunting had practically no effect on numbers of feral pigs and they therefore did not play a useful role in controlling wild pig numbers.

TABLE 3.10 Do you believe that Amateur Hunters have a Useful Role to play in helping control Wild Pigs? Responses

Answer	Frequency	Relative frequency %
Yes	87	86.1
No	9	8.9
No response	5	5.0
	101	100.0

The predominant view of this group, that amateur hunters had a useful role to play in controlling the population of wild pigs, contrasts with the *predominant* view of Pastures Protection Boards in New South Wales, as indicated by my earlier survey.[33] However, the majority of amateur hunters in the sample felt that amateur hunters could be more effectively organized to control wild pigs. Sixty-five hunters indicated that they could be more effectively organized, 33 believed that they could not, and four did not reply to this question.

Seventy-two suggestions were given as to how hunters could be more effectively organized. Greater co-operation between landowners and hunting organizations was suggested by 22 hunters; nine suggested the organization of pig drives, and seven recommended the advertising of hunting opportunities by landholders. Some of these suggestions may be worthy of consideration by policy-makers.

Compulsory Club Membership for Hunters and the Establishment of Hunting Reserves

On the question of whether all amateur hunters should be expected to be members of hunting clubs or associations, most respondents felt that they should not be, even though a considerable number felt otherwise. The distribution of their replies is given in Table 3.11. The main reasons given against compulsory membership were (a) that this would interfere with individual liberty (mentioned 22 times) and that (b) the high fees charged by clubs would be a deterrent to hunting by low income-earners (mentioned twice). The main reason given in favour of compulsory membership was that clubs would provide hunters with gun knowledge and control the irresponsible ('ratbag') element.

A majority (58 or 59 per cent) of respondents opposed the proposal to set aside government reserves for pighunting, 41 (40.6 per cent) favoured the proposal and two individuals did not answer the question. There are at present no such

TABLE 3.11 Should all Amateur Hunters be expected to be Members of Gun, Shooting or Hunting Clubs, Associations or Organizations? Responses

Response	Frequency	Relative frequency %
No	52	51.5
Yes	44	43.6
No answer	4	4.0
	101	100.0

reserves in Australia - unlike in some European countries, for example, Austria. Those opposing the establishment of reserves gave the following reasons:

(a) pigs cannot be confined to or contained in reserves (mentioned 13 times);

(b) not necessary because plenty of pigs are available elsewhere (on rangeland, etc.) (mentioned 10 times);

(c) dangerous to establish reserves because they will result in crowding by hunters and increase risks of accidental shooting of one another (mentioned 7 times);

(d) pigs are too damaging to consider having hunting reserves for them (mentioned twice);

(e) hunters would have to pay (mentioned once).

The main comment of those favouring the proposal was that reserves would help their sport (mentioned eight times). Four respondents also mentioned that National Parks should be open for the purpose of selective shooting. Such parks are not open to recreational hunters at present for selective hunting.

Members of the responding group were asked whether in principle they favoured a levy or tax on ammunition and firearms to be used for providing hunting reserves and managing wildlife for hunting purposes generally. A majority (56 or 55.4 per cent) were against this but 44 or 43.6 per cent favoured the proposal and one did not respond to this question. Eighty of the 101 respondents gave reasons for their answers. The most frequent reasons against the proposal were:

(a) taxes are high enough already (mentioned 24 times);

(b) the levy would be diverted by the government or politicians for other purposes (mentioned 14 times).

The most frequent comment (mentioned nine times) by those favouring the proposal was that this is done successfully overseas.

On each of the three proposals raised above, the opinion was divided, in the sense that there was not an overwhelming majority for or against the proposals.

Interest of Hunters in Pighunting and Number of Pighunters Known by Respondents

Respondents were asked to provide an estimate of the percentage of hunters in their district going wild pig hunting. Eighty provided an estimate. These estimates vary considerably between districts but averaged out at 54.8 per cent. The estimates given for country areas and capital cities were on average lowest (37 per cent), and those for large towns were highest (59 per cent). On the whole a considerable proportion of hunters seem to engage in the hunting of wild pigs, a result consistent with my observations on hunting magazines, as outlined in the previous section.

The respondents to the questionnaire on the whole appear well informed about pighunting, if the number of pighunters known by them is taken as an indicator. Ninety-seven respondents supplied us with the number of pighunters known by them. The average number was of the order of 65, 50 was the most frequently mentioned number, and the median number was 23. Half of the group of 97 knew at least 23 pighunters.

Wild Pigs and Tourism

Respondents were asked, 'Is there a prospect of wild pig hunting becoming a tourist attraction as with buffalo shooting in the Northern Territory?' (In the Northern Territory safaris are commercially organized for buffalo shooting.) The majority of respondents (52) felt that this was a possibility but 45 felt that it was not. The responses are summarized in Table 3.12. Seventy-seven of the 97 respondents commented. The most frequent comments of those saying 'yes' were:

(a) it is done at present in Australia (mentioned 14 times);

TABLE 3.12 Is there a Prospect of Wild Pig Hunting Becoming a Tourist Attraction? Responses

Response	Frequency	Relative frequency %
Yes	52	51.5
No	45	44.6
No answer	4	4.0

(b) the hunting of wild pigs is good sport (mentioned 13 times).

The most frequent comments of those saying that there was no prospect of wild pig hunting becoming a tourist attraction were:

(a) pigs are poor trophy animals (mentioned eight times);

(b) it would mean shooting with unknown shooters, and individuals would be worried or suspicious about this (mentioned three times).

Eating of Wild Pigs by Hunters

The majority of the respondents said that they ate wild pig. Of the 100 respondents to this question, 58 said they ate wild pig and 42 said that they did not. Forty of those saying that they did not eat wild pig gave their reasons. They most frequently mentioned the risk of contacting worms and diseases (29 times) as a reason. A few mentioned that they shot pigs in inaccessible regions and it was not worthwhile carrying them out.

Fifty-one of those saying that they ate wild pig (51 out of 58) said they were 'choosy' and all but one of these gave the basis or bases for their choice. Table 3.13 shows the frequency with which different considerations were mentioned. The most important consideration is the age of the pig. Other considerations include, for instance, 'pigs from areas which are known to be disease free'. In some instances more than one factor is taken into account - for example, young pig (see Fig. 3.16) on a good diet, young pig and personal inspection - but such combined considerations are in the minority.

Distance Travelled for Wild Pig Hunting and Areas where Hunting Occurs

Hunters of pigs were asked where they usually went wild pighunting and where they went on the last trip. This enabled us to estimate the approximate distance travelled, because we also knew the home address of the respondent. The usual return trip averaged about 1100 km, as did the average last trip. The details are set out in Table 3.14.

In terms of the costs of travel alone, hunters of wild pigs appear to spend a substantial amount per year. Allowing 5¢ per kilometre per year (in 1978-79), travel expenditure averages out to $55.68 per trip or, given the average of nine trips per year, approximately $495 annually. But travel costs are likely to be a little higher than this even if petrol alone is considered. However, since cars or jeeps may be shared by hunters we must be careful about generalizing from the above. Nevertheless the average overall expenditure figure given by pighunters

TABLE 3.13 Frequency of Factors Taken into Account in the Decision by Hunters to Eat Wild Pig

Factor or consideration	Times mentioned	Relative frequency %
Young pig (6 mths old or younger)	27) 42	34.6) 53.8
Porkers (yarded for a month or so)	15)	19.2)
Pig on good diet (for example, grain)	15	19.2
Personal inspection	11	14.1
Other considerations	10	12.8
	78	100.0

Fig. 3.16 These young feral porkers taken on 'Tipperary' station, N.S.W., are going to be eaten by the hunters. Most amateur hunters eat wild pig but many eat only young pigs or porkers.

TABLE 3.14 Estimated Return Distance Travelled (in kilometres) by Hunters of Wild Pigs

	Usual trip	Last trip
Mean	1113.6	1144.3

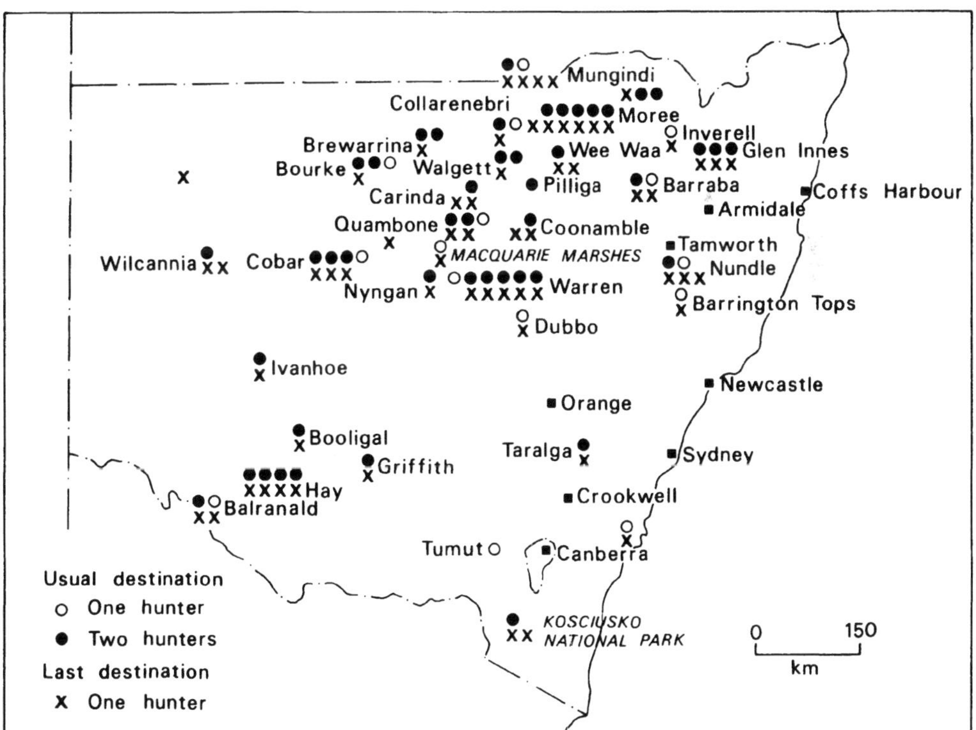

Fig. 3.17 Destinations in New South Wales of hunters of wild pigs, as indicated in the survey.

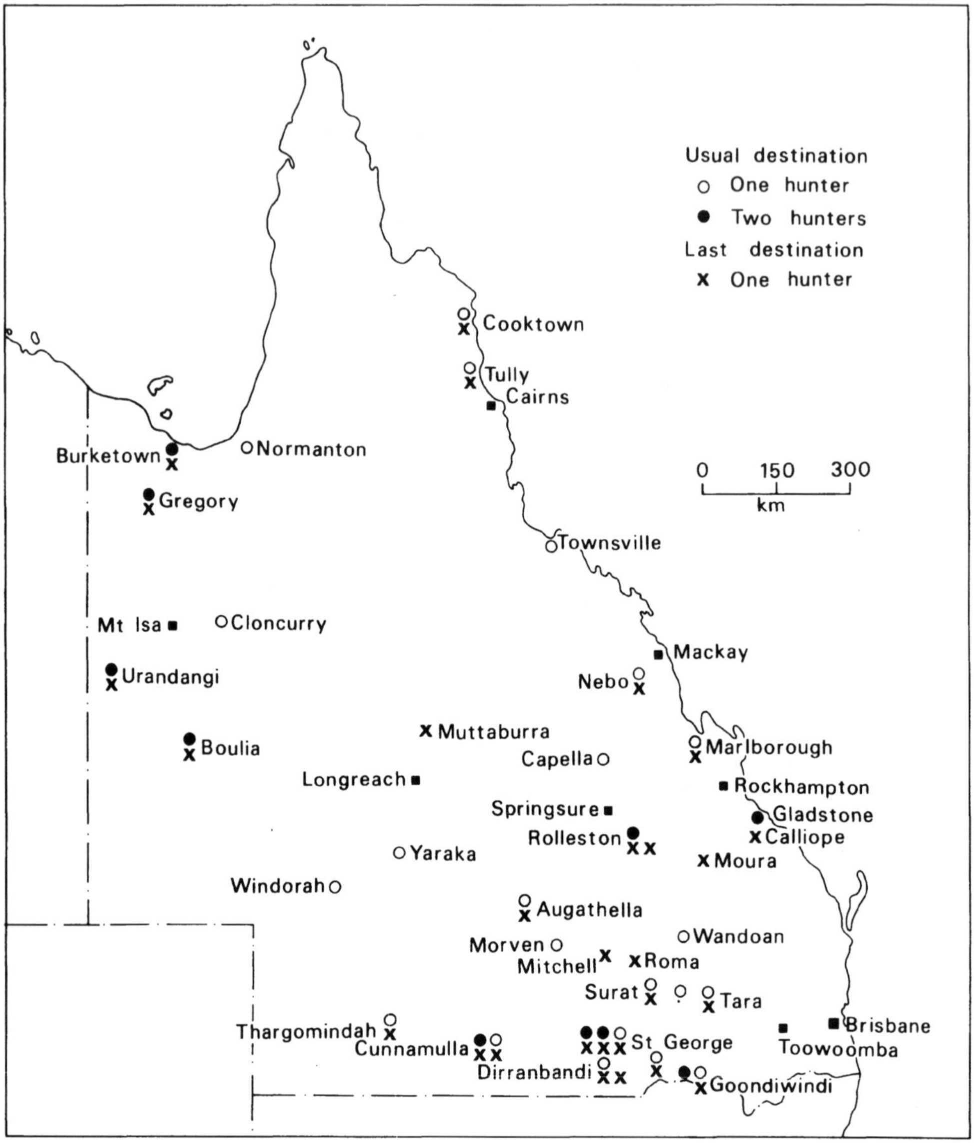

Fig. 3.18 Destination in Queensland of hunters of wild pigs, as indicated in the survey.

(see Table 3.8) of $447 per year does not appear inflated in the light of this consideration.

All the pighunters in the sample (except two who went shooting in Western Australia) went shooting in either New South Wales or Queensland. The places to which pighunters usually went and those to which they went on the last trip are shown in Figs 3.17 and 3.18.

In New South Wales, there is a concentration of hunting in the north-west of the state stretching from Cobar in the south, through the Macquarie Marshes to Mungindi on the Queensland border. These areas are known to contain considerable concentration of pigs.[34] In the south of the state the main concentration of wild pigs is in the Hay area. Pigs are available there in reasonable numbers and Hay is comparatively close for Victorians. Because, presumably, of the relative absence of wild pigs in Victoria, all Victorian hunters of wild pigs in this sample went hunting interstate.

The areas visited for hunting in Queensland are more dispersed on the whole than in New South Wales. This indicates that wild pigs are more widely spread in Queensland than in New South Wales in terms of worthwhile numbers for hunting purposes. However, the north-western area of concentration in New South Wales extends into Queensland, through to the area centred on St George. Results from this sample also indicate that there is more hunting of pigs on or near the coast in Queensland than in New South Wales. This would be consistent with greater concentrations of pigs in these areas than in New South Wales. Wild pigs are a problem throughout the sugar-growing areas of Queensland situated along the coast, as shown by the records kept by the Bureau of Sugar Experiment Stations on sugar cane losses due to wild pigs.[35]

Differences in Responses Between Hunting Club Members and Non-club Members

The degree of club membership of respondents varied between states. It was highest in Victoria (100 per cent) and smallest in New South Wales/Australian Capital Territory (59 per cent) - leaving Western Australia to one side because of the small number of respondents. In Queensland, the degree of club membership was 83 per cent. No significant conclusion can be drawn from this; the pattern possibly reflects the methods by which the respondents were obtained. Comparing the degree of club membership with the urbanization of the area where respondents live, for those in capital cities club membership was higher than average, and for those in non-capital cities it was lower than average. It was about average for the sample, in large country towns and in country areas.

Approximately the same proportion of club and non-club members in the sample took pigs by shooting but a higher proportion of those using a bow were club members. Bow-hunting may be more frequently a club sport.

Annual expenditure on pighunting by club members was greater than for non-club members. Expenditure per pig taken was also higher for club members, and the average number of their trips fewer. But the club members took more pigs per trip

TABLE 3.15 Number of Pig Trips per year, Kill, Costs and Distances Travelled by Members of Hunting Clubs and by Non-Members

	Item	Members		Non-members	
		Average or mean	Mode	Average or mean	Mode
1.	Trips per pighunter per year	8	2	11	6
2.	Number of pigs taken per year per pighunter	72.2	10	75.7	10
3.	Annual expenditure per pighunter on pighunting	$470.97	$500	$340.89	$500
4.	Cost per pig taken	$6.52	-	$4.50	-
5.	Return distance travelled by pig hunters				
	Usual trip	1210 km	-	878 km	-
	Last trip	1304 km	-	762 km	-

than non-members (nine compared to just under seven). Again this may reflect the location of club members as well as their enthusiasm and interest in hunting equipment itself. Figures for comparative costs of hunting and the frequency of trips are given in Table 3.15. Average distances travelled are also stated. These figures confirm that club members in the sample travel further than non-club members for hunting purposes.

A significantly greater proportion of club members than non-club members felt that all hunters should be expected to be members of shooting or hunting organizations. The comparative results are summarized in Table 3.16. Nevertheless 40 per cent of club members thought there should be no such compulsion.

A significantly different pattern of response was obtained from club members and non-club members as far as the establishment of reserves for wild pig hunting and the imposition of a tax/levy for the provision of game reserves are concerned. One half of club members were in favour of reserves for wild pig hunting and the majority of club members favoured the imposition of a levy for the creation of game hunting reserves. Non-club members, on the other hand, were overwhelmingly against these proposals. The results are summarized in Tables 3.17 and 3.18.

The responses of club members and non-club members differed on the question of whether wild pig hunting had a prospect of becoming a tourist attraction. The majority of club members (58.3 per cent) thought that there was a prospect of wild pig hunting becoming a tourist attraction whereas the majority of non-club members (58.6 per cent) thought otherwise.

TABLE 3.16 Should all Amateur Hunters be expected to be Members of Gun, Shooting or Hunting Clubs, Associations or Organizations? Comparative Responses

Responses	Club-members		Non-club members	
	Frequency	Relative frequency %	Frequency	Relative frequency %
Yes	40	55.5	4	13.8
No	29	40.3	24	82.8
No answer	3	4.2	1	3.4
	72	100.0	29	100.0

TABLE 3.17 Would You like to see the Government set aside Reserves for Wild Pig Hunting? Comparative Responses

Responses	Club members		Non-club members	
	Frequency	Relative frequency %	Frequency	Relative frequency %
No	36	50	22	75.9
Yes	35	49	6	20.7
No answer	1	1	1	3.4
	72	100	29	100.0

TABLE 3.18 Do You favour a Tax or Levy on Ammunition and Firearms to provide and manage Game Hunting Reserves? Comparative Responses

Responses	Club members		Non-club members	
	Frequency	Relative frequency %	Frequency	Relative frequency %
Yes	37	51.4	7	24.1
No	34	47.2	22	75.9
No answer	1	1.4	0	0
	72	100.0	29	100.0

Differences in Responses of Hunters from the Country, from Towns and from Cities

Table 3.19 sets out the number of trips taken annually by pighunters from capital cities, other major cities, large towns, and country areas. It also gives comparative data on the number of pigs taken, expenditure and distances travelled. The average number of trips taken per year is highest for pighunters in country areas. Expenditure per pig taken is on average around $5.00 for pighunters in capital cities, country areas and large towns. It is highest for non-capital cities with a population of 50 000 or more and averages $9.00 per pig taken. Hunters in this sample from capital cities on average are most efficient. Their *average* take of pigs per trip is greatest and their per unit expenditure per pig is least, even though the average distance travelled by this group is the greatest or almost so.

TABLE 3.19 Number of Pig Trips per Year, Kill, Costs and Distances Travelled by Degree or Urbanization of Hunter's Place of Residence

	Item	Capital city	Non-capital city (>50 000)	Large town	Country (<20 000)
1.	Average number of trips per pighunter per year	6	6	6	15
2.	Number of pigs taken per year per pighunter	76	58	64	86
3.	Average annual expenditure per pighunter on pighunting	$360	$529	$375	$439
4.	Average expenditure per pig taken	$4.75	$9.12	$5.87	$5.11
5.	Average return distance travelled by pighunters				
	Usual trip	1288 km	892 km	698 km	1302 km
	Last trip	1496 km	1250 km	680 km	936 km

Differences in Responses by Hunters from Different States

There were some differences between the states in the distribution of respondents by degree of urbanization. The highest proportion of hunters in country areas was in Queensland, the greatest proportion in capital cities was in Victoria and the greatest proportion in non-capital cities (over 50 000) was in New South Wales.

The relative frequency of bow-hunters in the group was greatest in Victoria and Queensland.

Expenditure and kill figures and travel distance figures, as set out in Table 3.20, indicate that annual expenditure per pighunter is roughly the same for New South Wales/Australian Capital Territory, Queensland and Victoria (around $400) and that average expenditure per pig taken is least for Queensland ($4.14). Average cost per pig taken is $6.93 for hunters from New South Wales/Australian Capital Territory and $5.68 for pighunters from Victoria. Mean distances travelled for pighunting are greatest for Victorian hunters, least for Queensland hunters and of intermediate amount for hunters from New South Wales and Australian Capital Territory. The Victorian distances are explained by the comparative lack of feral pigs in Victoria which means that Victorian hunters usually travel into New South Wales to hunt.

The breakdown of responses according to states also reveals that invitations from landholders are less likely to provide the initiative for pighunting in Queensland than elsewhere. Furthermore, most hunters in Victoria and in Queensland, unlike those in New South Wales, felt that there is a prospect of wild pig hunting becoming a tourist attraction. Indeed, more than 70 per cent of Victorian hunters thought this. The Victorian response probably reflects the relative scarcity of feral pigs in Victoria and the fact that hunters from Victoria usually travel interstate for pighunting purposes.

TABLE 3.20 Number of Pig Trips per Year, Kill, Costs and Distances Travelled by Hunters from N.S.W./A.C.T., Qld and Vic.

	Item	N.S.W./ A.C.T.	Qld	Vic.
1.	Average annual number of trips per pighunter per year	6	10	9
2.	Number of pigs taken per year per pighunter	57	104	78
3.	Average annual expenditure per pighunter on pighunting	$394	$430	$443
4.	Average expenditure per pig taken	$6.93	$4.14	$5.68
5.	Average return distance travelled by pighunters:			
	Usual trip	1006 km	912 km	1318 km
	Last trip	1092 km	828 km	1626 km

Conclusions from Survey

While one must be careful about generalizing from the results of the sample in this survey, the survey indicates that wild pig hunting is a popular sport and that individual hunters spend considerable sums of money on it annually. Furthermore, hunters are prepared to and do travel long distances to engage in the sport.

The majority of amateur hunters seem to believe that they play a useful role in controlling the numbers of feral pigs even though most believe that they could be more effectively organized for this purpose. More than half of the hunters in the sample eat wild pig and, while most are 'choosy', their criteria in general do not appear adequate to eliminate the risk of contacting parasites and diseases from feral pigs. Shooting seems to be the most common means used by amateur hunters to take pigs, followed by the use of the bow. While most pig trips are undertaken at the initiative of hunters themselves, in many cases trips are in response to invitations by landholders. The survey indicates that there is some demand amongst amateur shooters for game reserves and that a significant number of hunters believe that wild pigs have or are likely to have tourist value.

THE IMPACT OF AMATEUR HUNTERS ON PIG POPULATIONS

In Australia, amateur hunters kill a substantial number of feral pigs annually and make a significant contribution to the control of this Australian pest and thus reduce the extent of its damages to landholders. Nearly all respondents to the above survey were convinced of this. There is independent evidence from a survey of landholders in western New South Wales[36] indicating that the number of pigs taken by recreational hunters is about the same as that taken by agricultural landholders. But this latter survey, by Benson, probably underestimates the contribution of amateur hunters because it relies on information from landholders about kills by amateur hunters *known* to them.

It therefore may be interesting to do a 'back-of-the-envelope' calculation of the likely contribution of Australian amateur hunters to pig control. Assuming 100 000 hunters (the conservative figure suggested earlier) and that on average they kill five pigs each per year, amateur hunters would account for half a million pigs per year. The Benson survey indicated that in the western survey region of New South Wales amateur hunters accounted for about 150 000 pigs annually.[37] But this figure would have to be increased slightly to account for the whole of New South Wales (and for the likely underestimate mentioned) and the result might be doubled to take account of Queensland and Northern Territory. My estimate of about half a million pigs taken in Australia as a whole therefore does not seem unrealistic. I have estimated that the population of feral pigs in Australia (although it fluctuates) is around three million.[38] Hence one-sixth of the population (or roughly 15-20 per cent of the population) may be killed annually by amateur hunters. The actual figure could even range a little higher, but let us take the conservative side.

The effects of removing half a million pigs, however, is not to reduce the population by that much for a whole year. Replacement of adult pigs by younger ones takes place. But there is an *economic* gain, for *replacement of adult pigs by younger ones takes time* and younger pigs are less destructive than mature ones. Possibly for each adult removed the replacing pig causes half as much damage in the initial year of replacement.[39] Since amateur hunters remove mostly adults, the *effective* reduction in adult population for a year (as measured by the reduction in damage) is about half the number of pigs killed. From an economic point of view, amateur hunters therefore effectively remove a quarter of a million pigs or 7½ per cent of the estimated population annually. Assuming that each adult pig on average causes \$15 damage per year to landholders[40] the indirect benefit through pest control to landholders amounts to \$3.75 million annually. Thus amateur hunters not only place a high value on the opportunity of hunting wild pigs, but make a significant contribution towards controlling their population.

APPENDIX 3A

Market for Hunting Rights and a Socially Optimal Amount of Hunting

The economic theory of the market for hunting rights as outlined in this Appendix (which can be skipped by the less theoretically inclined) involves considerable simplification but is conceptually useful and clears up some possible misunderstandings. It shows, for example, that landholders may charge hunting fees even though they are net beneficiaries from hunting and it identifies circumstances in which a subsidy on hunting may be justified on economic efficiency grounds.

Demand for and Supply of Hunting Rights

The (net) demand for hunting rights depends on the difference between the overall demand for the activity and the (marginal) cost to hunters (of travel, ammunition, and so on) of engaging in hunting. For example, in Fig. A3.1 the curve GG might represent the demand for hunting rights all costs considered. If MC is the cost of engaging in hunting (travelling and so on) the (net) demand for hunting rights is shown by curve DD in Fig. A3.1 which is the difference between GG and MC. The demand for hunting rights will be higher other things equal the lower is the costs of engaging in hunting (MC) and the higher is the basic demand for hunting (GG). Other things equal, for instance, a rise in the price of petrol might be expected to lower the demand for hunting rights assuming that hunting is a normal good.

A supply curve of hunting rights can be constructed as follows:

(1) Let OP represent the net cost imposed on or net benefit received by landholders from each hunting visit. In technical language it is the net private marginal cost imposed or net private marginal benefit received from hunting activity. For example, if a net cost is imposed, the incidence of per-visit costs upon individual landholders might be as indicated by line PP' in Fig. A3.2. But a net benefit is possible, in which case the private net benefit curve would be below the origin. The line PP' indicates the price or fee at which landholders are just willing to supply hunting rights.

(2) The capacity of each property for providing hunting rights is limited. For instance, safety limits the number of hunters that can be on a property at one time. For simplicity, assume that there is a unique hunting capacity level on each property and that the market limit to the supply of hunting rights is as indicated by the vertical line CN_3. The supply curve of hunting rights is then

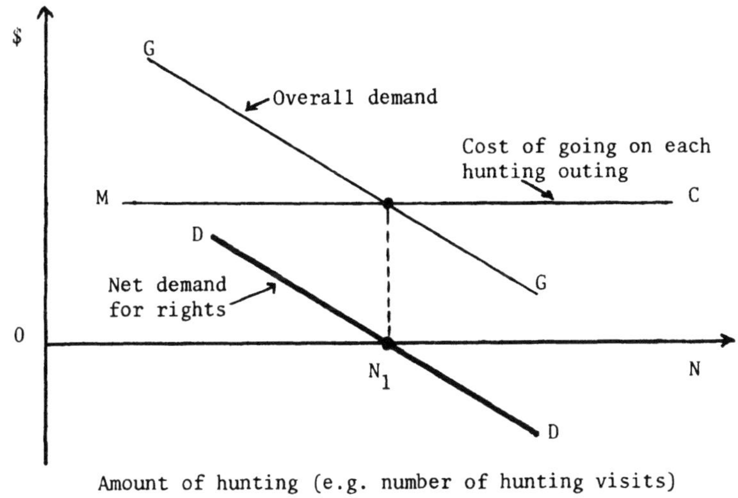

Fig. A3.1 The demand curve (DD) for hunting rights is derived from the difference between the overall demand for hunting (GG) and the (marginal) cost of going hunting (MC).

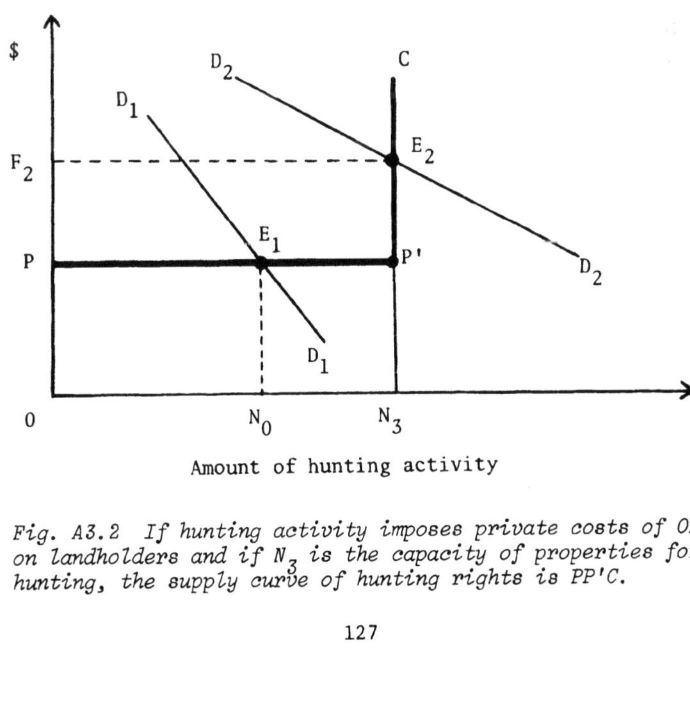

Fig. A3.2 If hunting activity imposes private costs of OP on landholders and if N_3 is the capacity of properties for hunting, the supply curve of hunting rights is PP'C.

the reversed L-shaped curve, PP'C, which may be regarded as a theoretical approximation to the actual position.

Fees for Hunting Rights Determined by Supply and Demand

The combined forces of supply and demand for hunting rights determine the level of fees charged for hunting rights. The equilibrium level of the fee is determined by the intersection of the demand curve for hunting rights and supply curve of these. In the case shown in Fig. A3.3, if demand is less than or equal to capacity, N_3, the fee is OP, equal to the net cost imposed by each hunting visit. For instance, this is so if the demand curve D_1D_1 applies. If demand exceeds capacity the equilibrium fee exceeds OP the net cost imposed on landholders and they earn a rent or surplus from the sale of hunting rights. For example, if demand is D_2D_2, the equilibrium fee is F_2, and the surplus or rent earned by landholders from the sale of hunting rights is $(F_2-P)N_3$.

But of course hunters need not impose a net (private) cost on landholders. Their net cost to landholders may be zero, in which case no fee is charged for

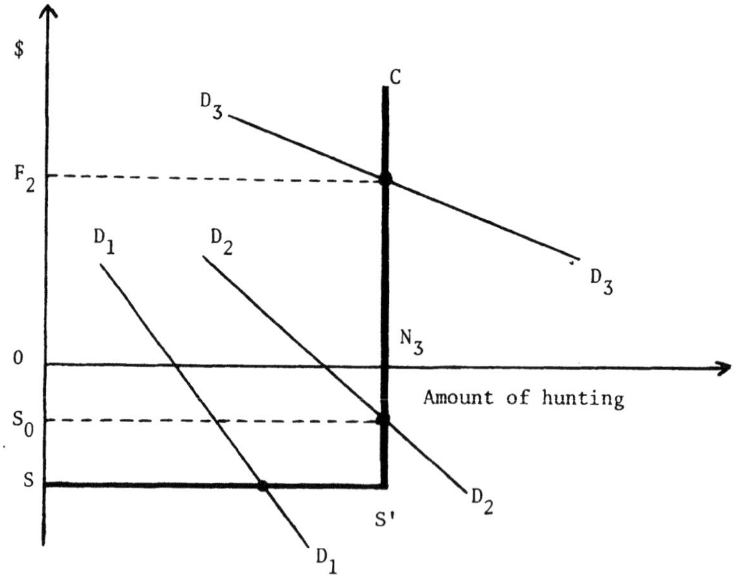

Fig. A3.3 If individual landholders obtain a net benefit from hunters through their contribution to pest control, they may subsidize hunters. But the subsidy depends on the level of demand. Indeed, if demand is high relative to capacity, hunting fees may be charged despite the net benefits obtained by landholders through pest control.

hunting rights until demand exceeds full capacity or, in view of the pest control benefits, the net benefit to individual landowners of hunters may be positive. In the latter case, individual landholders will subsidize hunters unless hunting demand equals or exceeds capacity if no fee is charged. The private subsidy from each landholder will equal the net benefit received by the landholder until demand is such as to utilize full capacity in which case the subsidy will be reduced to that level needed to ensure full capacity. If demand expands further so that it exceeds capacity at a zero change, a hunting fee emerges despite the fact that landholders are net beneficiaries from the activities of hunters even if fees are not charged. The possibilities are illustrated in Fig. A3.3. It is assumed that OS is the private benefit of each hunting visit so the supply curve of hunting rights is SS'C. If the demand curve for hunting rights is D_1D_1 (passing through SS') landholders find it optimal to pay a private subsidy on each hunting visit of OS. But if demand at this subsidy exceeds full capacity a smaller subsidy is dictated by supply and demand. For instance if demand is D_2D_2 the subsidy is reduced to OS_o. Whereas farmers received no surplus benefits before they now obtain benefit of $(S-S_o)N_3$ due for example to the pest control by hunters. Should demand exceed full capacity when hunting rights are free, a charge for hunting rights emerge and landholders obtain a double bounty - extra income and net benefits from pest control by hunters. For example, if demand is D_3D_3, a fee for hunting rights of F_2 emerges. Landholders obtain a surplus consisting of an income of F_3N_3 from hunters plus net benefits (through pest control) of SN_3 from hunting activities. Equity has nothing to do with the fee determined. It is determined by the impersonal forces of supply and demand. Incidentally, in some parts of Australia individual private landholders have and still do pay bounties to hunters taking pigs on their properties. For instance, this is done in some of the sugar cane fields of northern Queensland.

Divergence between the Social and the Private Costs or Benefits from Hunting

The social or collective costs or benefits of hunting (to landholders as a group or the community) may differ from private costs or benefits. When this occurs there is market failure in ensuring an optimal amount of hunting. In the case of pigs, for instance, a mobile pest species, an individual landholder only *appropriates a fraction* of the total economic gain from destruction of pigs on his property. Some of the pigs destroyed would have moved to other properties at least for some of their time. So the destruction benefits landholders who would have been saddled with these pigs at some time. Again the wider community may gain by the reduction in the population of a target species such as pigs, which are considered in a country such as Australia to be a danger to native species and to natural ecological and environmental conditions generally.

If the (net) social marginal benefit from hunting is greater than private benefit (or the social marginal cost of hunting is lower than the private marginal cost), the market for hunting rights will not encourage sufficient hunting activity to maximise social gains, unless hunting is at full capacity. This can be illustrated by Fig. A3.4. In the case shown, it is assumed that hunters impose a net private cost on landholders of OP so the supply curve is PP'C. But for landholders as a group (or the wider community), hunting of the target species may give a net benefit indicated by OW. This implies that if at the market price of hunting rights the amount of hunting is less than full capacity, hunting activity is smaller than socially optimal. For example, if the demand for hunting rights is D_1D_1 a fee of OP is charged for hunting and N_1 hunting visits occur. However, N_2 hunting visits are socially optimal (in the sense of ensuring that the additional value placed on hunting by hunters as indicated by their demand curve equals the marginal social

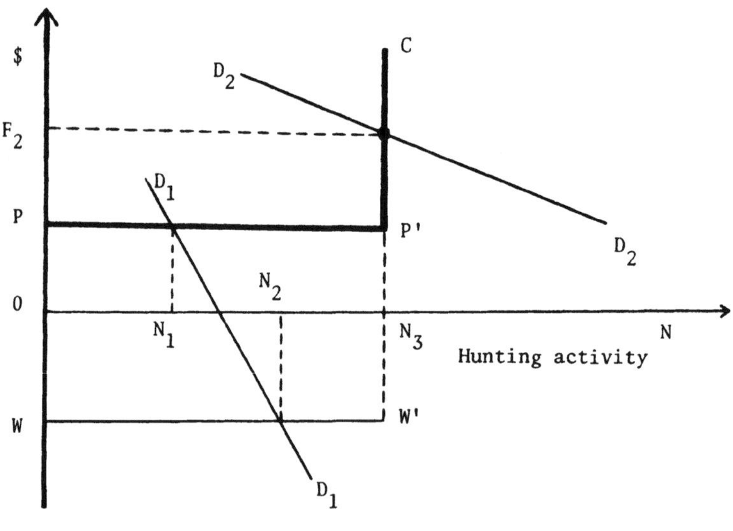

Fig. A3.4 It may be necessary to subsidize the supply of hunting rights if the (marginal) social benefits from hunting exceed (marginal) private benefits. In above instance, if private marginal costs on landholders are OP and social marginal benefit OW, a subsidy of PW corrects for the social undersupply of hunting rights if the demand for hunting rights is as shown by curve D_1D_1.

benefits from hunting). The deficiency in the market for hunting rights can be corrected by giving a subsidy of PW to landholders on the sale of hunting rights. The required subsidy, however, will be smaller if the demand curve for hunting rights passes through the dashed vertical P'W'. The required subsidy will be the one just sufficient to ensure full capacity and will be an amount equal to the distance between P' and the intersection point. No subsidy will be required to ensure social optimality of hunting if the demand curve for hunting rights ensures market equilibrium at full capacity. For instance, if the demand curve is D_2D_2 and the fee for hunting rights is F_2, hunting is at full capacity.

There may be a case for subsidizing the sale of pighunting rights in parts of Australia given that hunting is below capacity in some areas and the social benefit from controlling pig numbers exceeds the private gains of landholders.

However, theoretical completeness also requires account to be taken of the possibility that the social costs imposed by (or social benefits received from) hunters exceeds private costs imposed on (or benefits received by) landholders. This divergence could occur, for instance, because hunters endanger a species valued by the rest of the community and it may call for a tax on the issue of hunting rights. This is illustrated by the example in Fig. A3.5. Let OP represent the private costs imposed by hunters and OW the social costs. Given the demand D_1D_1 for hunting rights, a tax of PW ensures that hunting activity is brought into line with that level which equates the marginal value placed on hunting by hunters as

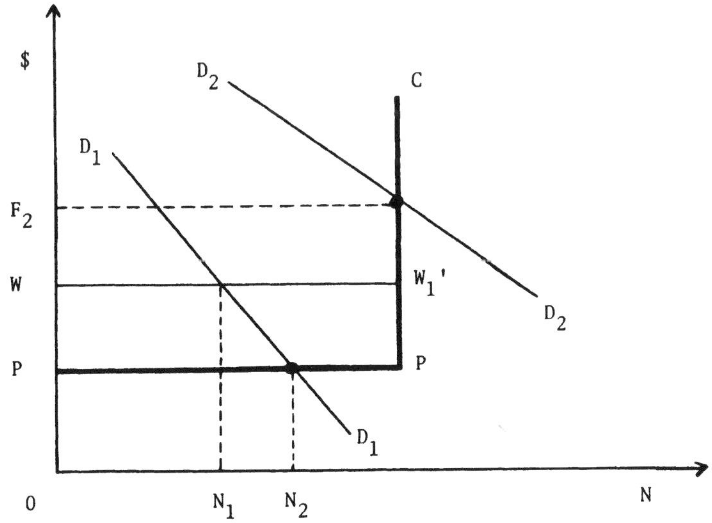

Fig. A3.5 If the social marginal costs imposed by hunters exceed the private marginal costs imposed by them on landholders, a tax on hunting rights may be called for.

indicated by D_1D_1 with the social marginal cost imposed by hunting. Hunting activity is reduced from the free market level N_2 to N_1. This level of tax equal to the divergence between private and social marginal costs is effective until capacity is reached in a free market. Once hunting capacity is utilized, no tax is required if the demand for hunting rights at a price equal to the full social marginal cost of hunting exceeds the full capacity level for hunting. For example, if the demand curve is D_2D_2, a social optimum is achieved in the market where the fees for hunting rights would be F_2. But if demand for hunting rights cuts the vertical P'W', a tax is still effective in ensuring a social optimum but a smaller one than PW is needed. The required tax will be equal to the difference between W' and the intersection point of the demand curve with P'W'.

It should be observed from Fig. 3.4 that if the demand for hunting is high and fully utilizing hunting capacity, no reduction in hunting may be called for from a social point of view even if social marginal costs imposed by hunters exceed the private marginal costs imposed on landholders by hunters. The optimal economic policy to pursue in relation to hunters is complex but the above simple theory identifies the main elements.

One simplification of the above theory is that the conditions are the same everywhere for all landholders but it can be modified to take account of differences. Spillovers or social benefits from hunting pigs may differ in different regions, so requiring variations in subsidies for pighunting in the different regions.[41]

CHAPTER 4

Landholders' Commercial Gain from Amateur Pighunting

Should landholders be allowed to make commercial gain or profit from animal pests? This is an issue not only with wild pigs but also with other pests, and there are different opinions about the effect of commercial use of a pest. Two broadly opposed views exist:

(a) Commercial use of a pest results in its conservation (or reduced control) by landholders, because they can profit from the pest population.

(b) The possibility of commercial gain results in greater harvesting or hunting pressure on a species, thus reducing its population.

Either effect is possible, depending upon the *identifiable* circumstances. Under some circumstances, commercial use leads to greater conservation of a pest, but under other circumstances - circumstances that seem to be more likely in the case of pigs - commercial use results in increased control of a pest population. Commercialization is an important strategy to consider in managing a pest.

Landholders have two main ways in which they can make commercial gain from wild pigs:

(a) Landholders may obtain income from fees charged to *amateur* pighunters for hunting, camping, accommodation, etc.

(b) By selling the rights to hunt pigs to commercial hunters, or by hunting pigs themselves and selling their harvest to meat producers, landholders can gain if wild pigs are *used in the meat trade.*

This chapter concentrates on the first possibility - an income from amateur pighunters' fees - and the next chapter deals with the second possibility.

The significance of a policy on commercial use of a pest can be seen from Australian and New Zealand experience. In Australia several organizations have opposed the use of feral pigs for commercial purposes, principally because of the fear that commercial use would result in reduced control of the pest. The opposition of the New South Wales Department of Agriculture to the commercial use of feral pigs was expressed in a letter to me in 1979:

> 'We are now introducing regulatory measures aimed at the eradication of feral pigs and can in no way condone any trade practice which would seek to traffic in a declared noxious animal.'[1]

The Department wishes to rely solely for control purposes on the noxious animal declaration, which imposes a legal obligation on landholders to take action against pigs on their property.[2]

By contrast, feral pigs in New Zealand have been used commercially for the game-meat export trade since at least 1967. This has proven to be a very effective control on the level of population of feral pigs[3] - a matter elaborated on in the next chapter. On the other hand, the commercial use of wild rabbits *Oryctolagus cuniculus*, a serious pest in New Zealand, has been banned on the grounds that it might result in less control of the rabbit population.

Policies dealing with the commercial use of animal populations often appear to be inconsistent. For instance, in New South Wales and other Australian states, the policies recognize that commercial exploitation of kangaroos for the meat and fur trades *can* threaten the populations of some species of kangaroos.[4] Similarly, commercial exploitation is recognized as a threat to other species, such as whales.[5] Hence opposition to the commercial use of feral pigs appears to be inconsistent. Why should commercialization lead to the preservation of feral pigs when it is believed to threaten the existence of other species such as kangaroos?

Four groups at least stand to benefit commercially, to some extent, from the existence of feral pigs:

(a) those selling ancillary commodities and services to hunters, such as ammunition, guns, petrol, camping equipment, etc.; as pointed out in the previous chapter, hunters spend a considerable amount on these items annually;

(b) landowners selling hunting rights and, in some instances tourist-type facilities, and landowners receiving the contribution of hunters in controlling the pig as a pest;

(c) tour operators using wild pigs as a tourist attraction;

(d) meat and game-meat processors and exporters able to utilize wild pigs.

The interests of the first group and, to a lesser extent, the third group are given particular attention in this chapter. (The fourth group is the centre of interest in the next chapter.) In the theoretical appendix to this chapter (Appendix 4B), the effect of hunting fees on the control of feral pigs is modelled, taking into account spillovers from amateur hunters. The model is of general interest and has application to the control of a wide range of pests.

As mentioned above, the charging of fees by landholders to hunters for the right to hunt wild pigs is one possibility for the commercial use of feral pigs. In Europe and North America the practice of charging for hunting rights is well developed.[6] In Australia the practice is not widespread, though it is established in south-western New South Wales for the hunting of feral pigs and occurs from the Victorian border north to the Lachlan River and its tributaries to a point on approximately the same latitude as Ivanhoe. Fees for pig hunting are reported as being charged in some parts of Queensland, but the practice does not appear to have been adopted in the central-west and north-west of New South Wales, where heavy pig infestations occur.

The payment of fees for pig hunting in south-western New South Wales can be explained by the area's proximity to Melbourne (and to Victoria generally) and the low number of feral pigs in Victoria. The main concentration of feral pigs in the south-west of New South Wales occurs near Hay, around the confluence of the Lachlan and Murrumbidgee Rivers, and around Willandra Creek slightly farther to the north. There is a big demand from amateur shooters from Victoria to use wild pig stocks in these areas, and the supply of pig-infested properties is relatively limited. In this supply-and-demand situation, a charge for hunting has been resorted to by many landholders as a rationing device.

A SURVEY OF LANDHOLDERS IN SOUTH-WESTERN NEW SOUTH WALES

The Hay district is a very suitable one for studying the charging of fees to amateur hunters for wild pig hunting. The town of Hay is 408 km from Melbourne, on the Murrumbidgee River, and can be reached by highway in 4-5 hours travelling time from Melbourne - so that weekend trips to this area are common. It is also close to the confluence of the Lachlan and Murrumbidgee Rivers, an area considered by the New South Wales Department of Agriculture to be one of high pig infestation[7] and a refuge for wild pigs.

Most properties in the area where pig infestation occurs are large and carry sheep or cattle. Most properties are subject to flooding, and many are subject to controlled flooding. A suitable habitat for pigs is provided by the cover obtained from lignum *Muehlenbeckia cunninghamii* (see Fig. 4.1) and the permanent water in swamps. The charging of fees for wild pig hunting has been common in the district.

On the advice of the veterinary officer in the district, 15 graziers were interviewed in July 1979. All except one (from the Gunbar area) had properties on the Lachlan River banks or its flood plain or on the Murrumbidgee River not far from the Lachlan. Most property holders from near Whealbah in the north to almost Balranald in the south had been contacted. Although the survey covered a large area, because the properties averaged 106 km^2 few graziers had to be contacted; together, the graziers contacted controlled 1600 km^2 of territory. A map of the survey area is shown in Figs 4.3 and 4.4.

The interviews were conducted verbally and as informally as possible, and the details written on the interview form (see Appendix 4A) immediately on leaving a property. (It was felt that a formal approach, with the survey form filled out in the presence of the respondents, would have inhibited their responses.)

Fig. 4.1 Lignum scrub forms a favourable pig habitat on the Lachlan River. Hunters may accidentally shoot stock in such scrub and can easily become lost in it.

Fig. 4.2 Feral pigs in the open, in the Hay district of south-western New South Wales

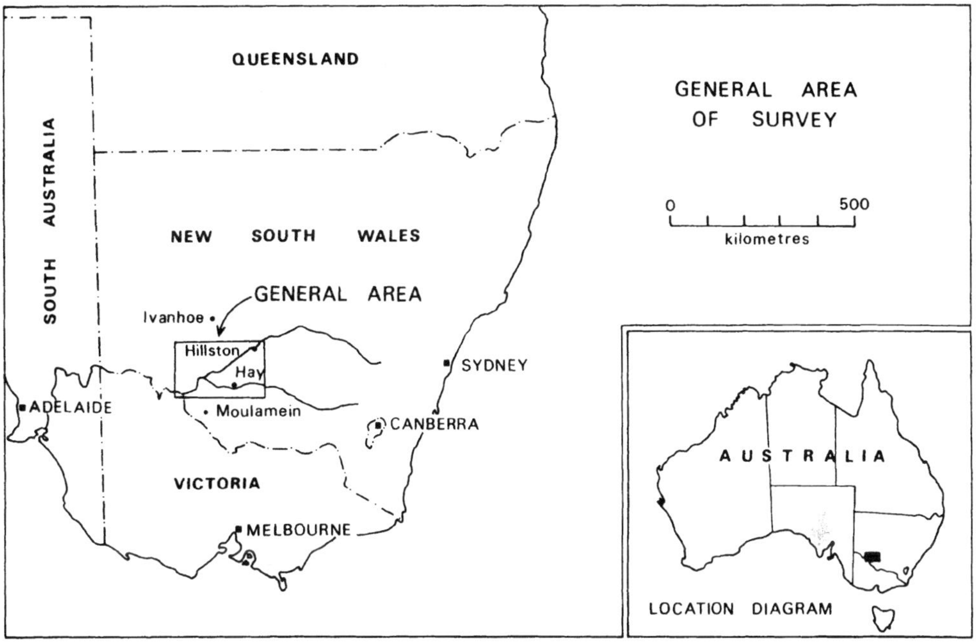

Fig. 4.3 Landowners in south-western New South Wales were surveyed to ascertain their attitudes to wild pig hunting on their properties.

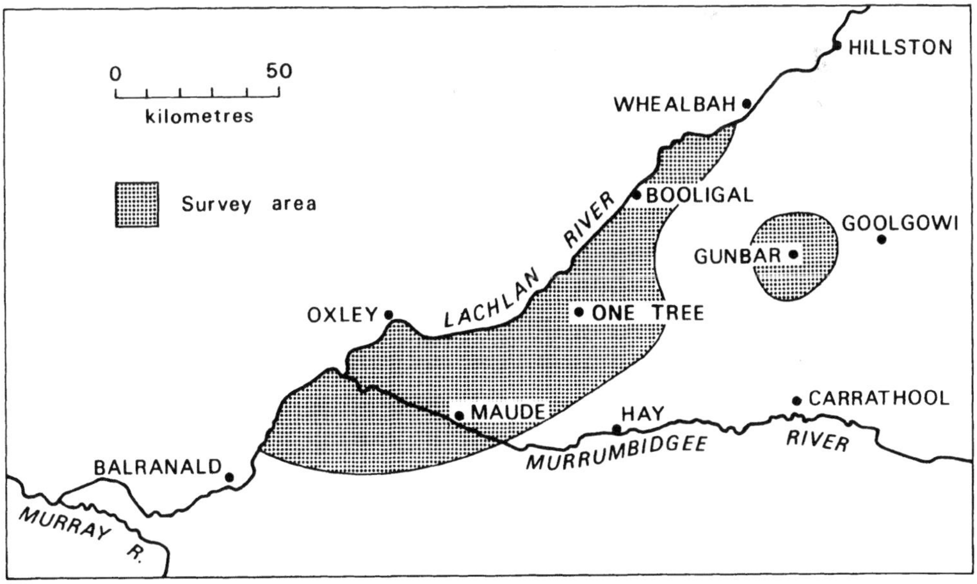

Fig. 4.4 The specific survey area in New South Wales

TABLE 4.1 Size of Properties in Interview Sample and Number of Pigs

Property	Acres	Hectares	km^2	Estimated minimum number of pigs
A	24 710	10 000	100.2	150
B	78 813	31 894	319.1	600
C*	9 885	4 000	40.2	300
D	35 000	14 164	141.7	200
E*	23 000	9 308	93.0	200
F	16 274	6 586	67.6	200
G*	19 300	7 811	78.2	200
H	25 000	10 117	101.3	200
I	7 000	2 833	28.2	100
J	17 000	6 880	68.9	400
K**	17 000	6 880	68.9	600
L	27 000	10 926	109.3	600
M	36 000	14 568	145.8	600
N	40 000	16 187	161.9	300
O**	18 000	7 284	73.1	300
Total	393 982	159.438	1 597.4	4 950
Average per property	26 265.5	10 629.2	106.493	330

* No fees charged, hunters allowed

** Hunters not allowed

Table 4.1 lists the sizes of properties and the estimated minimum number of pigs on each at the time of the survey. The graziers found it difficult to estimate precise numbers, but all pointed out that because of recent rain the pigs were more scattered and widespread than usual. The estimates of pig numbers should therefore be interpreted as giving merely an order of magnitude.

The Admission of Hunters to Properties and the Practice of Charging Hunting Fees

Of the 15 graziers and landholders interviewed, 13 admitted hunters to their properties for wild pig hunting and two did not. Of the 13 admitting hunters, 10 charged a fee. Let us consider the position, firstly, of graziers not admitting hunters; secondly, of landholders admitting pighunters for a hunting fee; and, thirdly, of those allowing pighunting free of charge.

The two graziers not admitting hunters did so for different reasons. One specialized in wool production and considered pigs to be a serious menace to his sheep flock. But in his view amateur hunters were not effective in controlling pigs; they missed too many pigs and scattered them, making it more difficult for him to control their numbers. He had his own control, using dogs - using scent dogs to track down pigs and using other dogs to hold them. The second grazier specialized in beef production. The manager did not reside on the property, and the owners feared that shooters would accidentally kill cattle if permitted on the property. The pigs were considered to be little or no economic threat to the cattle. No control measures were taken against pigs on this property, a matter complained about by some neighbours.

The graziers admitting hunters to their properties had a variety of reasons. In many cases, it was considered to be a cheap means of controlling the numbers of pigs, a source of additional income and a needed response to continual requests and pressure from hunters.

The 10 graziers charging fees did so for a variety of reasons. In almost half the cases the fees were considered necessary or desirable to ration the number of hunters coming on to the property. Before the introduction of charging, car loads of hunters would arrive on some weekends and pester owners to permit them to hunt, and the situation became worse as roads were improved. By charging hunters and allowing a limited number on and encouraging bookings, a grazier was more easily able to turn away casual shooters when it did not suit him to have additional shooters. A few graziers also claimed that 'the more undesirable shooter' would not pay and stayed away from the property when fees were charged.

With the introduction of fees, illegal hunting on the property was reported to be reduced, and the owner knew who was hunting on the property. Hunters legally on the property, having paid a fee for an area to shoot on, helped to keep poachers or illegal entrants away. They tended to enforce their hunting rights. In turn, this helped to reduce vandalism by undesirable hunters. The vandalism could take many forms:

(a) instances were related to me by graziers, and in one instance I was shown photographs of fences cut, gate locks shot out, gates damaged by ramming cars, outbuildings and water tanks damaged by shooting,

(b) and was told of the shooting of livestock, the theft of farm property, the lighting of fires, and the abandoning of pig hunting dogs, which then joined wild dog packs and/or savaged sheep.

Most of the landholders charging fees also mentioned that one reason for doing so was to help recompense them for any damage to their property that might occur as a result of hunting. None were covered by insurance for such damages. Also they pointed out that the fee helped to recompense them for extra effort on their part; they needed to give instructions to shooters (for example, where to shoot and other restrictions) and this took time. On occasions they were called on to find lost hunters or to pull out bogged vehicles owned by hunters.

One grazier related how he had tried different policies on the charging of hunting fees and the admission of hunters. At first, hunters were admitted without charge. But with improvements in roads and transport, the number of hunters increased, and damages caused on the property by hunters increased. He then decided to ban all hunters. Illegal entry occurred, however, and the damages continued. The owner did not know who was shooting on the property. He issued summonses against one group of hunters and obtained convictions but did not find this satisfactory. Reprisals were suspected - fires possibly deliberately started.

On the advice of an official of the Sporting Shooters' Association, he then decided to introduce a system of charging hunting fees and issuing permits. He found this to be the most satisfactory method of control. Some income was received to compensate for damages caused by hunters, the hunters paying for an area kept poachers away, and the owner obtained details of hunters permitted on the property. He would have preferred to have no hunters on the property but this could not be enforced, given the large size of the property, the fact that roads cut through it, the main road to it was sealed (enabling easy and speedy access) and there was considerable pressure by hunters for hunting. Restrictions could be more easily imposed on hunters allowed on the property, now that fees were charged, and this was considered by the owner to be a further advantage.

The three graziers allowing hunters but not charging fees had different reasons for not charging fees. One said that in the past he had charged fees but now, as a local government councillor, he was away too much and his ability to collect fees was therefore limited. His property was relatively inaccessible. The second grazier said that if he charged fees some unsuccessful pig hunters might kill cattle to get value for their money. The third did not charge because he believed he could thereby be more choosy about the type of hunter he permitted on his property. His property was well away from the main road and he limited hunters on the property to the group known to him as being reliable.

Landowners' Income from Hunting Fees

When fees are charged, the income received from hunting fees[8] appears to rise with the number of pigs on a property; this relationship is indicated in Fig. 4.5, which is based on Table 4.2. Fees on a daily equivalent basis (the daily fee, or

TABLE 4.2 Income Received from Feral Pigs by Landholders on Properties Charging Hunting Fees

Property	Number of Pigs (Min.est.)	Pigs/km^2 (Min.est.)	Fee on a daily basis $	Annual income from feral pigs $
A	150	1.5	5.00	790
B	600	2.0	5.00	3000
D	200	1.4	5.00	750
F	200	2.9	7.50	1080
H	200	2.0	10.00	3430
I*	100	3.5	1.00	160
J	400	5.8	15.00	1500
L	600	5.5	15.00 - 20.00	4000
M	600	4.1	20.00	3000
N	300	1.9	10.00	2000

* Exceptional property. Smaller than most with a considerable amount of area used for growing corn. Damages caused by pigs high.

half the weekend fee) indicated in Table 4.2 appear to rise with pig densities. The relationship is graphed in Fig. 4.6. On the whole, *properties with the highest densities of pigs charge the highest daily equivalent hunting fees.* The demand for hunting by hunters appears to be positively related to (that is, rises with) pig densities. (This relationship is used in the theory developed in Appendix 4B, which considers the implication for the control of pigs of the charging of fees for pig hunting.)

None of the landholders on the three properties with the greatest densities of pigs (and charging highest fees) were breeding sheep. They specialized in cattle production and, considering their product-mix, their properties were not greatly damaged by pigs. However, one grazier mentioned that pigs had forced his property out of sheep, and another grazier said that his lambing was done elsewhere to avoid pig predation. There appears to be a broad relationship between the level of the hunting fee charged and the damages anticipated from pigs on different properties. As a rule, the greater the anticipated damage from pigs, the lower the hunting fee, presumably to encourage hunters to control pig numbers. The fee was lowest on a property specializing in irrigated corn production (in one year, this property lost corn worth \$20 000 to \$30 000 to pigs); higher in the case of properties involved in sheep production; and highest on cattle properties. (The model outlined in Appendix 4B indicates that this pattern of behaviour is rational from the land-

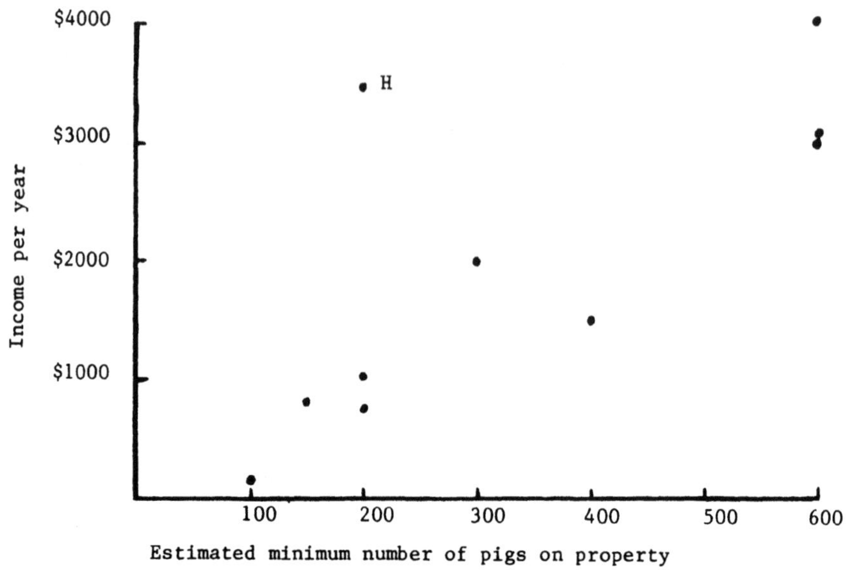

Fig. 4.5 Relationship between the annual income received by landholders from hunting fees and the number of pigs on the property. The relationship is positive. Property H was unusual, in that according to its owner the pig numbers were much lower than normal.

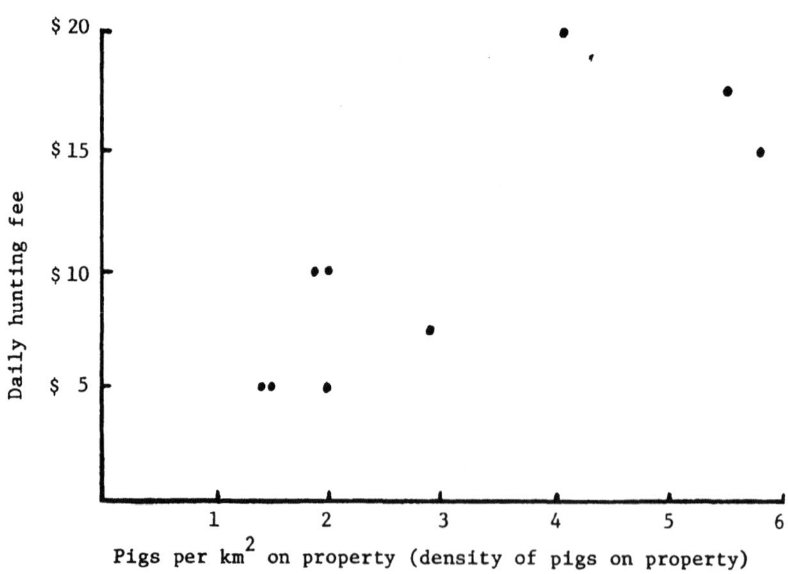

Fig. 4.6 Relationship between fees charged by landholders for pig hunting and density of pigs on the property. Fees tend to be highest on properties with the greatest pig densities.

owner's point of view).

It is noteworthy that hunting fees have not increased since the end of the rural recession of the last half of the 1960s that continued until 1972 and in real terms have therefore fallen. Because primary produce is now of greater value, the theory outlined in Appendix 4B predicts this result. It is now relatively more profitable to reduce pig numbers, in view of the higher value of the primary produce lost, and to encourage hunting. In at least two cases, farmers were increasing cropping on their properties; they therefore wished to reduce pig numbers and had kept their fees low with this in mind. Thus landholders tended to adjust their hunting fees rationally, taking into account their own economic gain, as their economic circumstances altered.

The fees charged in the survey area for wild pig hunting are low by world standards. In Texas, U.S.A., the fee for taking a European wild boar is $100-400 per pig, and $10-75 for a feral pig.[9] None of the graziers in the Hay district charged on the basis of number of pigs taken but, assuming an average of one pig taken per day, the comparable rates would range from $1 to $20 per pig.

Other Income of Landholders from Pig Hunters and Tourists

Two owners had diversified, to provide farm holidays on their properties; through Melbourne agents, cottages on these properties were let to holidaymakers. This type of business was considered to be marginally dependent on the presence of feral pigs. Camping sites were provided (at a fee) on another station and although these sites were mostly used by pig shooters, other people also used them. Another owner had intended to promote weekend 'safaris', using his own aeroplane, providing cabins on the property and supplying an 'outback package'; two groups had been taken, but he had not persisted with and promoted the safaris, presumably because, beef prices rose again and for him it became more profitable to attend to his cattle than to foster 'safaris'.

One novel method of earning income from less-successful hunters was observed in the Hay district. A grazier caught feral pigs for cross-breeding with white boars; the semi-domesticated feral pigs and their progeny were let to range and at night were attracted back to the sties with minimal amounts of grain (see Fig. 4.11). Less-successful hunters were offered small semi-domesticated feral pigs at $20 each and large ones at $40. The grazier could often earn more than $1000 annually from this source.

There is no doubt that the earning of income from hunting fees has been important to graziers in the Hay district. Of the 10 graziers charging hunting fees, at least four considered that the fees had been critical in keeping their properties viable during the 1968-72 rural recession. Eight of the 10 considered the income received from the hunters had been especially important to them during the recession. All those charging fees considered the extra income to be a useful bonus - except one, who set his fee at $1.00 per hunter merely to cover expenses associated with the exercise. One grazier, in particular, pointed out that his property was only marginally economically viable; the sole use of his land had been for wool production. He saw the hunting and camping enterprise on his property as a way to diversify and so reduce his risk and he had promoted it for this reason.

The Landholders' Reasons for allowing Hunting, and the Restrictions placed on Hunters

In all but three cases, control of pigs appeared to be an important reason for allowing hunters on to the property. In the three exceptional cases, control of wild pigs was not the main reason for admitting hunters. Additional income to be earned from the hunters was the prime consideration. One of these landholders had in fact *introduced* wild pigs to his area during the rural recession so as to earn a supplementary income from hunters. Since then, however, control had become his main concern, because he was growing more crops - crops are more susceptible to damage from pigs - and primary produce had increased in value. In the other two exceptional cases, little damage was experienced from pigs.

In the seven cases where control was important as a reason for admitting hunters and hunting fees were charged, income was also a significant consideration in six cases; only in the seventh case was income considered to be of no significance. As is pointed out in Appendix 4B, the earning of income from hunting fees and control of the number of wild pigs can be compatible. Furthermore, greater control of pigs may be achieved by charging fees to hunters than would occur if outside hunters were not allowed on properties and property-owners took their own most profitable control measures. *The commercial use of wild pigs can result in more control of pigs than their non-commercialization.*

All landholders charging fees pointed out that hunters would hunt on their properties, illegally if necessary. There is no practical way to keep hunters out. By charging fees it was possible to regulate the entry of hunters to the property. Hunters obtaining a hunting right kept away poachers, the numbers hunting were able to be controlled, knowledge was obtained about the hunter (his name, address, etc.), and a contractual arrangement could be established. Two landholders indicated that they felt it important, in principle, that shooters and city people be given access to country properties. In their view, all Australians had a 'right' to enjoy the countryside. One said that he wished to encourage hunting and good hunting habits and to give shooters a chance to learn these.

None of the property holders interviewed had specific insurance against risk from hunting (for example, accidental shooting of domestic stock) or expected hunters to take out such insurance. Those charging fees felt that the fees helped to cover their risks. A number of properties had public risk insurance.

To some extent, restrictions placed on hunters by property-holders helped to reduce risks from hunting. The following restrictions were noted on different properties:

(a) no shooting within a specified distance of the house and main buildings;

(b) shooting permitted for the party only in a specified area;

(c) limitations on size of firearms to be used.

Dogs brought by hunters were a contentious issue. On a few properties, dogs were not allowed at all and their use was restricted on others. In only one case was the use of dogs for pig-hunting actively encouraged, the case of the corn-grower who believed that they were most efficient in bringing pigs to bay. Sheep growers in particular were wary of dogs, because some had been left by hunters and had killed sheep. On one sheep station, dogs were not allowed and hunters were requested to shoot any dogs seen in their hunting area. The possibility that hunters would shoot livestock, damage property or steal it and create other risks was a problem for landholders and was likely to be taken into account by landholders when considering granting permission to hunt.

Many landholders took care to ascertain the identity and address of hunters. Car licences and, in some cases, valid shooters' licences had to be produced for this purpose. One landholder insisted on seeing valid New South Wales' Shooting Licences. Another required hunters to leave a deposit with him, to be collected at the end of their hunting period on the property. In some cases, hunters had to provide personal recommendations. One grazier said that if a hunter could produce membership of the Sporting Shooters' Association (a voluntary association of Australian shooters) he would automatically be allowed on the property.

All landholders regulated the number of shooters allowed on their property at any one time. Some allowed only one party at a time, but others allowed more. One owner, however, admitted that on long weekends too many parties may be allowed on his property to maintain safety. To improve safety, some owners specified different areas in which the parties could shoot; but not all did this and there was the problem that a hunter could not always identify his assigned area. Indeed, some owners had received complaints from neighbours about shooters 'accidentally' straying across property boundaries. In a few cases, the neighbours alleged that such trespassing had been encouraged by the property-owners' charging hunting fees.

During the summer months or in periods of fire danger, many property owners ban or restrict hunting to reduce their risk from fire. Hunters may accidentally start fires - from cigarettes or from campfires. One landholder expressed concern that with the introduction of the new New South Wales law allowing *public access* to rivers and streams it would be more difficult for him to control his risk because the Lachlan River flowed through his property. He foresaw a greater risk of fire and shooting of domestic stock.

No landholders surveyed in the Hay district placed a limit on the number or size of pigs taken by a hunter. However, in one case hunters were advised to take only boars and were allowed to take sows only as a last resort. This was a property on which pigs were not considered to be a pest even though they were present in large numbers. In another case, hunters were advised not to kill pigs wantonly. Hunters were asked that if they did not intend to consume the pigs themselves, they give them to the property-owner, so that he could supplement the diet of his poultry or use them in other ways. Two landholders charging fees expressed concern that their pig numbers had fallen and were worried about the possibility that if numbers fell too low their properties would no longer attract hunters. One grazier said that continual shooting on his property tended to scare pigs to the neighbouring property, where no control measures were taken. He systematically spelled areas from shooting, putting his domestic stock in the spelled areas. When these areas were eaten out and pig numbers built up, he shifted his stock and allowed shooters in; however, at the time of the survey he thought that he might need to spell the whole property from shooting for a while. Another said that shooters had been very effective in reducing numbers of wild pigs on his property and that he might have to limit numbers of shooters so that enough pigs remain to make the property attractive to shooters. He wanted to keep a reserve of shooters interested in hunting on the property, in case pig numbers should suddenly build up, and hunters were needed to exert more control. Furthermore, he did not want the hunting side of his business operations to collapse because prices for rural produce (for wool) could slump, and the hunting income was a useful stand-by. He also wished to continue with the hunting side of the business because he enjoyed social contact with the hunters and their families, many of whom had been coming for years. Several other graziers also mentioned the importance of social contact.

One grazier provided hunters with a written permit, a printed brochure on cautions and requirements (see Fig. 4.7) and a hand-drawn map of the specified hunting area covered by the permit. Most graziers were not so formal, although one was reputed to issue a lapel tag to authorized shooters. The tag had to be worn on the property so that the owner could more easily check whether a hunter had paid

CAUTIONS

(a) Snakes - all varieties. Snake bite outfit is essential.
(b) Broken bottles especially under water and in undergrowth.
(c) Treat wounded pigs carefully - can be very savage.
(d) Old sow with piglets if separated from young can be savage.
(e) Very easy to become bushed, make a land mark by sun or such like, before leaving camp.
(f) Communicate with other members of party frequently.
(g) Lock all camp gear away before leaving if unattended.
(h) Unauthorised shooters.

REQUIREMENTS

(a) All rubbish to be taken away from camp site.
(b) No shooting or entering areas other than those stipulated in the agreement.
(c) No fences to be damaged along with signs.
(d) Agreement to be produced on demand by police or game official.
(e) Portable cooking appliances preferred rather than open fire.
(f) No shooting of any other animal than those in the agreement.
(g) No spot lighting of any kind allowed at any time.
(h) Such restrictions as to lighting or making fires in the open as may exist from time to time are to be strictly adhered to.
(i) The prohibitions concerning all protected native flora and fauna are to be strictly observed.

Any of these requirements not adhered to will automatically lead to prosecution to the offending parties, it doesn't take much but a little care on your part for a successful holiday and shoot.

THANK YOU.

Fig. 4.7 Extract from brochure issued to campers and hunters by a property owner in the Hay district

for hunting rights. A few graziers collected data from hunters for analysis - number of pigs taken, etc.

Damages Caused by Pigs and Control Measures

Table 4.3 sets out the main agricultural and pastoral damages caused by feral pigs in the Hay district, as mentioned by property holders. Reported lamb losses ranged from zero or moderate to serious and *are correlated* with (that is, are closely associated with and rise with) the estimated density of pigs. Losses of lambs were serious in areas where pig densities were high. Property owners near the junction of the Lachlan and Murrumbidgee Rivers (an area of very high pig density) reported that they were unable to let their sheep *lamb* in the area because of predation by pigs. One grazier sent his sheep to Moulamein (130 km away) to lamb, and another respondent said he had been forced to give up sheep production because of pigs' killing lambs and relied on cattle for his income.

Crop losses from wild pigs can be considerable, although cropping is limited in the Hay district. Some crops are grown under irrigation. The Department of Agriculture agronomist at Hay said that pigs were especially damaging on irrigated crops in the district. Over the whole sown area in the district, he estimated the average loss of crop at 5 per cent, about equal to losses caused by galahs *Cacatua roseicapilla*. In areas where pigs are present, average losses to crops would be 8-10 per cent. In some years (particularly in dry years when there is little alternative feed for pigs) crop losses would average 10-30 per cent; and in good seasons of abundant feed, losses would be around 5-15 per cent.

Properties specializing in cattle production experienced little or no damage from pigs, according to their owners or managers. Some owners, however, expressed concern that pigs could carry exotic diseases, such as foot-and-mouth disease, if these happened to be introduced. Such a disease would devastate the Australian beef industry and halt much of current beef exports.

Thirteen of the landholders interviewed felt that feral pigs should continue to be classified as noxious in New South Wales, but most could not see how the regulation could be effectively enforced. One dissenter thought that a less rigid view was needed, although he felt that the feral pig should be regarded as a pest, and some control by landholders should be required. The classification 'noxious', however, places an eradication requirement on landholders and this was considered impractical and probably undesirable. The other dissenter opposed the noxious classification. Both of these landholders charged fees and both were worried about sheep predation by pigs. Only in one case did a landholder express a wish

TABLE 4.3 Damages Caused by Wild Pigs in the Hay District New South Wales, July 1979

Property	Type of Production	Damages by pigs complained of
A	Sheep, some cropping	Some damage to wheat, sorghum and other crops but some benefits since pigs eat mice. Since increasing cropping, is worried about damages from pigs, but because the season is good, expects little damage this year because pigs have alternative feed.
B	Sheep	Pigs only take weak lambs. Damage from pigs minimal.
C	Sheep	Little damage.
D	Sheep	Little damage. May kill some sheep.
E	Sheep, some cropping	Lambing losses but cannot be specific. Nests of pigs in sorghum cause some damage. Eat maize. Has had his maize harvester damaged by pigs not getting out of its path. Problem in crops.
F	Cattle, sheep, limited cropping	No great loss. Some lamb losses and some oats lost.
G	Cattle, sheep	Few pigs eat lambs but not a big problem.
H	Sheep	Lamb losses can be high. Forced into pig control by such losses.
I	Cropping, sheep, cattle	Lost $20 000 to $30 000 worth of corn in one year. Pigs knock down and eat corn. Problem in irrigated crops.
J	Cattle, sheep	Some lamb losses but mostly has cattle. Is growing wheat and is worried about losing it to pigs.
K	Cattle	No damage of significance.
L	Cattle, seasonally sheep	Cannot lamb on property because of pig predation. Lambing is done at Moulamein. Has a good deal of wheat in this year for the first time. Worried about damage of pigs to wheat. Will consider electric fencing of wheat crop.
M	Cattle	No damage to cattle but forced out of sheep by pig predation. Problem if foot-and-mouth disease enters the country.
N	Sheep	Lambing losses significant, rooting of ground by pigs provides a seed bed for Bathurst burr. (See Fig. 4.8) Significant in the spread and growth of Bathurst burr *Xanthium spinosum*.
O	Sheep	Sheep losses heavy especially in a dry season. Lambs and sheep of size killed. Lambing rate was as low as 20 per cent. Since active pig control with scent dogs and other dogs, lambing rate has been raised to 60 per cent.

to see the present regulation declaring the feral pig as a noxious animal fully enforced. On the whole, there would probably be considerable opposition by landholders to the full enforcement of the regulation in the Hay district, even if enforcement was practical.

By the way, it is not apparent that state authorities in New South Wales really expect all landholders to eliminate feral pigs on their properties. To date, no prosecutions have been made under the law declaring feral pigs noxious. In this respect, Howard's observation based on New Zealand experience seems relevant:

> 'Initially, with control programmes, public support can often be rallied more easily with extermination as the objective, but sooner or later a more realistic attitude must be developed or serious confusion results. However, unless responsible government specialists have the authority to decide at what level of infestation control must begin, an overall policy of extermination is often safer.'[10]

On the subject of control, eight of the 15 respondents believed the current position to be satisfactory as far as they were concerned. About five said that they would like to see more co-operation with neighbours in controlling pigs. Another pointed out that his property adjoined extremely large reed beds and swamps at the junction of the Lachlan and Murrumbidgee Rivers, a refuge for pigs and heavily infested; for control to be effective, it would need to occur in the reed beds - otherwise re-infestation would occur. One property owner, however, suggested that the regulations classifying feral pigs as noxious animals should be rigidly enforced and wild pigs decommercialized. He had suffered heavy corn losses due to pigs. In his opinion, those properties charging fees to hunters should be restricted to a minimum fee - for example, a couple of dollars per hunter to cover costs.

Nine of the responding landholders depended *solely* on amateur shooters for the control of pigs on their properties, and 12 or 13 depended *primarily* on amateur shooters. One of these graziers shot pigs, two others made limited use of electric fencing, and one had experimented with traps as a means of added control (see Fig. 4.9). One grazier who did not allow hunters on his property and was therefore *solely* responsible for the control of pigs on his property used dogs with good results: he had about 50 dogs for this purpose - scent dogs to track down pigs, and others to go in and hold pigs; he believed this to be the best method and more efficient than amateur hunters. On one property, no control measures of any kind were taken. Most of the landholders permitting amateur hunters on their properties believed that they were effective in controlling pig numbers, that the exercise was costless to them or nearly so, and that in most cases where fees were

Fig. 4.8 Effects of rooting by feral pigs. Such rooting provides a seed bed for the Bathurst burr, a weed pest.

Fig. 4.9 Three pigs caught in a trap at Hanna's property, 'Erin', near One Tree. Wild boars on this property have caused heavy lamb losses.

charged they yielded a profit.

None of the graziers were using poison on pigs, although one had in the past used SAP (Sayer-Allport phosphorus) and strychnine. Most were opposed to the use of 1080 (sodium monofluoroacetate) for the poisoning of pigs.

At the time of the survey, several landholders were seriously considering electric fencing of crops to keep pigs out. Even pig wire fencing (such as that shown on the trap in Fig. 4.9) would be economical in some cases. The Hay district agronomist gave me the example of one property in the district with 1000 hectares of irrigated maize having an expected yield of 6.8 tonnes to the hectare. With maize selling at $89 per tonne, and the loss of maize by pig predation being about 6-7 per cent, the loss would amount to about $40 000. The crop can be contract-fenced with pig wire and an electric wire offset provided for around $30 000. The cost of fencing to exclude pigs would therefore be recouped after one crop.

The Demand for Hunting Rights

Most of the hunters hunting in the Hay district came from Victoria, mainly from the Melbourne area. A few came from Griffith, N.S.W., or Canberra. Many properties had a stable clientele, and some national (ethnic) groups favoured particular properties. One landowner said that all of his clientele were of White Russian origin and that they had the first claim on hunting rights on his property. Several other national immigrant groups had priority on other properties, and some maintained overseas traditions associated with the wild pig.

None of the landholders felt there was a need to encourage more individuals to come hunting wild pigs in the district. Most felt that an excessive number of hunters wanted to come hunting on their properties. But most graziers thought it would be undesirable for them to raise their hunting fees, because unsuccessful hunters might be disgruntled, and a few graziers mentioned that they might therefore be more inclined to shoot stock. Landholders favoured the current position in which the demand for hunting rights exceeded the supply, because they could be more selective in allowing hunters on their property; they could exclude hunters who posed a risk to their properties.

The Use of Wild Pigs for Meat

Graziers reported that wild pigs taken by hunters were mostly eaten by the hunters and that this tendency had increased with the rise in meat prices. In some instances, hunters brought freezers or were loaned freezers by property owners to freeze the meat before taking it back to Victoria. Others came during winter and took the unfrozen meat back in the cooler weather.

As for other uses of feral pigs for meat, some graziers in the area caught feral pigs in sow and raised the litter for their own use as meat. Captured feral sows were sometimes crossed with a white boar (the progeny were mostly white) and the progeny raised for meat for the farm (see Fig. 4.10). However, feral pigs were reported to be 'poor doers' (that is, unthrifty) compared to domestic pigs and if well fed were inclined to add fat rather than meat. In the 1950s there had been a large operation in the Hay district to catch wild pigs using dogs, to de-worm them and hold them in sties for up to three months and then to sell them in Victoria; some had needed to be held for only six weeks (thus they were held for only a short period). In general, it does not appear economic to sty feral pigs for any length of time (in comparison to domestic types) because much of the food is converted into fat. In the rural recession, one grazier in the district had kept semi-feral pigs on range and would bring them to sties occasionally for feeding and for sale to local butchers.

This type of trade had become insignificant in the district but, as mentioned earlier, one grazier had 'domesticated' feral pigs and feral-domestic pig crosses to sell to hunters. These pigs were stied at night, given a little grain and allowed to roam by day at range (see Fig. 4.11). One property owner thought that feral pig meat should be made available to less-developed countries and deplored the possible waste in not making greater use of the meat. But in this respect, it should be observed that it is not costless to catch wild pigs and process and export them. Opportunity costs are involved, and for the same cost more effective forms of aid may be available.

The Husbanding of Feral Pigs

Only one grazier seemed to be husbanding feral pigs to any great extent. He had released white boars to upgrade the feral stock and requested hunters to take only boars. He had also taken wild pigs to sty, enticing them to the sties with semi-domesticated ferals, for example, with sows in heat. Another had followed a similar practice in the past and had been responsible for introducing feral and free-roaming pigs to his area for possible commercial gain; now he was anxious, because he had increased the growing of crops, to reduce their numbers as much as possible. Another grazier mentioned that a hunting friend had released a white boar on his property to upgrade the ferals. But the white boar disappeared quickly and was possibly killed by feral boars.

Fig. 4.10 The results of a white boar-feral sow cross. Petula, one family's pet pig, is in the centre. Some graziers rear such crosses to supply pork for home use.

Fig. 4.11 Some graziers rear stied feral pigs for domestic use but such pigs are reported to be 'bad doers' (unthrifty). One district grazier sells them to unsuccessful hunters.

As pointed out earlier, some graziers sometimes limited hunting pressure to ensure that pig numbers did not fall below a threshold required to make their properties attractive to hunters in the long term. One grazier also systematically spelled parts of his property to ensure a build-up of numbers of pigs in these before opening them to paying hunters. He managed wild pigs as a natural resource in order to increase his income from hunting fees.

Comments on the Charging of Hunting Fees

Some graziers charging hunting fees complained that their neighbours were excluding hunters and taking no control measures against wild pigs. The neighbours in turn complained that fee-chargers were husbanding pigs and increasing pig numbers. In one case already mentioned, a grazier suggested that wild pig hunting should be decommercialized to obtain effective control of the population of pigs. He suggested that the eradication intention involved in declaring feral pigs noxious be enforced and that landholders be limited to charging a nominal sum to hunters for the right to hunt pigs. He quoted the decommercialization of the rabbit in New Zealand as a worthy example to follow in achieving greater control.

However, as discussed in Appendix 4B, there may be *less* economic incentive to *control* feral pigs under decommercialization than under commercialization. In some areas, private control and even regional control of wild pigs can be *uneconomic* considered merely from the viewpoint of the landholders, or the optimal level of economic control may be on a small scale. Decommercialization could lead to less control of wild pig numbers, because, taking adverse spillovers or side-effects from hunters into account, graziers might admit fewer hunters and have little or no economic incentive to increase the number of pigs taken by them to compensate for the lower numbers taken by hunters. This is especially likely to be so on properties - such as cattle properties - where pigs cause little apparent damage. As mentioned earlier, the manager of one cattle property in the Hay district at the request of its overseas owners did not admit hunters, because even though pigs were present in large numbers they did little damage to the enterprise and it was anticipated that hunters would do more damage, by accidentally shooting cattle, for example. Other cattle properties admitted hunters, however, principally to obtain extra income. In the absence of worthwhile income, they might follow the example of the property just mentioned.

Most graziers who allowed paying hunters on to their properties in significant numbers believed that hunters were effective in controlling numbers of pigs. One owner who estimated that there were more than 300 pigs on the property had kept records indicating that approximately 2000 pigs had been taken by paying hunters in the previous 18 months. As he pointed out, if he had to buy only the ammunition to shoot this many pigs, it would have been expensive. Incidentally, as he had 100 visiting hunters per year, each hunter took on average about 10-15 pigs per visit.

TOURISM, SAFARIS AND WILD PIGS

There are groups apart from landholders interested in the possibility of tourist and hunting income from wild pigs. As well as those gaining from the sale of ammunition, guns, petrol, etc., some businessmen are interested in promoting tours and safaris, in which wild pigs play a part. Some tourists are mainly interested in seeing wild pigs and other forms of wildlife; in the survey, one grazier said that part of the reason for introducing wild pigs to his property was to give holidaying family groups staying in his holiday cottage a glimpse of them.

The organization of safaris or shooting expeditions for novices creates difficulties as far as safety is concerned. However, one group of professional shooters located in Hay (R. and M. Vivian) was planning to operate week-long game safaris and were hoping to attract experienced hunters from overseas. Game to be hunted included wild pigs and goats, and trophy tusks and horns were promised. The safari-side of the business was being planned to give greater diversification to their business and so reduce their risks. The group was planning to continue its operations in professional commercial hunting of various species, such as kangaroos.

There are other tourist operations in Australia drawing upon wild pigs as a resource, although not exclusively so. Outback Horizons, of Nailsworth, S.A., provides an outback safari service designed to cater for a wide range of groups. Their brochure indicates that they 'cater for all classes of hunters and observers - trophy hunters; tenderfoot hunters; lady hunters; bow hunters; non-hunting observers, photographers; muzzle loaders.'[11] For the safaris to the 'back of Bourke' (west of Bourke, N.S.W.), wild pigs, goats and foxes are often featured as attractions for hunters. WIMRAY Pty Ltd operate safari tours from Darwin, N.T., with the hunting area situated about 50 miles from Darwin. Pig shooting is one of the attractions. Indeed, the company reports that since it commenced operations customers have come mainly to hunt pigs, '...although we consider our main attractions to be buffaloes and barrumundi fishing.'[12]

The income received from hunting fees and from tours and safaris generated from the presence of wild pigs in Australia is small at present. But if international accessibility to Australia should increase, the demand by hunters could well rise. For instance, in Europe the demand for the hunting of pigs is high relative to wild pig stocks, and hunting in Australia could be encouraged as a part of an international tourism drive. A potential appears to exist which could be exploited with suitable promotion and back-up facilities in Australia.

CONCLUSIONS

The survey of landholders in the Hay district allowed us to understand the view of the landholders (who were mainly pastoralists) of the pig as pest and the pig as resource to be used commercially. The main commercial gain of landholders in the studied area from wild pigs was from hunting fees charged to amateur hunters. The research results show that landholders usually act rationally in their pig-control strategies, in their admission or exclusion of hunters from their properties, and in their charging of hunting fees. Landholders appear to be motivated by self-interest and in most cases this approximates to maximizing the economic gain (that is, net income) from their properties. The control strategy needed to maximize net income varies between properties.

The number of wild pigs on a property do appear to be related to the hunting pressures on the property. Indeed, many landholders surveyed believed that amateur hunters provided an effective means of controlling the number of feral pigs on their properties. On the other hand, the commercial use of feral pigs remains controversial in Australia, because some authorities fear that it will lead to reduced control of the pest. However, in some circumstances greater control of feral pig populations is obtained by their commercial use; this is examined further in Appendix 4B.

APPENDIX 4A

Name: ________________________________ Location: ______________________________

Size of property: _______ hectares. Estimated number of feral pigs: __________

Please tick appropriate boxes below

1. Are wild pigs hunted on your property by other than your family? [ ] Yes [ ] No
2. Do you charge a fee to hunters for hunting wild pigs on your property? [ ] Yes [ ] No

   If yes, what is this fee and how is it determined? _______________

   If no, why do you not charge? __________
3. How many shooters or hunters of wild pigs would visit your property per year:

   (a) as non-paying guests ______ (Number)

   (b) as paying guests ______ (Number)

   Where do they come from? __________
4. If you do charge for pig shooting, what would the average payment per customer be? $______
5. Would you mind indicating the gross income that you receive per year from hunters of wild pigs? $______

   How important is this extra income to you?

   [ ] necessary for economic viability of property

   [ ] necessary for economic viability of property in some years

   [ ] useful additional income

   [ ] other _________________
6. Do you obtain income from hunters of pigs from other than hunting fees? [ ] Yes [ ] No

   If yes, please specify __________
7. What is the main purpose of letting outside hunters on your property?

   [ ] to control the number of wild pigs

   [ ] to provide extra income

   [ ] other ________________

8. Do you have arrangements to insure against risks from hunting? [ ] Yes [ ] No

   If yes, please elaborate __________ .

9. Do you place any restrictions on hunters? __________

   Areas in which can shoot? ______ Number and size of pigs? ______

   (Bag limit) ______ Other ______

10. What are the main damages and extent of damages caused by pigs on your property? __________________________________

11. (a) Do you think that feral pigs should continue to be classified as noxious animals? [ ] Yes [ ] No

    (b) What should be done about controlling them? ________

    (c) What means of control do you consider are most economical?

    [ ] Shooting [ ] Poisoning [ ] Electric Fencing

    [ ] Fencing (other) [ ] Other (please specify) ________

    Do you use any of these means? [ ] Yes [ ] No

    If yes, please specify _______________

12. Do you think that a greater number of individuals can be encouraged to hunt wild pigs for a fee? [ ] Yes [ ] No

    Comment __________

13. Do you think that greater use could be made of wild pigs for meat? [ ] Yes [ ] No

    Comment __________

14. Do any landowners try to improve their feral pig stocks or enclose them? Comment __________

15. General comments on wild pig control or fees for hunting them __________

APPENDIX 4B

What impact is a landholder's charging of fees for hunting an animal pest, such as wild pigs, likely to have on the level of population of the pest? Must fees lead to an increase in the population because landholders try to conserve the pest to earn money? Certainly some public bodies believe, as was pointed out in Chapter 4, that conservation is to be expected. But the theory outlined here indicates that in certain circumstances greater control of a pest population is achieved when landholders are allowed to charge hunting fees than when they are not permitted to do so. While the theory is developed with specific reference to wild pigs, it is also relevant to other pest species in demand for hunting by amateurs.

It is assumed for theoretical purposes that each landholder aims to maximize his net profit from the operation of his property as a whole - a not unrealistic assumption in the light of the survey in the Hay district. The theory is developed in stages:

(a) The level of the hunting fee for maximizing income from hunters is determined.

(b) The benefit from reduced pig numbers (that is, a lowering of damage caused by pigs) is taken into account. The optimal fee must take this factor into account and is smaller than in case (a).

(c) The level of control of a pig population that would be achieved by a landholder acting alone is noted. This is useful for comparative purposes.

(d) Allowance is made for the fact that hunters may impose costs on a landholder - damage his property, require organizing by him, and so on.

For each of the possibilities (a), (b), (c) and (d), the corresponding levels of pig populations are determined and compared. Finally the question is taken up of whether the charging of hunting fees can promote the optimal management of wild pig populations from a social point of view.

Fees, Hunting and Pig Population Equilibrium on a Property

The level of hunting fees establishes the equilibrium number of hunting visits and the equilibrium size of the pig population on a property.

The number of pigs on a property, other things being equal, is assumed to be a decreasing function of the number of hunters using the property for hunting per unit of time. As was clear from the Hay district survey, the number of pigs on a

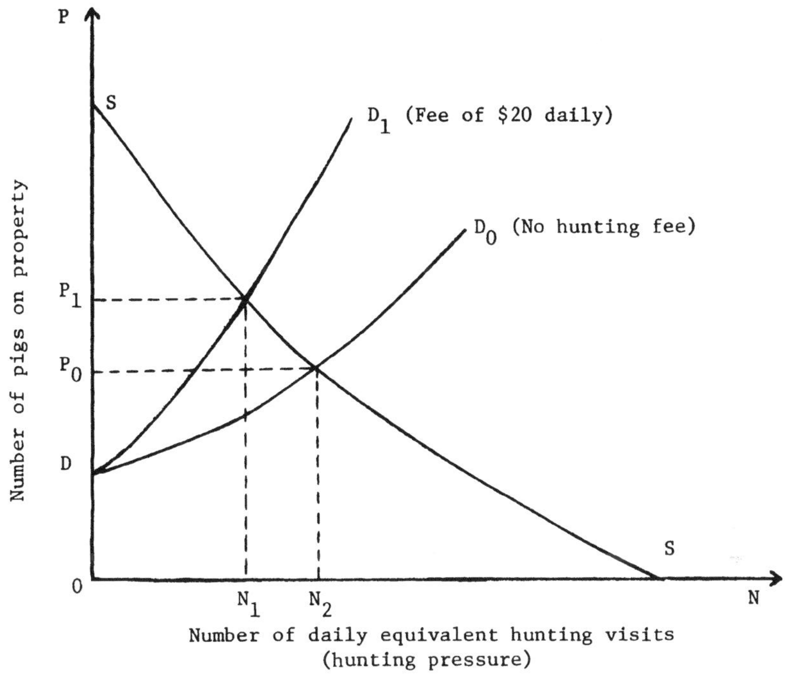

Fig. 4B.1 The equilibrium number of pigs on a property corresponds to the intersection of the pig supply curve, SS, and hunting demand curves such as those shown above. The equilibrium number of pigs can be expected to rise as the hunting fee is increased.

property declines with hunting pressure; an increasing number of pigs are taken and more are 'scared' to neighbouring properties as hunting pressure rises. A relationship of this kind is shown in Fig. 4B.1 by curve SS. SS, the pig supply curve, shows the equilibrium number of pigs corresponding to the hunting visits on the X-axis. Mathematically it is described by $P = g(N)$ where $\frac{dP}{dg} < 0$ and P represents the level of the pig population and N the number of hunting visits per period of time - for example, per year.

Now consider the demand for hunting visits. This is likely to depend on the supply of pigs and the level of the hunting fee. Other things being equal, the *demand to hunt* on a property can be expected to rise as the number or supply of pigs on the property increases. Thus at a daily hunting fee of $20, the annual demand for hunting on a property might be upward-sloping like DD_1 in Fig. 4B.1. In this case the equilibrium number of daily equivalent hunting visits per year (balancing the supply of pig stocks and the demand to use these) is N_1 and the equilibrium number of pigs is P_1. Assuming that the demand *as a function of price* for hunting is normal, the position of the demand curve in relation to pig numbers is likely to be closer to the horizontal axis in Fig. 4B.1 the lower the hunting fee. Thus if hunting is free, the demand curve might be as indicated by DD_0 and equilibrium established for P_0 pig stocks. This relationship implies that the equilibrium number of pigs maintained on the property rises with the level of the hunting fee.

Mathematically the demand relationship for hunting visits is of the form

$$N = h(F, P) \quad \text{where} \quad \frac{\partial N}{\partial P} > 0 \quad \text{and} \quad \frac{\partial N}{\partial F} < 0$$

and F represents the hunting fee. Solving the equations for the supply of pigs and demand for hunting visits, the equilibrium level of population of pigs, $\bar{P}$, can be expressed as a function of the hunting fee, for instance by $\bar{P} = \phi(F)$. $\bar{P}$ rises with F until the fee is sufficiently high to exclude all hunters and then $\bar{P}$ remains constant for higher fees.

The Income-Maximizing Hunting Fee and Variations in the Fee to allow for Pig Damages

Using the equilibria in Fig. 4B.1, the equilibrium total revenue for a property can be expressed as a function of its daily hunting fee. Thus in the example above at a fee of \$20 per day, equilibrium total revenue is \$20 × N_1. Thus in the absence of damages being caused by pigs on the property, the income maximizing fee is easily found, assuming for simplicity that the process of admitting hunters is costless. If the equilibrium relationship is as in Fig. 4B.2, the revenue-maximizing fee is F_2.

Looked at mathematically, the equilibrium number of hunters can be expressed as a function of the level of the hunting fee, that is

$$\bar{N} = \psi(F).$$

Hence the equilibrium level of total revenue is

$$R = \bar{N}F = \psi(F)F$$

and the revenue maximizing fee is found by maximizing this expression.

Bearing in mind that there is a one-to-one relationship between the level of hunting fees and the equilibrium number of pigs, marginal income from hunters can be expressed as a function of the equilibrium number of pigs on the property. Thus curve MM in Fig. 4B.3 might represent a landholder's equilibrium marginal income as a function of the equilibrium number of pigs, which in turn depends on the equilibrium number of hunters and the fee. The equilibrium number of pigs maximizing income *from hunting* is P_2 and corresponds to the daily fee of F_2 shown in Fig. 4B.2.

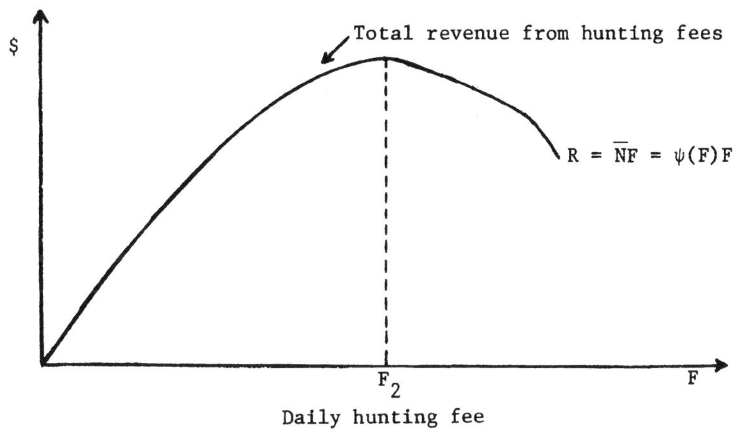

Fig. 4B.2 The total revenue curve above is based upon the equilibrium number of hunters as a function of the fee.

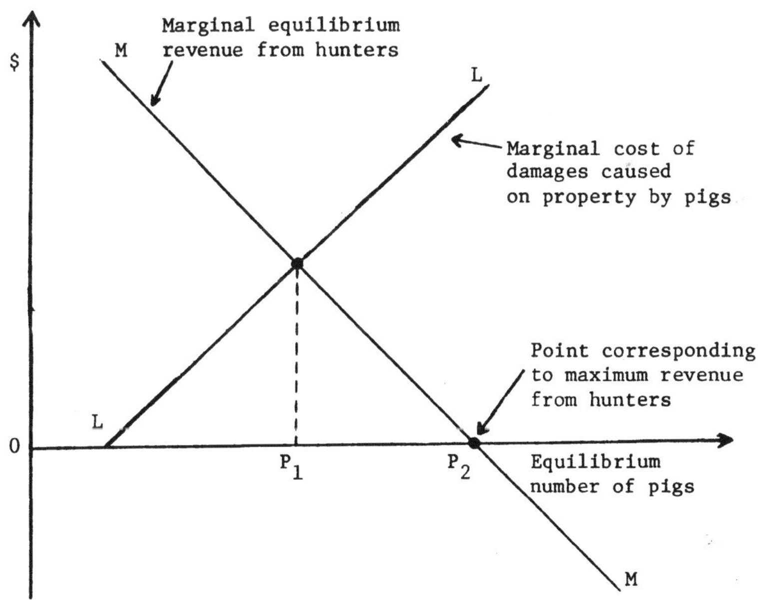

Fig. 4B.3 While pig population, P_2, corresponds to the fee, F_2, which maximises hunting revenue, the existence of marginal damages caused by pigs reduces the optimal number of pigs from the overall income-maximizing view of the property. In the case shown, P_1 is the optimal number of pigs when such damages are taken into account and corresponds to a fee lower than F_2.

However, the landholder must consider his profit from his operations as a whole. If pigs are causing damages on his property, he must take this into account. If the marginal cost of damages caused by pigs is as indicated by LL, the landholder's optimal equilibrium number of pigs will be P_1 and thus lower than that, P_2 maximizing his income from hunting fees. At P_1 the marginal cost of damages equals the marginal revenue from hunters and is achieved by charging a lower fee than F_2 which corresponds to P_2. Consequently, it is optimal to encourage a greater number of hunting visits when pigs are causing damage on the property. Indeed, the greater the marginal cost of damages, the lower is the optimal equilibrium number of pigs, the lower is the appropriate hunting fee, and the greater is the optimal number of hunting visits. These results accord with observations from the Hay district survey. Hunting fees tended to be lower on properties where pigs appeared to cause greater damage.

Comparative Control of the Pig Population by Landholders and Amateur Hunters

By adding a curve to Fig. 4B.3 to indicate the marginal cost to a landholder of controlling or reducing the number of pigs on his property by taking his own control measures, it is possible to compare the landholder's likely impact on the pig population with that of outside hunters. A landholder may exert less control over a pest population than when he permits hunters to hunt for a fee.

In Fig. 4B.4, the curve CC represents a landholder's marginal cost of controlling the number of pigs on his property taking his own measures. If LL continues to represent the marginal cost of damages caused by pigs on the property, then (assuming the unprofitability of a corner solution, that is no control) the most profitable level of control from the landholder's point of view is that for which the marginal cost of control equals the marginal cost of damages caused by pigs. In Fig. 4B.4 this occurs when the pig population is limited to P*. The lower the marginal cost of damages caused and the higher the marginal cost of control by landholders, the higher is the pig population tolerated by landholders.

Consider the position when landholders are able to charge hunters fees. If the marginal income received from charging hunters fees shown by the curve MM intersects LL (the marginal cost of damages curve) to the left of P*, greater control is exerted on pig numbers if paying hunters are allowed than would be exerted in the absence of outside hunters. For instance, in the case illustrated in Fig. 4B.4 and assuming that hunting fees are determined by taking into account overall gains to the landholder, including damages from pigs, P_1 is the optimal

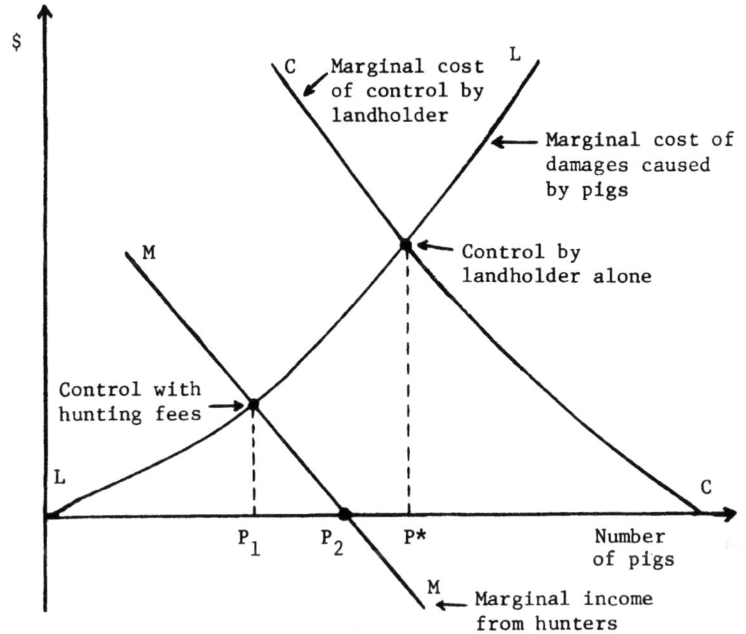

Fig. 4B.4 P is the level at which the pig population is held by a landholder in the absence of outside hunters. If MM is to the left of CC (intersects LL for a value of P less than P*) fee-paying hunters control the number of pigs to a greater extent than would occur in the absence of hunters.*

number of pigs and the corresponding hunting fee is readily determined in principle. Greater control of pig numbers is exerted by paying hunters than when the landholder takes only his own measures for controlling pigs. If the landholder 'goes it alone', pig numbers are merely limited to P*.

However, if MM intersects LL to the right of P* (the intersection point of CC and LL), less control of pig numbers is exerted when paying hunters are allowed than when the landholder goes it alone. So a different effect on the pig population occurs depending on the circumstances.

The lower the marginal costs of damages caused by pigs on a property and the higher costs to a landholder of limiting the number of pigs the more likely is the charging of fees for hunting to exert greater control on pig numbers than would occur in the absence of hunters. Of course, if hunting rights happen to be free and access free, even more control would be exerted on the population of pigs than if a hunting fee were charged. But account ought to be taken of costs imposed on landholders from hunting itself - for example, risks of the shooting of domestic stock.

Costs Imposed on Landholders by Hunters

Hunters may impose various costs on landholders. These include scrutinizing potential hunters, giving them permission and directions to hunt; possible shooting by hunters of domestic stock and of farm installations; and the risk that dogs used in pighunting are left, join feral packs and kill sheep. If the costs imposed by hunters are considered to be high enough by a landholder, no hunters may be permitted on to his property and he may go it alone in controlling pigs. In general, these costs operate towards restricting the number of hunters allowed on a property. But this reaction need not reduce the number of hunters allowed, to such an extent that control of pigs by paying hunters is less than if the landholder goes it alone. However, if it does, the landholder can be expected to find it profitable to exert extra control on pig numbers himself, to 'top up' the efforts of paying guests to bring pig numbers back to a more profitable level. In summary, no less control can be expected than if the landholder goes it alone; and indeed a little more control may be exerted, for his marginal control costs can be expected to be lower because his total needed effort is smaller in bringing pig numbers to a particular level after the reductions already achieved by hunters. Thus, even where paying hunters are not exerting control at the margin of pig numbers, greater control could be achieved as a result of their presence.

The effect of costs imposed by hunters can be seen to some extent from Fig. 4B.5. The independent variable is shown as the equilibrium number of hunting visits, which falls with increases in the hunting fee. The curve MRN_1 shows the marginal revenue that can be obtained from hunters. If revenue alone is considered, N_1 is the number of hunting visits which maximizes total revenue from hunting. But landholders obtain additional benefits from a reduction in this pest population through reduced damage to farm production. When these additional benefits are taken into account, the marginal overall benefit from hunters might be as indicated by the curve MR + MAB passing through N_3. Taking these additional benefits into account, the landholder's gains (if hunters impose no costs) are maximized by charging a fee which encourages N_3 hunting visits annually. But if hunters impose costs on a landholder, fewer visits will be optimal. The curve marked MC indicates the marginal costs imposed on landholders by hunters; the landholder *maximizes his net gain* by regulating the number of hunting visits, so that the marginal costs imposed by hunters are equal to the marginal overall benefits obtained by him. In the case shown, the optimal number of visits is N_2 and this implies a hunting fee higher than when no costs are imposed by hunters. If the marginal costs imposed by hunters are considered high enough, the hunting fee may well be set higher than

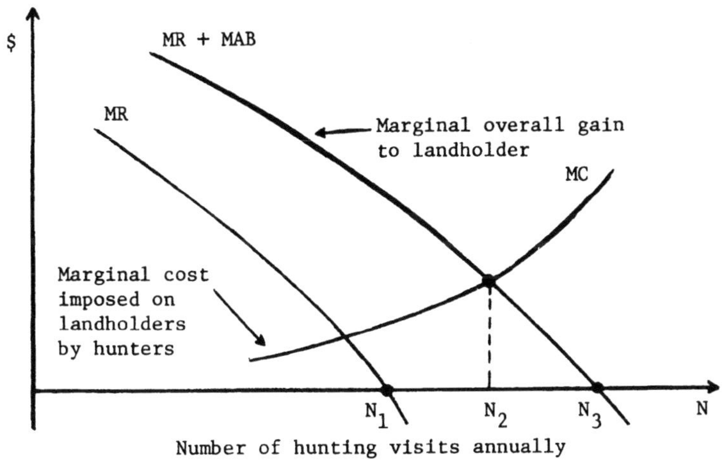

Fig. 4B.5 A landholder may find that hunting activity imposes costs on him. These costs cause him to limit the number of hunters. In the case shown, rather than permit N_3 hunting visits per year, the landholder sets a hunting fee which results in N_2 visits.

the level that maximizes total revenue from hunting: fewer than N_1 visits may be allowed. In the extreme case, an attempt may be made to exclude hunters completely, but as observed in Chapter 4 this is not always practical.

Social Optimality of Hunting Fees and Husbanding of Pests

The market mechanism works imperfectly in controlling a pest such as wild pigs because they are mobile, and the extent to which they are pests for different landholders varies. But even an imperfect market may be socially superior to none at all. Hunting fees for wild pigs allow the value of these to amateur hunters to be registered to some extent and encourages some conservation for hunting purposes because pigs are not perfectly mobile and their numbers are not independent of hunting activity on a property. Furthermore, in the absence of hunting fees many landholders may prefer to exclude hunters from their properties and not control pigs.

While landholders may exert greater control on pig numbers when they can charge hunting fees than when this is not allowed, they may not exert an optimal degree of control from a Kaldor-Hicks point of view. According to the welfare economics viewpoint of Kaldor and Hicks, an economic change is desirable if those who would gain from it *could* compensate those who lose and still be better off than before the change. This is likely to be so for at least two reasons:

(a) The demand curve for hunting on a particular property may slope slightly downward as a function of the hunting fee, as in the case of a monopolistically competitive firm. Each property has some relatively unique characteristics, and preferences appear to be built up for particular properties.

(b) Property owners do not take account of damages caused by pigs when they move to other properties. On a cattle property adjoining a sheep property, for example, there is likely to be undercontrol of pigs (from a Kaldor-Hicks point of view), even when hunting fees are charged. Pigs cause little damage to cattle properties but can cause considerable damage to sheep properties.

The possibility should also be considered that wildlife or pests, such as pigs, might be husbanded (for example, crops grown to increase their numbers) to gain from the sale of hunting rights. The profitability of doing this will depend on the alternatives and on the mobility of the pest.

Conclusion

The charging of fees to hunters for wild pig hunting can result in different circumstances in less, greater or the same control of pig populations as would occur if landholders merely took their own control measures, the overall income of the property being the goal in decision-making. Some of these different circumstances are identified above. The extent to which the alternative practices of charging and not charging fees for hunting are likely to promote the socially optimal regulation of a species still remains an open question. However, a likely effect of the abolition of fee charging for hunting on *some* properties would be a reduction in pig control measures and an increase in pig numbers.

CHAPTER 5

Commercial Use of Wild Pigs for Meat and Other Products

Although the wild pig is considered to be a pest by farmers, and by many conservationists in those countries where it has been introduced, it is also a useful exploitable resource. Apart from being an important asset for amateur (sporting) hunters in Australia and in other parts of the world, it can be harvested commercially for meat and other products. This is in fact done in a number of countries. Countries exporting the meat of the wild pig on a commercial basis include China, New Zealand and more recently Australia. Presently the main importing countries appear to be France, West Germany, United States, Japan, Netherlands and Sweden. The meat and other products of the wild pig are also traded within several other countries, for example in the U.S.S.R. and in Malaysia.

The commercial harvesting of wild pigs raises many questions about the management of this natural resource cum pest. For example: Will its commercial utilization lead to a reduction in the population of wild pigs or will commercial interests conserve the stock of wild pigs? How are the interests of farmers and other landholders, amateur hunters, and conservationists likely to be affected by the commercial utilization of wild pigs? What policies may be needed to accommodate these various interests? What are the likely economic problems of an industry harvesting wild pigs and what policies can be adopted to lessen these problems?

These questions will be taken up after aspects of the commercial industry are described. While reference will be made to the health and quality aspects of commercial exploitation of wild pigs in Australia, health aspects are discussed in more detail in Chapter 8. It should also be noted that the commercial value of wild pigs may extend beyond their direct supply of meat and related products from their carcases. The possibility must be considered of their providing a significant genetic reserve that may be used for improving varieties of domestic pigs in particular ways.

COMMERCIAL USE OF WILD PIGS OUTSIDE AUSTRALIA

The pork of the wild boar is regarded as a delicacy in most European countries, in America and in Asian countries amongst some non-Muslims. In these countries, it can often command a higher price than pork from domestic pigs, because the supply is relatively limited in relation to demand. Several countries import wild pig meat to satisfy their demands. These include West Germany, France, United States, Japan, Netherlands, Sweden, Hong Kong and Bahrain. However, in some countries commercial trade in the meat of the wild pig is based on local production or substantially on local production.

Diong examined aspects of the trade in wild pigmeat in Malaya in 1970.[1] At that time, the price per kati of cuts of wild pig in the local Kuala Lumpur market were about three-quarters that of similar cuts from the domestic pig.[2] Hence wild pig meat was at a discount on average, the discount varying with the food the pigs had been eating, the age of the animals and their sex. The current position is unclear. However, Diong states that wild pig meat is preferred by many Indians and Chinese and that a number of Chinese believe that this meat has therapeutic properties.

Diong reports that '...In cases when the harvest exceeds the local demand the meat is either exported to other states or used locally to make sausages, sweet meat, dried meat, meat floss and canned meat. In Ipoh (Perak) there are two factories making sausages and dried meat of the wild pig. The lard from domestic pigs is combined with the leaner meat of the wild pig for making sausages. In Kluang (Johore) two factories are known to make sausages from the meat of both wild and bearded pigs.'[3] Large amounts of wild pig meat are exported illegally to Singapore where they are used for canning and other purposes.

It is likely that similar types of commercial use occur in other countries of south-east Asia such as Thailand and Burma. Within China itself there is probably some trade in the meat of the wild pig but I have no details about this. However, China is exporting the meat of the wild boar to Sweden.[4] In Japan approximately 60 000 wild pigs (*Sus scrofa*) are harvested each year but only a small proportion of these are brought to market, most being directly consumed by the hunters themselves.[5]

Information about the commercial use of wild boar in the U.S.S.R. is sketchy. Commercial utilization occurs even though this seems to be on a smaller scale than at the beginning of this century. Sludskii reports that in Kazakhstan (Kazakh S.S.R.) commercial hunting of wild boar accounts for approximately 6000 head annually. These yield approximately 315 tonnes of meat, 63 tonnes of fat, 1575 square decimetres of hide and 3.15 tonnes of bristle.[6] In addition tusks are available for export.

A wide range of products is manufactured from wild boar in the U.S.S.R.[7] The most valuable product is the meat which may be used directly, salted or incorporated in smallgoods. Some of the fat may go into smallgoods and some of it is used for lard.

Much use is made of the hides of the wild boar. The hides of young pigs may be used to manufacture fancy goods or for chrome leather and linings in footwear manufacture. Heavy hides are used for soles, industrial leather goods and dressed leathers and in some places hunters' boots are cut from the tanned hides of large animals.

Bristles are also a valuable by-product. They are used in the production of brushes and footwear. While bristles used to be exported this is no longer so. Warm boots and mittens are made from the smaller 'downy' hairs ('pukh'). The tusks of the wild boar are exported for the manufacture of ivory goods.

Sludskii concludes that 'we do not find many wild animals, the hunting of which could give such a wide and valuable range of products as that yielded by boar hunting.'[8] He also takes up the question of husbanding the wild boar to increase its commercial value, a matter to which I shall give attention later in this chapter. Another possibility that he explores is the crossing of wild boar and domestic pigs to produce progeny that are superior to both types of pigs in some respects. This will also be discussed later in this chapter.

While *Sus scrofa* does not exist in the wild state in southern Africa (except possibly for a small colony of feral pigs in the Cape Province of South Africa) other members of the Suidae (pig) family do occur. These include the African bushpig *Potamochoerus porcus* and the warthog *Phacochoerus aethiopicus*. While an organization in the eastern Cape is reported to sell the meat of bushpig commercially, commercial operations are on a small scale. It also sells the skins, grinds up the bones for bone meal and supplies the tusks that are used as handles for various curios, such as bottle openers, and for necklaces and for jewellery. 'In Zululand, which is in the northern section of Natal, an organization called Bester Brothers prepares game meat commercially and it is possible to buy warthog polony, sausages and salami, most of which is exported to European countries. The supply of bushpig meat is too inconsistent to allow this kind of venture there.'[9]

In summarizing the position in South Africa, Darryl Mason of Kruger National Park has concluded:

> 'It seems to me that little commercial use has as yet been made of wild pigs, especially where cropping is undertaken as a management measure in conservation areas. Better use could be made of the potentially valuable meat, hides and tusks, as is the case with the cropping of elephants in Kruger National Park, and in the Luangwa Valley, Zambia.'[10]

Commercial Harvesting of Wild Pigs in New Zealand

The commercial harvesting of feral (wild) pigs in New Zealand is of special interest. Since the commercial utilization of feral pigs began, the wild pig has gone from being a serious pest in New Zealand to being only a minor pest. The feral pig was considered in earlier times to be a sufficiently serious pest for the New Zealand government to pay bounties on its destruction[11] but this is no longer considered to be necessary.

Apart from providing pointers to the likely effects of the commercial harvesting of feral pigs in Australia, the fortunes of the New Zealand industry have had an impact on the development of the Australian industry. As the availability of feral pigs in New Zealand has declined, New Zealand-backed firms have looked to Australia[12] to fill the widening gap in supplies. It has been New Zealand-backed firms in recent years that have successfully convinced the Australian Government to permit field dressing of game for export and have been the spearhead of the Australian industry.[13]

Dr Challies of the Forest Research Institute, New Zealand, has stated that currently in New Zealand 'feral pigs are generally of minor importance as a problem animal so consequently have attracted little effort from research organizations.

TABLE 5.1 New Zealand Exports of Feral Pig Meat 1967-79

Year	Quantity in tonnes	Price per kg $(NZ)	Deflated price* per kg $(NZ)
1967	339	0.95	0.87
1968	300	0.80	0.70
1969	442	0.68	0.58
1970	344	0.92	0.73
1971	407.5	1.00	0.73
1972	492.5	1.27	0.88
1973	732	1.43	0.90
1974	593	1.70	0.95
1975	367	1.64	0.84
1976	235	2.61	1.12
1977	252	4.00	1.44
1978	305	4.36	1.44
1979	210.5	4.30	1.20

Source: Information for all except the last column supplied by New Zealand Forest Service.

*Deflated by the New Zealand Consumer Price Index using 1965-66 as the base year. Hence prices are expressed in 1965-66 dollars. For further explanation, see C. A. Tisdell and J. T. Ward, *Economics in Our Society: New Zealand Edition*, Jacaranda Press, Auckland, 1981, pp. 252-5.

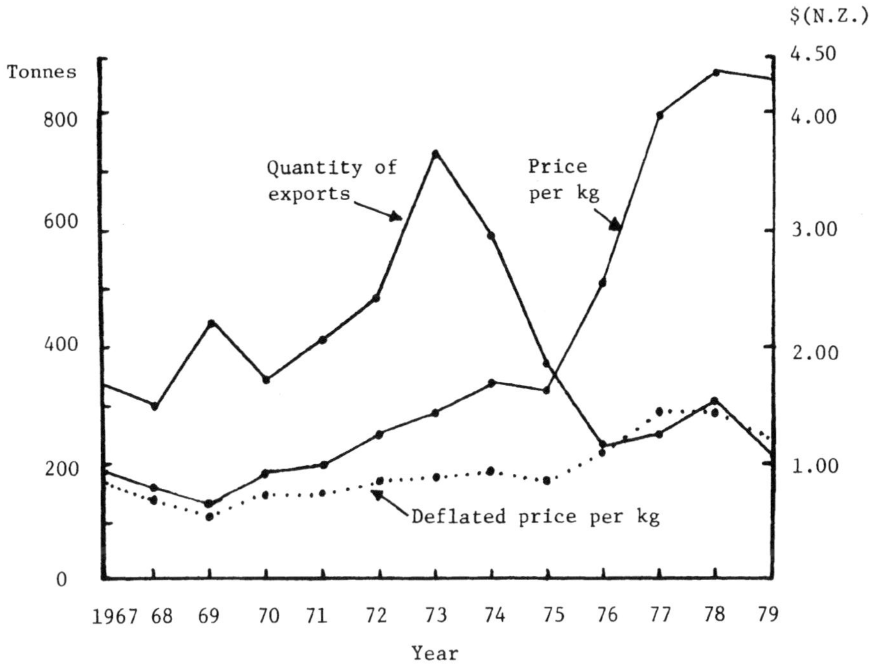

Fig. 5.1 Pattern 1967-79 of New Zealand's exports of feral pig meat and of prices received for these exports, 1967-79, based on Table 5.1.

They are locally a problem in production forests and on farm land but are usually easily controlled in these situations by either gun-and-dog or poison. Potentially they are a problem as a vector of diseases of domestic stock and would certainly come under close scrutiny if foot-and-mouth disease were to be established here. They have also been shown to adversely affect both flora and fauna in some reserves...'[14] He points out that the carcasses of feral pigs are processed by the game meat industry which is based primarily on deer.

Mr Fokerd, speaking for the Director-General of the New Zealand Forest Service, makes the interesting point that

> 'There is no doubt that the taking of feral pigs by the game meat industry for export has resulted in substantial reduction in the numbers of these animals right throughout the country. As previously explained, farmer complaints have almost entirely ceased. Farmers or more commonly commercial hunters working with property owners' approval have turned a former problem animal largely controlled by the state to monetary advantage.'[15]

All the feral pig meat harvested commercially in New Zealand is exported although sporting hunters naturally consume much of their own kill. The bulk of the meat is exported to France, Japan and the Federal Republic of Germany. Small quantities are sent to the United States of America, the Netherlands, Hong Kong and Bahrain.

Table 5.1 (see page 172) sets out the amount of feral pig meat exported from New Zealand, the average price per kilogram received and the deflated (real) price obtained per kilogram. These magnitudes are graphed in Figure 5.1 (see page 173).

From Table 5.1 and Figure 5.1, it can be seen that the price received per kilogram both in money terms and in real terms (that is, deflated or in terms of purchasing power) for New Zealand exports of feral pig meat has shown an upward trend since 1967, since commercial exploitation began. The quantity of exports peaked in 1973 and is now lower than in the earlier years of commercial exploitation. This is so even though the real price received for feral pig meat has risen significantly. The falling supply appears to be a result of a declining feral pig population. Commercial utilization has apparently caused a fall in these populations. This dwindling New Zealand supply seems to have been correlated with interest by New Zealand companies in developing exports of feral pig meat from Australia, an aspect that will be taken up in the next section.

COMMERCIAL UTILIZATION OF WILD PIGS IN AUSTRALIA

Commercial utilization of feral pigs on a significant scale in Australia began only in the 1980s and was stimulated by amendments at the end of 1979 to the Australian export meat regulations. These amendments allow the export for human consumption of game meat from field-shot animals. The industry is still developing. At present it is established in Queensland and expanding there and development could occur in the Northern Territory. The New South Wales government has remained opposed to a game industry based on field-shot animals in New South Wales on the basis that the development of such an industry could interfere with the eradication of feral pigs in the state. The New South Wales government believes that commercial utilization of this noxious animal in the state would lead to its husbanding. This issue will be discussed in some depth below.

Position before 1980

Before 1980 no systematic, widespread commercial use was made of feral pigs in Australia for meat or related products. Sporadic and localized use did occur with feral pigs being processed by meatworks in a similar way to domestic pigs and their products being put to similar uses. In essence, feral pigs passing the meat inspections were in most cases regarded as domestic pigs. No game-meat classification existed. In New South Wales and in relation to pig meat for consumption within Australia, this still seems to be the case.

In early 1979, I undertook a survey of the 29 abattoirs in New South Wales operating pig killing chains.[16] The purpose of the survey was to determine:

(a) whether or not abattoirs receive and process feral pigs and to what extent;

(b) the ways in which wild pigs are utilized;

(c) the commercial value of feral pig carcasses and skins and by-products;

(d) the incidence of disease and the rejection of feral pigs on this basis; and

(e) the prospects for the export of wild pig meat.

Ninety-three per cent (27 out of 29) of the abattoirs surveyed answered the questionnaire. The response rate was high. Of those replying only three said that they had slaughtered feral pigs in the previous 12-18 months or pigs that from their characteristics appear to be feral. This means that 11 per cent of responding abattoirs slaughter what appear to be feral pigs. Thus the practice of slaughtering feral-type pigs at abattoirs in New South Wales does not appear to be common.

It is possible that the proportion of New South Wales abattoirs receiving feral pigs is underestimated in this survey, even though the underestimate may not be considerable. An underestimate is possible (a) because it can be difficult to identify feral pigs, and (b) in some quarters there appears to be a stigma associated with the use of feral pigs for human consumption.

Abattoir managers were asked to comment on the possible accuracy of the situation indicated by the survey. Opinion was divided. Some managers thought that the survey underestimated the proportion of abattoirs receiving feral pigs, whereas others thought that the result was 'about right'. I independently obtained direct evidence in writing from a senior manager of one abattoir which claimed in the survey not to have slaughtered feral pigs in the previous 18 months that the abattoir is slaughtering feral pigs. If my survey results are adjusted for this, at least four abattoirs in New South Wales (14 per cent of those with pig-killing chains) slaughter feral pigs. In all probability the actual figure is somewhat higher. Data on rejection of pigs for sparganosis (a parasitic infection commonly associated with feral pigs) by abattoirs could provide an independent indicator of the extent and frequency with which feral pigs or pigs running under similar conditions are received by abattoirs. But I was unable to obtain N.S.W. sparganosis-rejection data on an individual abattoir basis.

Certainly the number of abattoirs utilizing feral pigs in Queensland seems higher than in New South Wales. In Queensland at least seven abattoirs appeared to be processing feral pigs in the mid 1970s.[17]

Well-fed (e.g. grain-fed) feral pigs that have recently escaped from domesticity are difficult or impossible to distinguish physically from domestic pigs.

Feral pigs caught in traps at a field day at Nundle, N.S.W. (and shown in Fig. 5.2, for example), had the characteristics of domestic pigs and were in excellent condition.[18] Reports of feral pigs of this excellent physical nature are not uncommon from grain-growing areas.[19]

It appears that feral pigs are influenced physically by their environment and their parent stock and may vary in conformation and appearance from those that are virtually indistinguishable from domestic pigs to those exhibiting 'stock' feral pig characteristics. According to respondents to the survey the stock characteristics of feral pigs are:

Long snout
Coarse hair
High back-bone
High thick-set forequarters
Light hindquarters
Coloured (usually black or black/tan)
Excitable, flighty and difficult to handle.

Most of the respondents receiving feral pigs agreed that it is possible for a feral pig to pass as a domestic pig. Some commented that this depends on how long the feral pigs have been stied and on recent interbreeding with domestic stock.

It is also possible that abattoir managers discourage the receival of feral pigs because the cost of processing them is higher in relation to their value than domestic pigs. In some areas the slaughtering charges for feral pigs are higher than for domestic pigs and at some abattoirs managers insist on freezing the carcasses for at least 48 hours. One slaughtering superintendent said, 'The

Fig. 5.2 The feral pigs caught in the above trap in New South Wales are difficult or impossible to distinguish from domestic pigs.

slaughter of feral pigs is not encouraged at our abattoir. However, during very slack times a lot of works welcome any type of slaughterings to keep their works going instead of issuing stand-down notices.'

All responding abattoirs handling wild pigs reported that the practice exists of catching feral pigs, holding and fattening them for slaughter but that this does not occur on a large scale. Examples of this practice were given in the previous chapter.

Feral pigs slaughtered at responding abattoirs were all used for human meat consumption. Hides were not used as only the bristles and not hides are removed for meat consumption in Australia. Skinned pigs are not presentable as joints and must as a rule be used in canned meat, sausages, pressed ham, etc. Some meat and bone meal was produced as a by-product.

The three abattoirs admitting to the processing of feral pigs (let's call them A, B and C) reported that the following individuals bring feral pigs to their abattoirs for slaughter: (A) amateur hunters; (B) farmers and graziers, amateur hunters (and a local smallgoods processor buys these pigs from amateur hunters); (C) commercial operators.

Respondents said that none of their feral pig meat was used in the pet food trade but that it was all used for human consumption. While at some centres the meat was principally privately consumed (by hunters, for example) in at least one centre 80 per cent of feral pig meat processed for human consumption entered the commercial trade.

Three abattoirs gave figures for the average dressed weight of feral pig carcasses at their works. These figures suggest that the *average* dressed weight of a feral porker at about 30 kg is 10-20 per cent less than that of a domestic porker and that of a feral baconer at about 45 kg is on average 12-24 per cent less than for a domestic baconer.[20] Assuming that feral pork passing through the abattoirs is sold at the same price as domestic, the price for feral pigs would be less than for domestic pigs. One abattoir reported to me independently of the survey that '...on the rare occasions that feral pigs pass through the abattoir, they can be sold for human consumption, if they are classified as fit for humans. In terms of the economic viability, once released the feral pork meat is sold at the same price as ordinary pork.' It is possible too that wild pig meat if sold as a specialty item (as wild boar, for example) could command a premium over domestic pig meat.

The economics of utilizing wild pigs for human consumption depends to some extent on the incidence of rejection of wild pigs because of disease and parasitic infection and the extra cost of measures to reduce health risks, such as extended periods of freezing of feral pig meat. Abattoirs slaughtering feral pigs did not reject any pigs prior to slaughter. Estimates for rejection of feral pigs because of disease, parasitic infection or other reasons after slaughter ranged from ½ per cent at one abattoir to 25 per cent at another.

Even allowing for a further reduction in value for higher rejection rates for feral pigs at abattoirs, feral pigs on current meat prices are valuable for meat for human consumption. Assuming that rejected feral pigs cover slaughtering costs from the meat and bone meal or bone fertilizer produced from them and given the *most pessimistic* rejection rate, feral pigs per kilogram would be at a discount of 25 per cent. This estimate is based on estimated dressed weights and assumes that the feral pork is traded as if it is from a domesticated animal.

Rejected pigs are as a rule transferred by meatworks to digesters which produce meat and bone meal or fertilizer. The main reasons for rejecting feral pigs were reported to be:

(a) the presence of sparganosis (a parasitic infection) or the occurrence of other parasitic infections;
(b) emaciation; and
(c) septic injuries.

Sparganosis is especially difficult to detect and is widespread in feral pig populations but fortunately can be destroyed by either freezing the meat for 24 hours or by cooking it well.[21] A number of meatworks insist on freezing feral pig carcasses for 48 hours to guard against the possibility that some consumers may not cook the meat well and become infected with the parasitic cysts from spargana. As a rule infection does not have serious health consequences. If spargana in insufficiently cooked or frozen pork are eaten by humans, they 'migrate from the gastro-intestinal tract of men to become resident in the loose fibrous tissue under the skin or between the muscles of the stomach and the nearby surrounding regions. Patients usually become aware of a small lump in this region, which can vary in size up to 3 cm and often moves about as the parasite migrates.'[22] However, provided that the pork is cooked well, there is no risk of humans being infected with parasites from eating feral pig meat.[23]

The possibility of using feral pigs directly in blood and bone, meat meal or in tallow was also explored. One rendering works[24] indicated that it could pay approximately 4 cents a kilogram for pigs for such purposes delivered dead within approximately 10 hours of death. A trapper or hunter would therefore receive only \$2.00 for a 50-kg pig delivered for this purpose and \$3-4 for a larger pig. Delivery would appear to be economic only if a number of pigs can be carted to the rendering works over a fairly short distance and the points of pick-up of pig carcasses are easily accessible by wheeled vehicle. But prices may rise as meat becomes shorter in supply. Meat meal is now about \$300 per tonne. It is used in pet food and stock food.

The prospects for using feral pigs in pet food was also explored. As mentioned earlier, all responding abattoirs indicated that feral pig meat from their abattoirs is not used for pet food.

Enquiries with companies processing pet food indicate that feral pig meat can technically be used in canned pet food. Although it has been suggested to me that pork fat can cause rancidity in pet food, the fat can be drawn off in the cooking process and used for tallow. I have been unable to determine the extent to which feral pig meat is used in canned pet food by pet food processors, but the amount of use is probably small. The manager of one manufacturer's research centre reported in relation to the health risk of using feral pig meat: 'Where feral meats are used in canned food I know of no problem re disease provided normal canning temperatures are used which will definitely destroy both parasites and bacterial pathogens.' The manager also reported that feral pigs are suitable for the production of meat meal provided that they are not infected with anthrax. The production of meat meal normally involves cooking the pulverized meat and bones and drawing off the tallow. All parasites are killed by adequate cooking and most pathogens are also destroyed. In the case of anthrax there is, however, a danger of it forming heat-resistant spores. As mentioned earlier, however, the current value of pigs for meat meal and tallow appears to be low and operations involved in collecting feral pigs for this purpose are likely to be economic only in special circumstances. But meat meal is a major ingredient in dry dog foods and in stock food, so a substantial market exists.

One large abattoir independently stated that it uses some pork in its meat meal. Some feral pigs find their way into this end-use after being rejected for human consumption. No evidence has been received, however, to suggest that it is economic at present to take feral pigs for the specific purpose of producing meat meal.

All abattoirs slaughtering feral pigs said that feral pig skins are of no commercial value. Enquiries with skin buyers indicate that the cost of producing pig skin hide in Australia is likely to be about $2 per square foot whereas imported pig hide from India currently costs $1.10 per square foot.[25] At current comparative costs it is uneconomic to produce pig hides in Australia or to kill feral pigs just for their skins.

In the light of the subsequent development of an export market for game meat, it is interesting to note that in early 1979 all responding abattoirs felt that there were no prospects for the export of wild pig meat or for the export of skins. On the comparative costs quoted earlier, there is clearly no prospect for export of skins.

At the time, however, I suggested that the prediction about Australian exports of feral pig meat could prove to be inaccurate and that there were prospects of game-meat exports being developed.[26] I also suggested that the economic operations of a game-meat industry in Australia would be enhanced by allowing pigs to be used for game meat to be slaughtered in the field and eviscerated. I went on to point out that if pigs were placed in chillers (but not frozen) and taken to a central point for inspection, further processing and possible freezing, economies of scale would be achieved and the considerable cost of transporting and handling live animals avoided.

I also pointed out that the economics of the operation would depend too on whether existing meatworks handled game meat or special works were established for this purpose. If special meatworks were established, regularity of supplies of feral animals is likely to be a factor influencing the commercial viability of the meatworks.

The establishment and holding of export markets in game meat will depend upon the maintenance of hygiene requirements. If an export market were established all exporters may suffer if there is laxity by a single exporter. Government supervision therefore appears to be necessary. In the light of these observations and predictions, it is interesting to trace the subsequent development of Australian exports of feral pork as game meat.

Position from 1980 onwards

Before 1980, commercial utilization of feral pig meat in Australia was very limited. However, in November 1979 the Commonwealth Minister for Primary Industry approved amendments to the export regulations to allow meat from field-shot animals to be exported. In Australia exports are under the control of the Commonwealth Government and it can regulate conditions of production by using this power. However, production can also be controlled within a state by the state government and regulations controlling the sale of meat within a state are exclusively a matter for each state government. Thus the meat processors in each state are, if they engage in export, subject to regulations both by their state government and by the Commonwealth Government. A state government has the power to prevent or control the commercial utilization of game such as feral pigs even if exports are permitted by the Commonwealth Government. It is possible for the New South Wales Government to ban the commercial use of field-shot feral pigs even if these are only intended for export and for the Queensland Government to allow this activity but not to permit the sale of the meat from these animals for human consumption in Queensland.

The amendments to the Exports (Meat) Regulations in November 1979 were intended in the first instance to facilitate the export of game meat from feral pigs and kangaroos. The Australian industry has already developed to utilize these animals, and other species (such as feral goats) are likely to be added in the future.

Game animals are defined as 'those which have not been husbanded or confined in any way and which have been shot in their free state.'[27] The amendments require shot animals to be bled and partially eviscerated in the field. They may then be delivered directly to a registered game preparation establishment or temporarily to a chiller (either mobile or in a depot). Chilling must meet various requirements. If delivered to a chiller the carcasses must reach the game preparation establishment within 72 hours of killing of the animals.

Game preparation establishments have to be approved and registered by the Australian Bureau of Animal Health and must meet various requirements affecting hygiene. After the carcasses are delivered to a game preparation establishment they are inspected by Australian Bureau of Animal Health inspectors and condemned carcasses rejected for human consumption, washed and trimmed and completely eviscerated. The regulations require that postmortem inspection on the carcass and remaining viscera of wild pig shall be carried out to the extent possible in accordance with the requirements laid down for domestic pigs.[28] A number of specific hygiene requirements and precautions apply. These include the condition that

'Game meat derived from wild pigs shall have been:

(i) held at a temperature not higher than -18°C for not less than 30 days; or

(ii) held at a temperature not higher than -28°C for not less than 12 days.'

This condition appears to be more than adequate to guard against the possibility of live spargana remaining in the flesh of the pig and infecting humans with the parasite as a result of their eating undercooked meat.

It is interesting to note that '...the amendments make a clear distinction between the operation of a registered meat establishment (that is, an establishment registered to slaughter and/or prepare meat from domestic or captive animals) and that of a registered game preparation establishment. Carcasses or meat derived from game animals will not be permitted entry into a registered meat establishment except in the fully frozen and wrapped form for cold storage.'[29] Some feral pigs not shot in the field may as in the past continue to be processed through registered meat establishments. Technically feral pigs caught in traps are not classed as game but this distinction is likely to be difficult to police.

Let us now consider the structure of the industry and the economic nature of its operation. Currently the only company exporting feral pig meat from Australia is Game Management Pty Ltd but other companies are considering operations. Game Management is equally owned by a New Zealand-based company, T. J. Edmonds Pty Ltd, and by D. Wild Pty Ltd. In New Zealand T. J. Edmonds exports venison and some feral pig meat. Whereas in New Zealand venison is the backbone of the game-meat industry, in Australia it is likely that kangaroo meat will play a similar role. Game Management is exporting kangaroo meat for human consumption as well as feral pig meat and expects to add feral goat meat to its range of exports. Although other companies have commenced game-meat operations in other states, Queensland appears to be the only state where feral pig meat is being processed as wild game on any scale. Southern Game Meat Pty Ltd in Sydney advised me in August 1981 that it concentrates mainly on exports of kangaroo meat but a few field-shot feral pigs are being processed for export to Germany.

Game Management has a game preparation establishment in Brisbane and is planning a second establishment at Goondiwindi in south-west Queensland. The latter establishment would be well placed not only to receive feral pigs from this area but also from north-west New South Wales. The company is reported to have 60 hunters, each of whom is a licensed kangaroo shooter. Thus hunters jointly take kangaroos and pigs. They are paid on a piece rate. The rate is said to be 45¢ per kilogram based on the weight of partially eviscerated pigs[31] delivered to chillers or the preparation establishment.

While it has been reported that Game Management expects to export 4000 tonnes of feral pig meat annually to Europe,[32] this may be an overestimate. Given that 4800 feral pigs are processed monthly by the company[33] exports would, on the basis of average dressed weights quoted earlier, be closer to 2000 tonnes per annum. Nevertheless these are considerable exports. The more conservative estimate amounts to about ten times New Zealand's exports of feral pig meat in 1979 and three times its peak exports of feral pig meat and, as discussed below, there is further scope for expansion of the industry.

The main destinations of Australian exports are the Federal Republic of Germany, France and Japan. The price per kilogram received for Australian exports fluctuates but has been recently in the range \$3-5.[34] Taking \$4 per kilogram as an average, Australian exports on current operations would be worth approximately \$8 million annually, assuming 2000 tonnes of exports per year. In Europe the meat retails at a premium compared to domestic pork and is reported to sell at about \$10 per kilogram.

ECONOMIC POTENTIAL OF AUSTRALIAN FERAL PIG MEAT INDUSTRY

The rapid rate of increase of feral pigs under Australian conditions implies that Australia has a large harvestable resource. While the level of feral pig population in Australia fluctuates, it is estimated to be between 5 and 10 million head.[35] Under stationary ecological conditions a population is reckoned to be able to sustain a 60 per cent continuous annual harvest.[36] However, even if one takes the conservative population estimate of 5 million pigs and a 50 per cent annual harvest rate, the population would be sustainable given the rapid rate of increase of pig populations. Thus in theory, it would seem that 2½ million feral pigs could be harvested in Australia annually without depleting the population.

Currently a considerable number of feral pigs are taken annually by amateur (sporting) hunters, possibly up to half a million as indicated in Chapter 3. The Australian game industry is at present taking only about 60 000 head of feral pigs annually. This harvest could be expanded considerably - for example, tenfold - without seriously depleting feral pig stocks. A commercial annual harvest of half a million feral pigs annually could be easily sustained (along with a sporting harvest of equal amount) and, on current export prices, would yield feral pig meat valued at about $50 million annually. Whether this increased supply would depress export prices would have to be considered, for a reduction in price would lower the value of a large harvest.

The maximum sustainable harvest of feral pigs in Australia is unlikely to be reached. This is because colonies of pigs exist in inaccessible and remote places where harvest under any forseeable economic conditions is likely to be unprofitable.

There are said to be 1 million pigs in the area that Game Management Pty Ltd draws its supply from.[37] Since its annual harvest is 60 000 this amounts to only 6 per cent of the population and is well below the sustainable harvesting rate.

The fact that feral pigs can be jointly harvested and processed along with other game species assists the economic operation of this industry. Presently kangaroo is the main wild species jointly harvested, but scope exists for harvesting other feral animals jointly, such as feral goat in Queensland and New South Wales, and buffalo in the Northern Territory. Joint harvesting and processing allow overhead costs to be spread more widely and consequently ensure lower costs per unit of production. Furthermore, joint production may help to smooth out fluctuations in supplies due, for example, to changing seasonal conditions. The pig population is subject to greater fluctuations but that of the kangaroo is more stable.

The economics of the industry is assisted marginally by its sales of by-products - for example, meat that is rejected for human consumption but which can be used in pet food, meat meal or blood and bone. However, game meat for human consumption is the mainstay of the industry.

The economic growth of the industry would be facilitated if in addition to the export market, the domestic market for game meat could be widened. Already kangaroo meat is sold for human consumption in South Australia but other Australian states have yet to follow this practice. As far as I am aware, other game meat is not available for sale for domestic consumption in Australia because of regulations by state governments.

ECONOMIC AND SOCIAL BENEFITS AND CONFLICTS IN THE GAME-MEAT INDUSTRY

A number of benefits and several conflicts arise from development of the game-meat industry, especially its utilization of feral pigs. While the Queensland Government sees (as did the New Zealand Government) commercialization as a way to control vertebrate pests, the New South Wales Government has remained suspicious of this proposition and has held to the idea that the commercial use of feral pigs could encourage their preservation rather than reduction in their numbers.[38] The arguments both ways are complex.

In New Zealand the commercial use of feral pigs has led, it seems, to a reduction in their numbers. The impact of commercial harvesting on feral pig numbers is likely to vary with circumstances. The following are some of the possibilities:

(a) If common-access to properties is allowed to enable pigs to be taken they are virtually common-property, and commercial use (as in the case of whales) is likely to lead to a reduction in their populations. Harvesters have no incentive to husband or conserve a species under this circumstance.[39] Harvesting, however, tends to be inefficient from an economic viewpoint. For example, the cost of harvesting is higher than it need be because the same areas may be worked over by a number of people.

(b) Landowners are in a position to prevent common-access by hunters. Although they do not have a property right in wildlife (it is either common-property as in the case of the feral pig, or state-property as in the case of the kangaroo), landholders have some powers to prevent trespass on their land and thus stop the taking of wildlife on their land. Hence, although technically the wildlife is not owned by them they may be able to charge a fee or price to hunters for harvesting on their land and may profit thereby.[40] In this regard a comment by a New Zealand correspondent is relevant. He points out that in New Zealand, 'Theoretically landowners obtain no income from selling of shooting rights as under the Wild Animal Control Act 1977 wild animals are the property of the Crown. Nonetheless, it is common practice for landowners to exact some tribute by various means from commercial hunters. Similar tribute is generally not asked by amateur or sportsmen hunters although access to private land is not always readily given now that animals have a high monetary value. Hunting on Crown-owned land is by permit, free of charge. Hunting of wild pigs continues to be a popular sport, and in many areas there are still sufficient numbers of wild pigs for sportsmen to follow this pursuit.'[41]

If demand for pigs exceeds supply at zero price (or after costs imposed on landowners by hunting are taken into account) a payment to landowners for allowing hunters is likely to emerge. The emergence of this payment or an increase in this payment is on the face of it evidence of increased hunting pressure.[42] While the landholder may reduce his pest control activities against feral pigs to benefit from their sale, the net destruction of feral pigs on his property may rise. The greater the mobility of the game animals and the greater the extent to which their presence on a property tends to be independent of hunting on that property because of mobility of the animals, the less restrictive is any landholder likely to be about harvesting on his property, and the less likely is he to husband or conserve game. Nevertheless, his interest may no longer be in eradicating the game completely (if it is a pest) and he may want to keep sufficient numbers on his property to make the property a continuing attraction to hunters, as explained in the previous chapter.

(c) The distribution of hunting rights to game processors is likely to have an impact on conservation. Where professional hunters have exclusive rights of hunting in particular areas and especially where a game processor has exclusive rights to harvesting in a region,[43] the processor has an incentive (and to some extent so do professional hunters, depending on the size of their area) to conserve their stock of feral pigs or other game. Even though more pigs may be taken than before commercialization, the structure of the industry tends to be such as to prevent harvesting on a scale that drives the species to extinction.[44] Thus more pigs *could* be left than is socially optimal taking into account their damages. On the other hand, to allow many processors in an area could add to the costs of harvesting because overheads are then more thinly spread and a monopoly may be required to make it worthwhile for a company to bear the risks of establishing an industry in a new region. A large investment is required in processing works and other facilities.

Additional benefits and conflicts from a game industry are as follows:

(a) The granting of exclusive rights to a processor in a region means that effectively hunters (and indirectly landholders) are tied in their rewards to a single buyer (a monopsonist). This can mean that the payments received by them are lower than in a more competitive situation.[45] This may help to conserve species but reduces the incomes of professional hunters and the royalty payments to landholders. Social conflict is liable to occur over such payments.

(b) Amateur or sporting hunters can be expected to lose as a result of the development of a commercial game industry. To the extent that the industry creates an extra demand for game: (i) they may find that they have to pay fees or higher fees for hunting rights, or (ii) that they are excluded from more properties because landholders prefer to deal exclusively with regular professional hunters, or (iii) less game is available for them. As a result of this there is a conflict. It may lead, for example, to political demands. For instance, sporting hunters might demand exclusive access under permit to Crown Lands free of charge or demand more state hunting preserves, or ask for access to state forests or national parks where this is denied.

(c) The question of hunting in national parks and state forests in Australia is considered in the next two chapters. Cropping of animals is allowed in a number of parks in Africa[46] but this has not been done in Australia as yet. It could be opposed by conservationists. However, there remains the difficult problem of deciding what to do about exotic animals such as feral pigs in Australian national parks and forests. Conservationists need to address themselves to this issue. There could be greater pressure from sporting hunters to use such areas in the future as the game industry expands, and one can even foresee the possibility that processors would be interested in cropping suitable open areas of this type.

(d) The game processing industry can assist employment in country regions of high unemployment, in Goondiwindi, for example.

Even though the game-meat industry has only recently established itself, there are already signs of some conflict between it and sporting hunters in Australia. For example, I recently received a letter making the following point:

> 'I am involved in a considerable amount of hunting with the bow and arrow in south-east Queensland where I am currently living. Due to the impact of game-meat hunting for overseas consumer markets plus unfavourable climatic conditions, the wild pig has been considerably affected. It has become apparent that a considerable knowledge of their ...habits...is needed.'[47]

Already the commercial use of deer in Australia has led to some conflict between amateur hunters and commercial users.[48] Some political accommodation of the interests of commercial and sporting hunters in Australia is likely to be needed in the future.

The fact must also be faced that the economic status of a pest can change. Its value may alter from negative to positive so that it goes from a species worthy of eradication to one that is useful to conserve and harvest as an economic resource. The feral pig may change its status in parts of Australia in this way as it is utilized commercially for game meat. Although farmers may continue to suffer damages from feral pigs, they may obtain some compensation by selling hunting rights, and the value of pigs for game meat may outweigh the damages caused by them. When this is so, feral pigs become a valuable economic resource.

The question still remains of whether farmers are likely to be compensated adequately for tolerating an animal that may cause losses to them. In France, farmers are indemnified for damages caused by game animals, including wild boar, from a fund to which hunters contribute.[49] Possibly a similar fund could be considered in Australia. However, in the case of the kangaroo, it might be argued that the government as well as processors should contribute to the fund. The reason for government contribution is that some measures to conserve the kangaroo (restrictions on shooting it) are taken by the government because the kangaroo is regarded as a public good.[50]

FURTHER ECONOMIC POSSIBILITIES OF THE WILD PIG

The possibilities of landholders earning income from sporting hunters by selling hunting rights to wild pigs and from tourism were discussed in the previous two chapters. These commercial aspects are capable of greater development in Australia and the commercial use of feral pigs for game meat is likely to result in the practice spreading of charging for amateur hunting rights.

At a more general level, wild and feral pig populations provide a potentially important gene pool that can be used to improve domestic pigs in specific ways - for example, to increase their resistance to particular diseases or to improve their ability to utilize material of low food value.

The comments by Sludskii are interesting in this regard.[51] He points out that experiments have been conducted in Kazakh S.S.R. of the Soviet Union in which wild boar have been mated with sows of the Large White breed. The resulting offspring after several crosses have been found to display characteristics superior to that of the Large White breed. The death rate among such pigs is found to be lower, weight gains on low-concentration food are greater, they are better able to utilize summer pastures in south-eastern Kazakhstan, and they have no burns or ill effects from sun and heat unlike the Large White breed. Sludskii concludes that

> 'It is obvious that the wild boar is not only a valuable hunting animal, but it may also be widely used for hybridization with domestic pigs, in the first place with the purpose of obtaining hybrids of a commercial type.'[52]

In areas where domestic pigs continue to be run on range in Australia, it is possible that systematic crossing of selected feral pigs and domestic pigs could produce a more productive variety.

The most profitable commercial use of some land areas, such as marshy areas, may be to run feral or wild pigs and harvest these. The feral pigs could be upgraded by culling and crossing and as Sludskii points out[53] suitable crops and other environmental changes can be made to improve the quality and volume of the harvest. The increased use of feral animals and kangaroos in Australia could make it economically worthwhile to husband these species in areas where their presence constitutes the most valuable commercial use of land. In game or hunting reserves, of course, special attention needs to be given to the management and husbanding of game species.

While the husbanding of feral species, such as feral pigs, in *some* areas could give the greatest commercial gain from the land, I am not necessarily advocating it. The systematic development of 'super-feral' animals and their husbanding in Australia would possibly threaten native species to a greater extent than in the past through increased competition and habitat change. While, for example, feral pigs could make the most profitable commercial use of wetlands, man's systematic husbanding of them in this ecological niche in Australia is likely to have serious consequences for native wildlife. However, if this wildlife is bound to disappear in any case even if feral animals are not husbanded, there may be a logical case for husbanding feral animals for commercial gain. The possible impact of feral pigs on wildlife is considered in the next two chapters.

CHAPTER 6

Pigs in National Parks and as a Threat to Wildlife

The introduction of pigs *Sus scrofa* (and other exotic feral animals) to Australia, New Zealand and islands in the southern Pacific threatens some native species, has disrupted the ecological balance and may change the pattern of evolution of species. They pose a special danger on islands where specialized species have evolved in the absence of significant predators.[1] This is evident in the south Pacific - for example, on Lord Howe Island, the islands of New Zealand and to a lesser extent Australia where, for instance, birds commonly nest on the ground and many are flightless.

However, surprisingly little is known about the impact of feral pigs upon wildlife in Australia. Frith, for example, points out that for Australia 'There has been no study of the effect of feral pig populations on wildlife. However, that these large, omnivorous and numerous animals undoubtedly destroy many ground-nesting birds and their eggs, small mammals and reptiles. The effect of their rooting and wallowing on the habitat of many small animals is severe.'[2] While the Commonwealth Scientific and Industrial Research Organization (CSIRO) is carrying out a long-term study of the combined ecological impact of water buffaloes *Bubalis bubalis* and feral pigs in the Northern Territory, detailed findings will not be available until many years hence.[3] Some research is also being carried out by the New South Wales Department of Agriculture and National Parks and Wildlife Service into the attractiveness to non-target species of poisoned baits intended for feral pigs.[4] Nevertheless there continues to be a paucity of information about the impact of feral pigs and of measures to control their numbers upon Australian wildlife.

Given the current state of information, it is difficult to see how rational policy decisions can be made about the control of feral pigs. This is especially so in national parks that aim to conserve native species or natural environments. But it is useful to bring together the fragmentary information that exists on this subject. It is hoped that this chapter and the next, which deals with feral pigs in forests, will make a small contribution to this task.

In order to add to my knowledge about the impact of feral pigs on wildlife and natural environments, I undertook five surveys on this subject. Field officers in the national parks of New South Wales, Queensland and Western Australia and forest officers in New South Wales and Queensland were asked to provide information about feral pigs in their parks and forests. This chapter and the next draws heavily on the responses to these surveys.

But before looking at the specific findings from these surveys, it is worthwhile considering the following broad questions: What is the general impact of feral pigs on wildlife? By what means do feral pigs compete with other wildlife and how serious is this competition in Australia? What factors ought to be taken into account in deciding to control the population of feral pigs in national parks?

An Overview of the Impact of Feral Pigs on Wildlife in Australia

Because of the relative isolation of Australia from other continents, a unique set of fauna evolved here. Of the 16 orders of land mammals, only four had representatives in Australia before the coming of European man. Marsupials were represented in many forms, many of which were unique, and unique species were to be found amongst other orders of animals. The introduction of foreign species by European man and his own activities have had serious consequences for some of the native species. Just what effect the feral pig has had is uncertain, but Harry Butler has epitomized the concern of many Australians as follows:

> 'When the Wet comes the whole Top End (Northern Australia) becomes an enormous lake as the rivers flow out over the surrounding plains. The huge flocks of water birds are one of the spectacular sights of this fascinating area. But it's the buffalo that's become the symbol of the Top End: an introduced animal that destroys thousands of acres through erosion and silting of waterholes. Wild pigs do the same, so these animals must be controlled if we are to save our native wildlife - like the magpie geese [*Anseranas semipalmata*] totally dependent on these waterholes in the dry season.'[5]

The Northern Territory Board of Inquiry into feral animals concluded (in relation to the Northern Territory) that

> 'The feral pig has few friends. Throughout most of its range the pig overlaps with the buffalo, and while some of the environmental damage it causes may be supplementary it also produces its own undesirable effect... Generally pigs tend to concentrate near watercourses and billabongs but during the wet season they range further through the forest country as evidenced by pig holes frequently encountered in this terrain. (Amongst other things they eat Pandanus palm.) The Board considers that whilst the pig can be extremely damaging to vulnerable plant communities such as rain forest, in stopping natural regeneration, the buffalo has far greater environmental impact.'[6]

But in the regions of Queensland and New South Wales where the buffalo is not present the pig may have the greatest environmental impact of any introduced animal.

The state of scientific work on the ecological consequences of the pig in

Australia can be gauged from the following statement by the CSIRO Division of Wildlife Research on its research:

> 'Preliminary work has concluded on the feral pig in the monsoonal wetlands, including Kapalga, and has provided information on food, breeding, diseases and parasites. This work has pointed to an apparent interaction between the pig and the buffalo, to predation on young pigs by dingoes, and to a need to define the pigs' habitat quantitatively. As the pig is largely nocturnal the problems of studying its distribution and abundance are particularly difficult.'[7]

In New South Wales, J. Giles was engaged in the 1970s on extensive ecological and biological research into feral pigs in two localities, one at Yantabulla near Hungerford in far western New South Wales and the other in the Macquarie Marshes near Warren in the central western New South Wales. He and his colleagues found that

> 'Green vegetative material and roots formed the bulk of their diet at all sites. The most common herbs identified were *Medicago, Trigonella, Marsilea, Portulaca, Calandrinia* and *Tetragonia* spp. In the marsh habitat the rhizomes of *Typha* (cumbungi) and *Phragmites* (bullrushes) and parts of *Paspalum paspaloides* were prominent in the diet. Around the crop growing areas of Moree, grain was common in pig stomachs. Seeds and fruit of acacias and grasses were important items of diet when they were available.
>
> A variety of fauna were in the diet including frogs, earthworms, insects and carcasses of sheep, cattle, kangaroos and pigs. There was a notable lack of bird remains in the stomachs.
>
> Frogs were common in the stomachs of pigs collected in swamps and floodwaters, particularly *Litoria peronii, Litoria caerulea, Lymnodynastes fletcheri* and *Lymnodynastes tasmaniensus*.'[8]

The wide range of food eaten by feral pigs in Australia and the extent of their threat to native wildlife is also evident from Pullar's 1950 article in which he indicates that the diet of the feral pig includes the following:

> 'Vegetable foods include grasses, leaves, fruits, nuts, berries and edible roots such as yams and water lilies. In some areas the roots of the nut grass (*Cuperus rotundus*), bullrush or cumbungie (*Typha augustifolia* var. Brownii and *Typha nuehleri*) and lignum (*Muehlenbeckia cunninghamii*) appear to be the staple diet.
>
> Animal foods include grubs, beetles and worms, obtained by rooting; crabs, fish and clams on the sea shore; yabbies, fish and fresh water mussels from inland waters. They break up decaying logs and stumps to obtain witchety grubs. They also hunt and kill small mammals, reptiles and amphibians, and are

responsible for the destruction of large numbers of eggs of ground nesting birds.

Carrion, if available, appears to form a large part of their diet and decaying carcasses in infested country are soon consumed.'[9]

In the Australian natural environment pigs are a particular problem because:

(a) Their natural predators such as wolves and tigers[10] are absent, and dingoes while they may have some impact on the populations of feral pigs do not seem to have a decisive effect and in any case the dingo *Canis familiaris dingo* is destroyed as a pest. Hunters in Australia are also in insufficient total numbers compared to Europe to reduce the stock of feral pigs drastically.

(b) Before settlement by Europeans of Australia, no fauna existed in Australia with a penchant for wallowing and rooting in marshes, swamps and billabongs, as pigs do. It is true that a number of native animals burrow, dig and scratch - for example, the wombats Vombatidae and the bandicoots Peramelidae - but they do not carry out these activities in wetlands and bogs. It is possibly in these areas that pigs cause greatest damage to habitat and pose the greatest threat to some other forms of wildlife. For example, the brolga *Grus rubicundus* (see Figure 6.1) is likely to be adversely affected by food competition from pigs, habitat destruction and predation on its eggs and chicks.

But let us look more closely at possible implications of the presence of wild pigs for native fauna.

Impact on Fauna

It is difficult to assess broadly the impact of feral pigs on Australian natural fauna given the vacuum of knowledge on this subject, but the following are some of the mechanisms that may operate to the detriment of native fauna:[11]

(a) food competition from the pig;

(b) habitat destruction by pigs, making native animals more susceptible to other predators, eliminating nesting sites and shelters or ruining food and water;

(c) direct predation by pigs.

It is of course possible that a few native species may benefit from the presence of the feral pig - for instance, the dingo may find it an additional useful source of food - but this appears to be the exception.

The pig is omnivorous, and this helps to explain its successfulness as a colonizer of new environments, but it must have shade and ready access to water for its survival. If the pig happened to be more restricted in its diet, it would be easier to pinpoint the native species likely to be threatened by it. Nevertheless it is clear that species dependent upon wetlands, marshes, billabongs and

Fig. 6.1 Brolga (or native companion) Grus rubicundus. *Feral pigs may compete with this bird for food, destroy its habitat and eat its eggs and young.*

watercourses bear the main brunt of the presence of the feral pig in Australia, since these areas are its preferred habitat, and it damages them by wallowing and rooting and eats whatever eggs and small animals it can catch or root out of the ground. Let us look more specifically at the possible interaction of feral pigs with groups of native animals.

Birds.[12] About 30 per cent of Australian bird species nest on or in the ground, and the eggs and chicks of these birds are prone to be eaten by pigs and other introduced mammals such as foxes. However, it seems likely that the pig is an *opportunistic* feeder on eggs and chicks. It is not clear that they are sought out by pigs. Even so, considerable damage could be done in areas where pigs are dense - for example, around waterholes.

Pigs may also disturb the nests of ground-nesting birds - for example, the nest of the brush turkey *Alectura lathami* - and destroy suitable ground nesting sites, by rooting out tall vegetation around billabongs and watercourses, for instancc.

Apart from this, food competition with a number of waterbirds can be severe, especially during dry seasons when both pigs and some of the waterbird species become entirely dependent on scattered waterholes and swamps for food. The pig in these areas feeds on lilies, bulbous plants, frogs and green matter that is often the food source of waterbirds. Conceivably some of these species of birds might be seriously threatened by competition from the pig, for example, the brolga.

I wrote to Professor Barrett about the possible impact of feral pigs on birds in Australia. A part of his reply may be of interest in the light of the above. He said:

> 'I can certainly imagine pigs could have an adverse impact by predation of nests and also, in the case of magpie geese, by competition for food e.g. eleocharis bulbs around swamp edges. I personally believe the potential conservation problem is sufficient to warrant some intensive ecological studies.'[13]

Mammals.[14] Small mammals dependent upon the edges of lagoons can be expected to be reduced in numbers by the presence of feral pigs. The pigs destroy their habitat and if the animals are shallow burrowers may be dug out and eaten by pigs. Pigs seem to consider Murinae (rats and mice) to be a treat.

Apart from direct predation, food competition and habitat destruction poses problems for some of the smaller mammals. We might, for example, expect the population of the false water rat *Xeromys myoides*, which nests in a mound of leaves and mud on the edge of swamps, and the Australian water rat *Hydromus chrysogaster*, which burrows in the banks of streams and lagoons, to be reduced.

The range of many of the bandicoots (Peramelidae family) has been reduced by the presence of feral pigs, feral goats and other introduced species. Habitat destruction and food competition have all played a part.

In rainforests it is conceivable that the rufous rat-kangaroo *Aepyprymnus rufescens* may be affected by habitat destruction and food competition from pigs. These small herbivores build nests from material in the forest floor. Another adversely affected dweller of the rainforest could be the musk rat-kangaroo *Hypsiprymnodon moschatus*. This small rat-sized animal is the most primitive member of its family, forming a connecting link between possums and kangaroos. It appears to be uncommon in most of its limited range in Queensland. 'It runs on all feet instead of leaping and lives upon insects, tubers and other vegetable matter which it obtains by scratching among the litter of the rainforest floor'[15] so some food competition from pigs cannot be ruled out.

Reptiles.[16] Feral pigs opportunistically eat small reptiles, are reported to eat the eggs of crocodile species (for example, *Crocodylus johnstoni* which has been regarded as an endangered species), possibly eat the eggs of other species including turtles, and there is some overlap of their diet with that of some species of lizards - goannas (Varanidae family) eat carrion as do pigs, and some lizards eat snails as will pigs. Predation on tortoises possibly cannot be ruled out. There may also be some food competition with snakes for frogs; and pigs will eat snakes! Indeed, in *The Australian Encyclopedia* it is enthusiastically reported that 'any economic value possessed by feral pigs is limited to the eating of carrion, the killing of snakes and the provision of a useful alternative source of meat in outback areas.'[17] Pullar also believes that they have a major impact on snake populations. He says:

> 'Pigs kill and eat large numbers of venomous and other small and medium sized snakes, and in some districts have been responsible for their virtual extinction. Pigs are not naturally immune to snake bite, but depend on a tough resistant integument with a thick layer of subcutaneous adipose tissue, and like the mongoose, on relatively quicker muscular movements.'[18]

Amphibians. Frogs are the only Australian representatives of modern amphibians, and number of Australian species are unique. Feral pigs appear to be especially fond of frogs as food, as the study by Giles mentioned earlier indicates. Wood and Barrett have pointed out that in the United States '... Portions of birds, small mammals, fawns, piglets, frogs, snakes, salamanders and turtles have all been found in pig stomachs, along with a larger number of invertebrates, particularly earthworms',[19] and in correspondence with me, Barrett has said 'Here in California I am particularly concerned about the potential for newly introduced pigs to further endanger several more amphibian species.' Not only may predation of pigs on frogs be significant but rooting by pigs may damage their breeding sites, as observed by Gorham in Fiji.[20]

Fish.[21] It is possible that feral pig activity could interfere with the breeding activity of native fish that spawn in still freshwater areas such as billabongs. While pigs will eat fish, predation is unlikely to be important.

Conclusion. This brief outline indicates that feral pigs in Australia may reduce the populations of a wide range of species. Species of birds, mammals, reptiles, amphibians and fish can be reduced or threatened by the presence of feral pigs. This possibility and others raise several questions about the desirability of controlling feral pigs in national parks in order to conserve native wildlife.[22]

DETERMINING THE OPTIMAL LEVEL AND METHODS OF CONTROL OF PIGS IN NATIONAL PARKS

To determine the socially optimal level or best methods of control of feral pigs in national parks is not easy. But is clear that a rational decision requires the costs and benefits of control of feral pigs to be specified.

Social Benefits and Costs of Control

The social *benefits* of control of feral pigs in national parks might be grouped as:

(a) benefits obtained within the park - such as less damage to natural environments, continued conservation of native species in the park[23], reduced damage to walking tracks, and so on;

(b) benefits obtained in the area adjoining the park - for example, less agricultural damage caused by pigs straying from the park;

(c) spillover benefits of wider nature - for instance, from the conservation of a greater diversity of species, or by eliminating feral pigs as a potential disease reservoir for the infection of domestic livestock, or by reducing erosion and siltation of dams.

It is doubtful whether improved public relations with adjoining landholders should be regarded as a social benefit, even though it is an important political consideration in the management of many parks and forests and for other government authorities instigating feral pig control programmes. However, we might add:

(d) the demonstration or induced control benefit as a social benefit - adjoining landholders may be more likely to take beneficial action to control feral pigs if the park does have a control programme.

Now consider the social *cost* of control. It is necessary to take into account for each possible method of control or mixture of methods the following:

(a) the actual costs incurred by the park authority - this should include actual outlays or expenses for control, such as costs of labour and materials;

(b) the effectiveness of the method in reducing the population of feral pigs;

(c) the adverse side-effects of the method used - this should include an allowance for the adverse side-effects of the method:

 (i) in the park - for example, upon non-target species in the park, upon visitors to the park;

 (ii) and unfavourable spillovers outside the park - for example, the use of swine fever would kill domestic pigs.

Thus, for example, if poisoning of pigs in a park were being considered it would be necessary to take into account the benefits as indicated above, the actual costs or

TABLE 6.1 General Factors to Consider in Determining the Control of Feral Pigs in National Parks

Benefits of control:	Within the park
	In areas adjoining the park
	Global or more widespread benefits
	Induced control by other landholders
	Improved public relations (?)
Costs of control:	Direct cost to authority in relation to effectiveness of the method in reducing the pig population
	Adverse side-effects of the control in the park, e.g. on non-target species
	Adverse side-effects beyond the park, e.g. the use of a disease for biological control would damage the domestic pig industry

outlays by the park authority in purchasing and using the poison and the bait proposed to carry it, the effectiveness of the poison in reducing feral pig populations, the susceptibility of non-target species to the poisoning programme, and the value of non-target animals lost. As will be discussed in Chapter 9, species differ in their susceptibility to various poisons and baits. For example, dingoes and dogs show little tolerance to 1080 (sodium monofluoroacetate) and are especially at risk if meat baits are used. Birds appear to be very tolerant of this poison. To avoid damage to meat-eaters, it might be decided, for instance, to spread the poison in grain bait.

Table 6.1 summarizes the general factors that need to be considered in controlling feral pigs (or, for that matter, any pest) in a park.

Socially Optimal Technique of Control

The socially optimal technique for any desired degree of control of feral pigs is the one that maximizes the net social benefit for that degree of control - that is, the one that maximizes social benefits less social costs for that degree of control. The optimal technique (or mixture of control techniques) may alter as the desired level of control changes. This can be illustrated as in Fig. 6.2.

The curve EFG represents the hypothetical net social benefits from the technique of shooting, and the curve ABCD represents that from poisoning feral pigs. Shooting

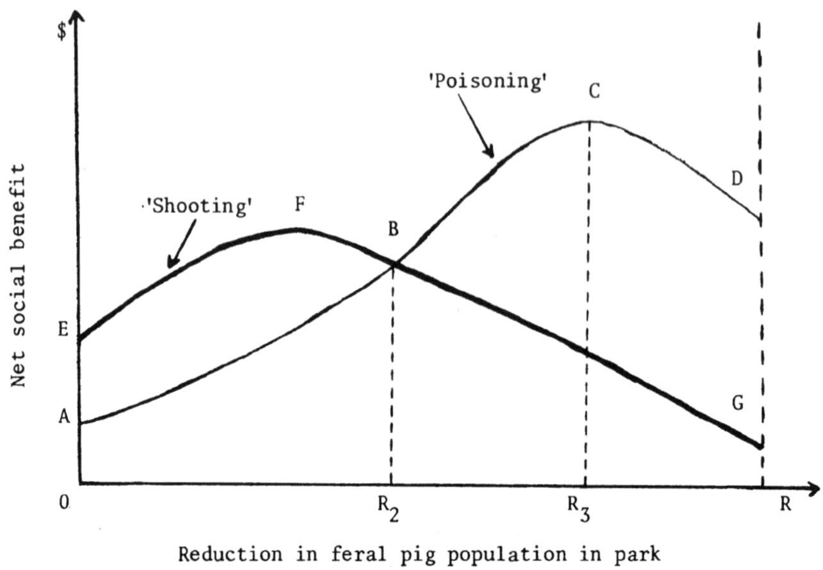

Fig. 6.2 The maximum net social benefit at different levels of feral pig control is indicated by the composite curve EFBCD. 'Shooting' is optimal for a desired level of control of less than R_3 and poisoning for a greater level of control. Net social benefit is maximized when the pig population is reduced by R_3.

is the socially preferred technique if a reduction of less than R_2 is required in the feral pig population, but poisoning is preferable if a greater reduction is desired. However, given that only these two techniques are available, the socially optimal pig control policy is to reduce the park population of pigs by R_3, using poisoning in this instance. Let us look at a little more closely the socially worthwhile and optimal levels of control.

Socially Worthwhile and Optimal Levels of Control of Feral Pigs in Parks

Any level of control of pigs (or other pests in parks) is only socially worthwhile if the social benefits from that control exceed the social costs. Assuming that techniques are chosen so as to derive social optimality, four possibilities are worth specifying:

(a) No control of the pig population is optimal. At all possible levels of reduction in the population, social costs exceed the social benefits of control expressed in money terms. For example, this is the case if the net social benefit curve of control is as indicated by that marked (a) in Fig. 6.3. This case is likely when the adverse side-effects of all available control methods are high, the spillovers from pigs in the park are small, and no particular species is endangered by the presence of pigs.

(b) Control is only socially beneficial if it is carried on beyond at least a minimum threshold scale and only worthwhile if kept within limits. This case is illustrated by the net social benefit curve (b) in Fig. 6.3.

(c) Control is socially worthwhile at all possible levels. This case is illustrated by curve (c) in Fig. 6.3.

(d) Only complete eradication in the park and exclusion of feral pigs is worthwhile. This is not illustrated in Fig. 6.3, but it could be optimal policy in which the benefits from control are very high and the fecundity and mobility of pigs is an important consideration.

For net social benefit curves (b) and (c) in Fig. 6.3, the optimal level of reduction of the feral pig population in the park is R_2.

The net social benefits (NSB) of controlling feral pigs and the nature of the

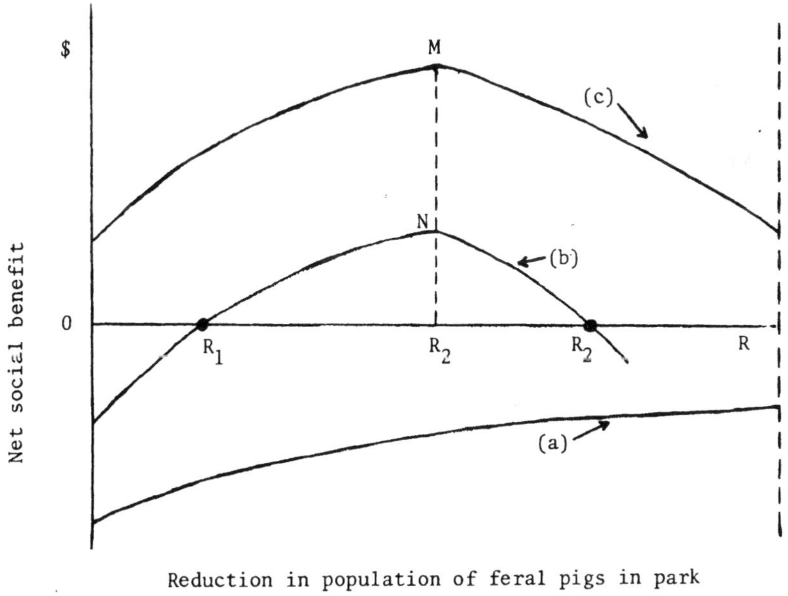

Fig. 6.3 Some different possible relationships shown by curves (a) (b) and (c) between net social benefits and the reduction in feral pigs in a park. As explained in the text, the socially worthwhile level of control of the pest varies with the nature of these curves.

NSB function are likely to vary between parks and regions. In some parks no control of feral pigs may be socially optimal; in others complete eradication and exclusion of pigs could be worthwhile; and in still others different types of intermediate degrees of control may be socially optimal, depending upon the nature of the relevant NSB function. State-wide plans or priorities for control of feral pigs need to be drawn up with this non-uniformity in mind.

None of the above is intended to suggest that it is an easy or an unambiguous task in practice to estimate net social benefit functions for the control of feral pigs or for any other pest in a national park. But rational social behaviour requires us to think along the above lines. Difficult problems are not solved by ignoring them. Furthermore, even if we cannot perfectly solve such problems, exploring the issues and considerations involved in a solution may help us to improve our imperfect policy decisions. It was with this and the above framework in mind that I surveyed officers in the national parks and forests of Australian states.

SURVEYS OF WILD PIGS IN NATIONAL PARKS IN AUSTRALIA

During 1979 and 1980 district officers in the national parks and forests of New South Wales and Queensland and in the national parks of Western Australia were surveyed to obtain information about feral pigs in their parks and forests. A summary of the results for the surveys of the national parks is given in this chapter, and results from the survey of forests is covered in the next chapter.

The questionnaires were designed to provide information on the following:

- The size and density of wild pig populations in the parks.
- The damages caused by and benefits of wild pigs in parks, including damage to flora and fauna.
- Other damages or ill-effects from pigs in parks, such as increased erosion and the potential of wild pigs for carrying diseases and parasites.
- The straying of wild pigs from the parks, the frequency of complaints about this and the type of damage caused when straying.
- Control measures taken against feral pigs in parks, the cost of these measures and their effectiveness.
- Control measures recommended by National Parks and Wildlife Service officers.
- The relative seriousness of feral pigs as pests in comparison to other feral and introduced mammals in national parks.

The surveys indicated the presence of more than 40 000 wild pigs in the national parks and nature reserves of New South Wales, Queensland and Western Australia. In most parks where they were present, feral pigs were reported to damage native flora and fauna and to have other ill-effects such as increasing the rate of erosion. The straying of pigs and complaints about this were common. There are more complaints about this feral animal straying from parks than any other in Australia.

New South Wales is the only state in which the National Parks and Wildlife Service is attempting on a significant scale to control the population of feral pigs in parks, but limited control is attempted in Queensland and Western Australia. In the eastern states shooting, poisoning and trapping are employed as control measures. Existing control measures have limited impact on the level of feral pig populations in Australian parks.

In Queensland parks, feral pigs rank as the most serious pest, closely followed by cats.

In New South Wales parks, feral pigs are ranked as a serious pest; cats and rabbits are considered to be slightly greater pests.

In the Western Australian parks in which they occur, feral pigs are considered to be a serious pest.

One must exercise caution in drawing conclusions from the results of these surveys. In many instances the information available to officers is limited. For example, no systematic counts of feral animal populations in the parks have been undertaken so the estimates are based upon the experience and 'guesswork' of officers. But the surveys do provide a good deal of information that is worthwhile investigating further and scientifically, and they point to the considerable gaps in our existing knowledge about the ecological consequences of feral pigs in Australia.

Let us consider the main findings from the surveys of National Parks and Wildlife Service officers in New South Wales, Western Australia and Queensland respectively.

RESULTS FROM A SURVEY OF WILD PIGS IN NATIONAL PARKS IN NEW SOUTH WALES

As outlined in Chapter 2, feral pigs in New South Wales cause considerable damage to farming and grazing properties. They damage crops and pastures, help spread some types of noxious weeds, damage fences and watering points, promote erosion, are frequently responsible for lamb predation and are potential carriers of exotic diseases, such as foot-and-mouth disease, which would have a disastrous impact on the Australian livestock industry if introduced. In an earlier survey by me of

Pastures Protection Boards in New South Wales (see Chapter 2), several Boards expressed concern about the presence of wild pigs in national parks and their straying to private land. In part, this influenced me to undertake a survey of N.P.W.S. officers in New South Wales to gather data on the presence of feral pigs and other feral and introduced mammals in national parks and nature reserves in New South Wales, their straying from such areas, and complaints about it. Throughout the world, complaints from agriculturalists and graziers about wild animals straying from adjoining national parks are common. But this spillover is not the only matter considered. Feral pigs damage native flora and fauna and cause other damages in parks and are therefore considered to be a pest in such areas. This section reports on the opinions of the N.P.W.S. District Officers about the type and extent of such damages occurring within their areas and the seriousness of the threat to native flora and fauna posed by feral and introduced mammals. It also provides information about measures being taken in parks and nature reserves to control the population of feral pigs. While the survey concentrated on feral pigs, extra comparative data were collected on other feral and introduced mammals.

Survey forms[24] were sent by post to N.P.W.S. District Officers in New South Wales in May 1979, with a covering letter, a copy of my permit for the survey from the N.S.W. National Parks and Wildlife Service, and a post-paid reply envelope. Forms were forwarded to 23 District Offices and 22 completed survey forms were received. The response rate was therefore 95 per cent.

Presence and Population of Pigs in N.S.W. Parks

Pigs are widespread in national parks and reserves in New South Wales. Seventeen (or 81 per cent) of responding Districts reported pigs to be present. Ignoring rabbits, pigs are more numerous than any other feral animal in national parks. Respondents estimated there to be more than 13 000 feral pigs in areas under their control. This compares to an estimate of 10 000 + for feral goats. In most districts (Lower Darling and Mootwingee are important exceptions) pig populations exceed those for goats.

Table 6.2 ranks Districts by the total number of pigs reported to be present in national parks and nature reserves. The distribution of numbers between Districts is very uneven. The Coonabarabran District accounts for more than half the total number of pigs reported, the Upper Hunter District accounts for 16 per cent (mostly in Barrington Tops area), and considerable numbers are present in the Moree-Narrabri, Tibooburra and Griffith Districts.

Considerable variation in numbers of feral pigs and their densities occur between parks in the same District (and of course within parks) and this is hidden

TABLE 6.2 N.S.W. National Parks and Wildlife Service Districts Ranked by the Number of Pigs Reported*

District		Area of parks and nature reserves containing pigs (hectares)	Number of pigs	Density in parks and nature reserves containing pigs (hectares/pig)**
Coonabarabran	(W)	105 670	7 050	15
Upper Hunter	(T)	22 000	2 101	11
Moree-Narrabri	(W)	32 332	1 050	31
Tibooburra	(W)	1 113 951	1 030	1 081
Griffith	(W)	28 524	1 028	28
Kinchega	(W)	44 000	300	147
Lismore	(C)	10 467	200	52
Blue Mountains	(T)	267 308	200	1 337
Narooma	(C)	156 050	170	918
Morton-Nowra	(C)	2 000	150	13
Glen Innes	(T)	6 050	70	86
Hawkesbury	(C)	7 890	40	197
Mootwingee	(W)	34 183	35	977
Grafton	(C)	3 902	20	195
Lower Darling	(W)	58 000	6	9 666

(C) Coast (T) Tablelands (W) Western Slopes/Plains

*Five Districts with pigs did not give complete pig numbers. Morton-Nowra and Lower Darling provided figures for some areas, and these have been included. Kosciusko, Bathurst and Mootwingee gave no figures but numbers were considered to be low. Pigs were reported not to be present in the North Metropolitan, Taree, Eden and Maitland Districts. The reply from Armidale was not received in time for processing.

**Based on areas where pigs were present and estimates of pig numbers were received.

in the crude figures given in Table 6.2. One nature reserve, Macquarie Marshes[25] (Coonabarabran District) (see Fig. 6.4) accounts for 5 000 pigs, or approximately 38 per cent of the total reported from all districts. Barrington Tops (Hunter), Pilliga (Coonabarabran), Warrumbungle (Coonabarabran), Nocoleche (Tibooburra), Mt Kaputar (Moree) and Willandra (Griffith) account for considerable numbers of pigs. Seven parks and nature reserves account for almost 90 per cent of all numbers of pigs reported as being present in national parks and nature reserves in New South Wales.

Fig. 6.4 The Macquarie Marshes near Warren, in the Coonabarabran N.P.W.S. District of New South Wales, contains one of the greatest concentration of feral pigs in the state. This aerial photograph shows damage done to the marsh by the rooting of feral pigs.

Fig. 6.5 gives a general view of the spatial distribution of the number of pigs in parks.[26] The main population of pigs in parks occurs from north of the Macquarie River through to the Queensland border. There is a substantial population along the Upper Hunter and towards the lower end of the Lachlan River. These areas as a whole are regarded by the N.S.W. Department of Agriculture as ones of heavy pig infestation.

If national parks and reserves are ranked by density of pigs reported to be present, it can be seen that most of the parks or reserves with high populations of pigs have high densities. For instance, Macquarie Marshes and Barrington Tops have high densities. But some small areas such as Goonawarra (Griffith) and Illawong (Narooma) have high densities even though their total pig population is low. Nocoleche has a considerable population of pigs but a reasonably low density, since it is a very large park.

Fig. 6.6 provides a spatial view of pig densities in national parks and reserves in New South Wales.[27] These densities are high in most parks and reserves north of the Macquarie River and in isolated nature reserves in the south-west of New South Wales. On the Tablelands, the Barrington Tops area reports high densities.[28] On the Coast Illawong (Narooma), Bungawalbin (Lismore) and Murramarang

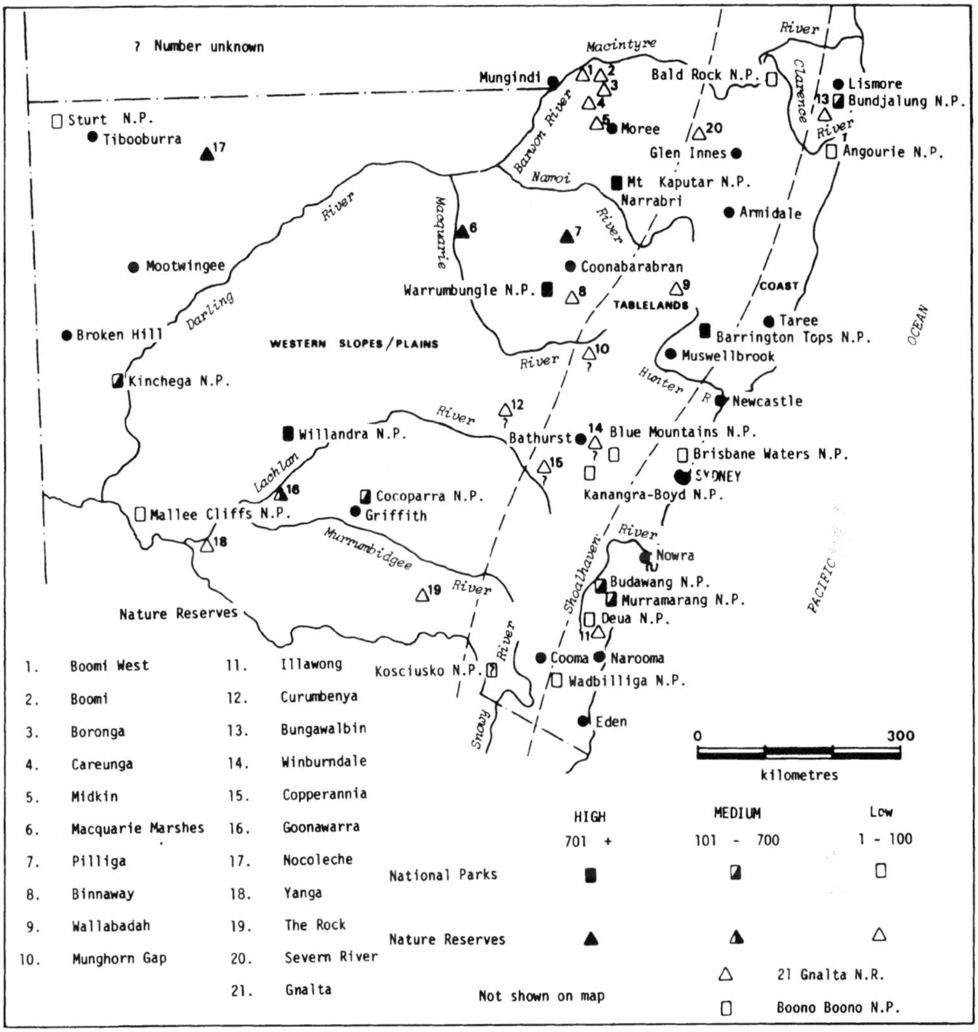

Fig. 6.5 Population levels of feral pigs in national parks and nature reserves in New South Wales, as reported by N.P.W.S. officers. The lowest populations are in the lower eastern section of the state.

National Park (Nowra) report high densities. Isolated coastal pockets have high densities.

Damages Caused by and Benefits from Wild Pigs in N.S.W. Parks

For the purposes of this report, New South Wales has been divided into three regions:

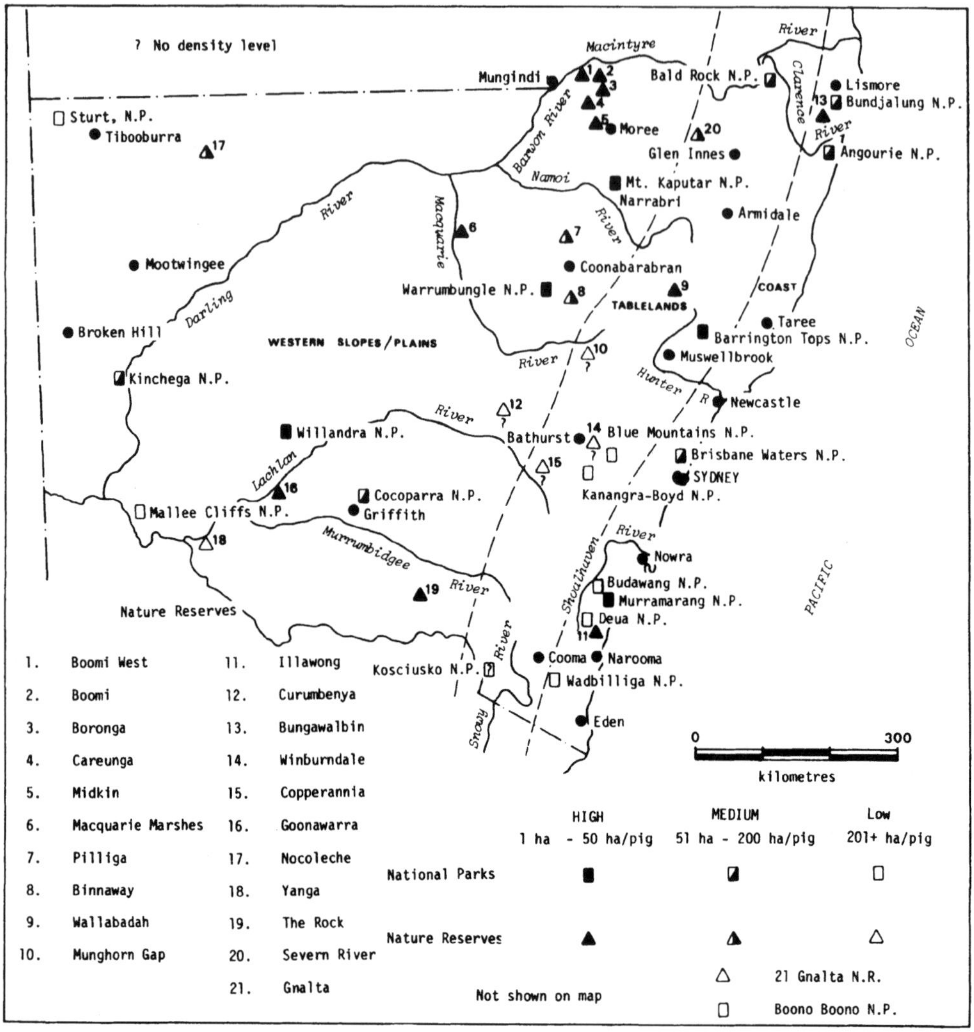

Fig. 6.6 Densities of feral pigs in national parks and nature reserves in New South Wales, as calculated from data supplied by N.P.W.S. officers.

the Coast (C); the Tablelands (T); and the Western Slopes and Plains (W); as indicated approximately in Fig. 6.5. The Coast is taken to include a region extending back from the sea for a distance of approximately 80-100 kilometres. The Tablelands consists of those areas on and closely adjoining the Great Dividing Range (the main catchment area of the river systems) and therefore includes the eastern slopes of this range. The Western Slopes and Plains covers all areas farther to the west.[29]

Fauna and flora damage. Table 6.3 provides information from districts

TABLE 6.3 Number of Districts Containing Wild Pigs in Parks Reporting that Pigs Damage Native Flora and Fauna (N.S.W.)

Region	Yes, damage occurs No.	%	No damage occurs No.	%	Total districts with pigs No.	%
Coastal	5	100			5	100
Tablelands	5	100			5	100
Western Slopes & Plains	6	86	1	14	7	100
All regions	16	94%	1	6%	17	100%

containing wild pigs of the frequency with which damage to native flora and fauna from pigs occurs. In all cases but one, damage is reported as occurring. The district reporting no damage was Mootwingee which had few pigs.

Flora. Table 6.4 lists the reported types of damage to flora caused by pigs and Table 6.5 sets out these data for fauna. The reported damages are likely to be the most obvious, because most District Offices are not in a position to do detailed observation and many indicate that they cannot be sure of the extent and types of all damage caused by feral pigs. This damage in parks and reserves warrants closer ecological research.

The most frequently mentioned complaint was that from rooting and soil disturbance causing damage to vegetation (see Fig. 6.7). The second most commonly mentioned nuisance as far as flora is concerned is that pigs help to spread and propagate noxious and introduced weeds. Their rooting provides ideal seed beds for burrs, thistles and other weeds. They also spread a number of weeds as a result of eating their fruits (for example, blackberry), but other animals do this also. This was as frequent as the complaint that pigs damage ground cover species by rooting and eating. On the South Coast both Morton-Nowra and Narooma reported extensive damage to burrawangs *Macrozamia*. For instance, Narooma reported: 'Burrawangs are a good food source for the pig - high in starch. The pig completely destroys these plants.' Interestingly one District Officer mentioned the possibility that pigs may help spread soil fungi. Some of these are responsible for the death of trees in Australia, and this subject will be considered in depth in the next chapter dealing with forests.

Fauna. As for damage to fauna, it was most frequently stated that pigs tended to reduce the population of ground dwelling birds and small mammals. This can occur

TABLE 6.4 Types of Damage to Flora Reported: Frequency Reported by Districts with Feral Pigs (N.S.W.)

Damage	Coast	Tablelands	Western* Slopes & Plains	Total frequency
Rooting and soil disturbance causing vegetation damage, roots, bulbs	4	3	3	10
Selective grazing of ground cover species		3	4	7
Spread of noxious weed and introduced plants (weeds, thistles)	2	2	3	7
Rooting around bogs, creeks, waterholes causing vegetation loss		1	3	4
Impact on swamps and moist areas (damage to sphagnum swamps)		3		3
Native grasses and sedges damaged	2			2
Trampling of vegetation			2	2
Uproot and destroy burrawangs (*Macrozamia* spp.)	2			2
Possible spread of root fungi			1	1
Erosion problems resulting in damage to flora			1	1
Disturb vegetation in colonized sand areas near beaches	1			1

*One Western Slopes & Plains District reported that damage caused to flora in parks in the district was unknown.

TABLE 6.5 Types of Damage to Fauna Reported: Frequency Reported by Districts with Feral Pigs (N.S.W.)

Damage	Coast*	Tablelands**	Western Slopes & Plains	Total frequency
Predation on native ground dwelling species	2		3	5
Competition for food		1	1	2
Damage to fauna habitat			2	2
Possible damage to clutches of emu eggs and other eggs			2	2

*Four of the five Coastal Districts reporting pigs stated they did not know the extent of damage or threat to the native fauna in their parks.

**One of the Tablelands Districts reporting pigs did not know the extent of damage or threat to native fauna in the park. Two other Districts reported some damage but gave no specific examples as they were not aware of the exact damage.

Fig. 6.7 Rooting by feral pigs amongst snow grass Poa siebarana *on the edge of the 'Pol Blue' sphagnum swamp, Barrington Tops, N.S.W. Pigs also do considerable damage to the swamps.*

Fig. 6.8 The habitat of a small mammal (possibly a swamp rat) on the border of a swampy area. Wild pigs may destroy such habitats and consume these mammals or their young.

by destruction of habitat (see Fig. 6.8), competition for food, by pigs eating the eggs of birds, and by their consuming small animals. In several Districts, District Officers indicated that they knew damages occurred but had insufficient information to specify these precisely. On the whole, knowledge about this subject is limited. The following is a sample of the more specific answers:

Kinchega N.P.: Fauna is damaged by pigs as a result of:

'(1) Competition for available food;

(2) Destruction (rooting and trampling) of lake edge habitat;

(3) Possible consumption of eggs of shore-breeding birds, e.g. plovers, dotterels.' (Charadriidae family)

Coonabarabran: 'As pigs are generally agreed to be omnivorous, they must have some effect on ground nesting birds and small animals. Boggy marsh and fern-type turfs are turned over by the pigs which is an action that the plants in these areas have not previously experienced.'

Blue Mountains N.P.: 'Competition for food and resources, that is, indirect effect particularly with macropods.' (Macropodidae family: rat-kangaroos, wallabies, kangaroos)

In many districts where pigs are present they are reported to cause or add to erosion in parks. This was reported in 41 per cent of the Districts having pigs in parks (see Table 6.6). The following comments indicate the range of the seriousness of the problem:

Lismore: 'Not serious at this stage as populations are low and inhabit relatively flat country.'

Upper Hunter: 'Swamps are becoming silted in the Barrington Tops area and sphagnum moss swamps are badly affected (see Fig. 6.9). This is serious since the area is an important water catchment and "natural" water storage area for the Hunter and Manning Rivers.'

Kosciusko: 'Degree of seriousness unknown.'

Moree: 'Soil wash is occurring due to natural sphagnum swamps on rocky plateaus being rooted up. As a result of rooting dehydration of soil and destruction of plant communities occurs in these bogs.'

Coonabarabran: 'Mild erosion is occurring along creek banks as a result of activities by pigs.'

Other damages caused by pigs in parks. Officers in the majority of districts with pigs in parks reported damages from pigs other than induced erosion and damage to native flora and fauna. Nine reported such damages, one suggested that no such damages occurred in his parks and six officers were uncertain about whether additional damage was caused. The main types of additional damage complained about are listed in Table 6.7. The two most common types of additional damage mentioned were

TABLE 6.6 Number of Districts Containing Wild Pigs in Parks Reporting Pig-Induced Erosion or its Absence in Parks (N.S.W.)

Region	Yes, induced erosion No.	%	No induced erosion No.	%	Do not know No.	%
Coast	1	20	3	60	1	20
Tablelands	3	60	2	40	-	-
Western Slopes & Plains	3	43	4	57	-	-
TOTAL	7	41%	9	53%	1	6%

TABLE 6.7 Types of Damage Reported Other Than Erosion and Damage to Native Flora and Fauna: Frequency Reported by Districts with Feral Pigs (N.S.W.)

Damage	Coast	Tablelands	Western Slopes & Plains	Total frequency
Spread of noxious weeds		2	2	4
Minor damage to boundary fencing		1	2	3
Danger to park staff and visitors			1	1
Digging up fire trails making them impassable at times			1	1
Pollute waterholes by wallowing			1	1
Increase numbers of prey species (e.g. dingoes)		1		1
Visual damage very evident	1			1
Unpleasant odour	1			1
Changes in vegetation, hydrology and water quality in small streams	1			1

Fig. 6.9 'Pol Blue' sphagnum swamp in the Barrington Tops, N.S.W. Mountain cord rush Restio australis *in the foreground; and under the line of trees, at the back, Spanish broom* Teline monspessulana, *which provides ideal cover for pigs. Feral pigs utilize this swamp during the summer, damaging it extensively, but move to lower ground during the winter.*

Fig. 6.10 Spanish broom Teline monspessulana *forming a dense shrublayer on the edge of a forest roadway in alpine woodland, Barrington Tops State Forest, N.S.W. This exotic plant pest provides pigs with shelter and their rooting may assist in its regeneration.*

that feral pigs help spread noxious weeds (see Fig. 6.10) and cause minor damage to the boundary fencing of parks.

Potential and problem of wild pigs carrying diseases and parasites. Almost half of the responding District Officers said that the potential of wild pigs for carrying diseases and parasites posed a problem for parks in their area. The responses are summarised in Table 6.8. N.P.W.S. District Officers seem to be especially aware of the possibility of feral pigs carrying diseases contagious to domestic livestock. The following comments were amongst those received:

Morton-Nowra(C): 'Potentially the chance of major disease is always present in colonizing fauna, particularly pigs.'

Glen Innes (T): 'Fouling of drinking water by pathogens is a risk.'

Raymond Terrace (C): 'Not directly a problem but have a moral obligation to be aware and control. I don't believe that much of the wildlife would be threatened by diseases carried by pigs.'

Coonabarabran (W): 'Major distributor of foot-and-mouth disease and of swine fever if an outbreak occurs. While the impact on native fauna is not known, the potential may be devastating. Kangaroos are known to be susceptible to foot-and-mouth disease.'

Kinchega (W): 'Moral obligation to surrounding landholders to control in case of contagious diseases entering the country. The extent of transmission of diseases from pigs to native fauna is not known.'

Lower Darling (W): 'All pigs carry disease which affect domestic animals. Pose a problem to everyone in the area.'

TABLE 6.8 Does the potential of wild pigs for carrying diseases and parasites pose a problem for your park(s)?
Frequency of District Responses*

Region	Yes	No	Do not know	No answer
Coast	3	1	3	2
Tablelands	3	1	1	
Western Slopes & Plains	4	2	1	
	10 (48%)	4 (19%)	5 (24%)	2 (9%)

*Districts with or without pigs could answer this question

Tibooburra (W): 'Ability to carry foot-and-mouth disease poses a risk.'

Moree-Narrabri (W): 'Problem since parks and reserves located near major stock routes.'

Kosciusko (T): 'Potential for foot-and-mouth disease spread is always a problem with wild species.'

Benefits. All Districts with feral pigs, except Kosciusko, Moree-Narrabri and Coonabarabran, reported no benefits from wild pigs in their parks. The benefits reported from the three Districts, even though none regarded pigs as a net benefit, were:

Kosciusko: 'provide a non-native prey species for wild dogs.'

Moree-Narrabri: 'minimal value in turning turf and in burial of seed in forest community possibly promoting germination.'

Coonabarabran: 'may open up huge reed beds in Macquarie Marshes Nature Reserve, allowing for greater diversity in the number and species browsing the marsh.'

Pigs compared to other Feral Animals as a Threat to Native Flora and Fauna

N.P.W.S. Officers from 17 Districts (81% of respondents) said that other feral animals pose as great (or greater) threat than feral pigs to native flora and fauna in New South Wales. Table 6.9 sets out the frequency with which Districts mentioned particular feral animals as posing as great a threat to native flora and fauna as pigs. Some Districts mentioned more than one species as posing such a threat. Feral cats were the most frequently mentioned (see Fig. 6.11) followed by feral goats, rabbits and feral dogs. Some regional differences are apparent. On the Western Slopes and Plains, feral goats and rabbits are most frequently mentioned as posing a threat, but the frequency with which they are mentioned is much lower on the Tablelands and still lower on the Coast. On the Coast, rabbits are not mentioned, except by Narooma where they were ranked highest. After feral cats, feral dogs were most frequently mentioned on the Coast as posing as great a threat as pigs.

Further information about the comparative threat of feral pigs and other introduced animals to native flora and fauna is given later, but it is worthwhile considering some comments here. Comments from N.P.W.S. Officers about other animals posing as great a threat to native flora and fauna as pigs included:

Coonabarabran: 'Feral cat - aggressive predator of birds and small mammals. Goat - aggressive competitor for habitat and food resources of the brushtailed rock wallaby *Petrogale pencillata* in Warrumbungle National Park; damages flora.

TABLE 6.9 Frequency with which Districts mentioned Particular Feral Animals as Posing as Great a Threat to Native Flora and Fauna as Pigs (N.S.W.)

	Coast	Tablelands	Western Slopes & Plains	Total frequency
Feral cats *Felis catus*	5	3	4	12
Feral goats *Capra hircus*	1	2	6	9
Rabbits *Oryctolagus cuniculus*		1	6	7
Feral dogs *Canis familiaris*	3	2		5
Foxes *Vulpes vulpes*		1	3	4
Horses *Equus caballus*		2		2
Cattle *Bos taurus*		1		1
Dingoes *Canis familiaris dingo*		1		1

Fig. 6.11 The feral cat Felis catus *is regarded by N.P.W.S. Officers as the most serious threat to native wildlife in New South Wales. This one was caught by an R.S.P.C.A. officer near Moree, N.S.W. Its size can be appreciated in relation to the standard cat-carrying cage. It was reputed to have 'savaged a dog and attacked a man - its weight - almost 20 kg'. (Source: Royal Society for the Prevention of Cruelty to Animals,* Animals 18(1979): 1(March), 19).

Rabbit - competitor with macropods for feed only noticeable during the dry season.'

Lower Darling: 'Large numbers of goats damage smaller shrubs and trees. Rabbits prevent regeneration of young trees, etc. Cats and foxes affect population of bird and small animals.'

Griffith: 'Goats graze shrubs and trees and use rock shelters which otherwise may be used by native fauna ...'

Kinchega: 'I would rank as follows: (1) rabbits, (2) cats, (3) pigs, (4) foxes. Reasons: Rabbits occur in large numbers, have a high breeding potential and are widespread and difficult to control. Cats have a more direct effect on fauna, are efficient hunters and are difficult to control. Pigs are dependent on water and are easier to locate and destroy.'

Kosciusko: 'Brumbies - erosion ...'

Morton-Nowra: 'The feral cat is probably the most abundant introduced predator in most parks and reserves within the district. Since their distribution is not closely linked with rabbit populations, it is assumed that their diet is native fauna.'

Straying of Feral Pigs from Parks in N.S.W.

Throughout the world farmers and graziers frequently complain of damages from animals straying from neighbouring national parks. In Australia complaints about feral pigs straying from national parks and nature reserves are common. As will be pointed out in a later comparative section of this chapter, complaints about feral pigs straying from national parks in New South Wales are more frequent than for any other feral or introduced mammal.

Approximately two-thirds of Districts with pigs in their parks received complaints about damages caused by pigs straying from their parks, although most indicated that they received few complaints. On the whole there appears to be less complaint the farther west one goes in the state. In areas of more intensive agriculture, such as the Coast, complaints are most frequent (see Fig. 6.12).

The groups complaining are reported to be mostly neighbouring farmers and graziers. But in one District it was mentioned that residents of villages adjoining parks complain, and in another case that 'ill-informed outside pressure groups complain.'

The reported frequency by Districts of types of damages from pigs straying from parks and complained about are listed in Table 6.10. The largest number of districts reported that they receive complaints about damages to fences caused by feral pigs. Complaints about damages to crops rank next. Several District Officers

Fig. 6.12 Pasture dug out on the Newstead property of D. & J. McDonald at Woodburn, north coast of N.S.W., by feral pigs from the adjoining Broadwater National Park.

mentioned that the National Parks and Wildlife Service is accused of harbouring pigs.

Measures Taken to Control Feral Pigs in N.S.W. Parks

There is widespread misunderstanding amongst farmers and graziers about the extent to which the National Parks and Wildlife Service in New South Wales is attempting to control wild pigs in its holdings. Many farmers and graziers appear to believe that attempts at control of feral pigs by the National Parks and Wildlife Service are the exception rather than the rule. However, at present attempts at controlling feral pigs in parks are widespread, and control activity appears to have increased in comparison with earlier years. Only three of the 17 Districts reporting pigs to be present were not taking control measures. The overall position is summarized in Table 6.11. Those Districts taking no control gave the following reasons:

Narooma (C): 'Narooma district has only been manned for a period of one year. Staff shortage and other higher priority work does not leave time to enter into a pig control programme at this stage.'

TABLE 6.10 Types of Damage from Pigs Straying from Parks Complained About: Frequency Reported (N.S.W.)

Damage	Coast	Tablelands	Western Slopes & Plains	Total frequency
Damage to fencing		2	2	4
Damage to crops		1	2	3
Pigs present, park accused of harbouring pigs	2			2
Damage to gardens and lawns	1			1
Robbing garbage receptacles, domestic and caravan park	1			1
Newly planted sugar cane areas	1			1
Damage to market gardens			1	1
Pigs enter neighbouring grazing and farming land			1	1
Threat to lambing flocks			1	1
Comment by 'ill-informed pressure groups'			1	1
Possible cross-breeding with domestic stud stock			1	1
Damage to pasture		1	1	2

TABLE 6.11 Number of Districts Containing Wild Pigs in Parks Taking Measures to Control the Pigs (N.S.W.)

Region	Control measures taken	No control measures taken
Coast	4	1
Tablelands	4	1
Western Slopes & Plains	6	1
	14 (82%)	3 (18%)

Mootwingee (W): 'No programme initiated as yet. Trapping will commence this year.'

Blue Mountains (T): 'No, because actual existence of pigs in the parks has not yet been confirmed by Service personnel.'

Methods employed. The most mentioned methods of control of feral pigs by districts were shooting and trapping (see Table 6.12). Eleven of the 14 Districts controlling pigs adopted these methods. Half of the Districts attempting control were using 1080 poison (sodium monofluoroacetate) (in grain pellets) but none of these was on the Coast. Poisoning appears to be less effective on the Coast because of the greater rainfall in the area, the difficulty of getting pigs to take the bait and the greater risk of accidental poisoning of domestic animals, such as domestic dogs eating poisoned pig carcasses, in view of the extent to which the area is settled.

Effectiveness of methods employed. Comments made about the effectiveness of the different control measures are summarized in Table 6.13. In general, District Officers appeared to have reservations about the effectiveness of trapping, especially on the Coast. The amount of labour involved in manning traps was a problem. The most favourable comments about trapping were from the Western Slopes and Plains. As far as poisoning with 1080 is concerned, the most favourable reactions, once again, were from the West. Some reservations have been expressed on the Tablelands about poisoning and 1080 is not used on the Coast. I shall return to the effectiveness of the different methods later, when costs are related to the number of pigs killed, but it might be mentioned here that actual cost figures confirm the general view of District Officers that trapping is costly per pig taken.

Most District Officers (13 out of 17) having pigs in their District felt that there should be more control of pigs in parks under their control. Only three thought that this was unnecessary and one did not know.

TABLE 6.12 Frequency with which Different Types of Control Measures were Mentioned by Districts as having been Adopted to Control Pigs (N.S.W.)

Method	Coast	Tablelands	Western Slopes & Plains	All regions
Shooting	3	3	5	11
Trapping	3	3	5	11
Poisoning		3	4	7
Helicopter shooting			1	1

TABLE 6.13 Effectiveness of Measures taken to Control Pigs in Parks: Frequency Commented on by N.P.W.S. Officers (N.S.W.)

Comment	Coast	Tablelands	Western Slopes & Plains	All regions
Shooting				
No comment	3		2	5
Fair to good		1	1	2
Shooting & dogging more effective than trapping in dense scrub & steep country		1		1
Population declining		1		1
Very effective control (pigs eliminated in study area Willandra)			1	1
Eliminates those pigs seen			1	1
Trapping				
No comment			2	2
Only effective in short term unless continued action is taken	1		1	2
Ineffective: too time-consuming manning pig traps	1			1
Adjoining landholder upsets critical control time with his shooting & trapping	1			1
Ineffective due to low trap numbers and lack of concerted effort	1			1
Shooters broke up groups and disturbed feeding		1		1
Population declining		1		1
Traps working but not as effective as shooting in dense scrub and steep country		1		1
Very effective control in study area Willandra			1	1
Reasonably effective			1	1
Poisoning				
No comment			1	1
Poisoning at refuse tips *would* significantly reduce numbers *but* Shoalhaven Shire Council will not permit poisoning	1			1

Cont.

Table 6.13 *Cont.*

Comment	Coast	Tablelands	Western Slopes & Plains	All regions
Poisoning (cont.)				
Fair to good		1		1
Population declining		1		1
Shooters broke up feeding groups		1		1
Cheapest, most efficient and effective means of control			1	1
Effective control in study area at Willandra			1	1
Control reasonable at Cocoparra for isolated pig population			1	1
Effective only if funding continues			1	1
Temporarily ineffective due to flood and rain (few weeks)			1	1
Helicopter Shooting				
Very effective. Future use on Kinchega planned for hot weather when pigs are on waterways			1	1

Special difficulties in controlling feral pigs in parks. As for special difficulties encountered in controlling feral pigs in parks, the most frequently mentioned problem was lack of staff members, followed by limitations imposed by terrain, difficulty of access to areas containing pigs, lack of co-operation from neighbours, limited finance and the distance of some parks from headquarters. Some of the other individual difficulties set out in Table 6.14 are of interest. Reinvasion or possible reinvasion of pigs from neighbouring properties after control in parks was mentioned directly or indirectly in three cases. This problem is indirectly implied in the District where the absence of a pig-proof boundary is mentioned and in the case where difficulties arise because the park abuts a forest.

What should be done to control feral pigs in parks? A wide range of suggestions were received from District Officers about what they felt should be done to control wild pigs in national parks. These are summarized in Table 6.15. The most frequently mentioned suggestions were: control should be in co-operation with neighbours; control work must be monitored to judge its effectiveness; funds should be adequate to allow year-round control using current systems; and pig population and dietary analysis should be undertaken.

TABLE 6.14 What special difficulties are there in controlling pigs in your Park? Frequency of Difficulties Mentioned by Districts with Pigs (N.S.W.)

Difficulties	Coast	Tablelands	Western Slopes & Plains	All regions
Staff numbers limited	3	2	2	7
Terrain	2	2	2	6
Access to pig areas	1	2	1	4
Lack of co-operation from neighbours		1	2	3
Limited finance	1	1		2
Distance of park from headquarters		1	1	2
Concentration of pigs around villages	1			1
Climate	1			1
Danger to public (shooting, poison)	1			1
Knowledge of their whereabouts		1		1
Lack of time			1	1
No pig proof boundary			1	1
Reinvasion rapid after control			1	1
Large area adjacent to state forest			1	1
Large park area			1	1
Not possible to use 1080 as regulations state all carcasses to be recovered and buried. This is impossible.			1	1
Lake edge susceptible to damage by vehicles used for poisoning and trapping control			1	1

The Opening of National Parks for Hunting

Since some amateur shooters in an earlier survey (see Chapter 3) said that they would like to be permitted to shoot pigs in national parks at selected times of the year, the views of District Officers were canvassed about this possibility. Nearly all District Officers were opposed to allowing such entry. Of the 21 Districts reporting pigs, 18 opposed such entry, two favoured it and one District was undecided.

TABLE 6.15 What should be done to control wild pigs in national parks? Suggestions Received from N.P.W.S. Officers (N.S.W.)

Suggestions	Frequency
Systematic trapping programmes should be expanded	3
Control in co-operation with neighbours	3
Work must be monitored to judge effectiveness	2
Current system of control will work if financial allocation would allow for year-round control	2
Population counts and dietary analysis should precede control	2
Each park should be analyzed to establish pig status and control problems so that control techniques can be developed	1
Pig control should be carried out by park staff	1
Continue current programme	1
Multi-measure control programmes	1
Guaranteed finances to allow for long-term planning and staff increases	1
Hunters or responsible shooters to be permitted at optimum shooting times	1
Long-term management aimed at eradication by baiting, trapping and shooting	1
Combination of trapping and poisoning most important	1
1080 poison most suitable for Moree-Narrabri area	1
Systematic poisoning programmes	1
Parks should become part of an eradication area systematically poisoned with 1080	1
Dogs, shooting used for isolated pig numbers	1

The main reasons given by District Officers for and against open shooting of pigs in parks are summarized in Table 6.16. Opposition was principally based on the following views: shooting cannot result in eradication of pigs; shooters are difficult to control; shooting is a danger to visitors to parks and to fauna other than pigs; parks are for passive recreation; and shooters may damage park areas by driving and movement. One of two Districts favouring open hunting suggested that a shooting drive against pigs in conjunction with neighbouring properties would decimate the pig population and the other District qualified its response by saying that 'it depends on who the shooters are but perhaps a local gun club could assist, in a trouble spot, on request.'

TABLE 6.16 Main Reasons Given by N.P.W.S. Officers (N.S.W.) for or against Open Shooting or Hunting of Wild Pigs in Parks

Reasons	Frequency
Against open hunting	
Shooting reduces numbers but does not eradicate	7
Difficult to control shooters	6
Shooting: danger to visitors	5
Parks are for passive recreation	3
Shooting: danger to other fauna	3
Shooters damage park areas by driving and movement	2
Allowing hunters on park areas forces you to authorize entry against regulations	2
Park management problem: pig control	1
Opening an area to periodic culling would create a situation for political pressure to maintain a stock (breeding) of feral game for continued recreational culling	1
High cost of issuing permits and patrolling compared to existing control methods	1
Would attract undesirable park users	1
Might be thought that parks are becoming game reserves	1
Outside trophy hunters are not responsible people for eradication purposes	1
Offence under the Act to carry firearms	1
For open hunting	
One or two drives carried out in known pig populations in conjunction with drives on adjacent properties would almost decimate the population	1
Depends on who shooters are. Perhaps a local gun club could assist, in a trouble spot, on request	1

Expenditure by National Parks and Wildlife Service (N.S.W.) on Feral Pig Control: Costs and Effectiveness[30]

In the 12 months prior to the survey, the annual expenditure on wild pig control by park authorities amounted to more than $18 000. Table 6.17 sets out control measures taken against wild pigs in N.P.W.S. Districts and cost per pig taken in the 12 months prior to the survey, where sufficient information had been provided by respondents. Annual expenditure on wild pig control by Districts ranged from $147 to $4 000, with the highest expenditure tending to be on the Western Slopes and

Plains. The Districts providing adequate information (ten) spent between them \$18 621 on wild pig control in the 12 months' period immediately preceding the survey and took an estimated 1 785 pigs. This works out at an average cost of \$11.43 per pig taken.

However, costs vary considerably from District to District and according to method used. Evidence is fragmentary for the Tablelands, but it appears that cost per pig taken is lowest in the west. The average expenditure per pig taken on the Western Slopes and Plains was \$7.55. On the Tablelands expenditure per pig taken was \$27.40, considerably higher than on the Western Slopes and Plains because of the terrain and other difficulties. On the Coast the average expenditure per pig taken was \$78.50.

Cost-effectiveness of different methods of control. Evidence about the cost per pig taken by the various methods used by park authorities is fragmentary but points towards trapping as being most costly per pig taken, followed by poisoning, and then shooting. Overall, shooting appears to be least costly. Table 6.17 sets out the costs in dollars per pig taken by different methods for Districts where it has been possible to calculate the cost per pig taken by individual methods. But it would be misleading to conclude from this that shooting is necessarily the most economical method. The most economical method is likely to vary with circumstances and it may be best to use multiple methods. Furthermore, the shooting figures are likely to refer to the fact that a ranger may carry a gun on his normal rounds and shoot any pigs incidentally seen. The cost per pig killed in a systematic shooting programme would be much higher than for casual shooting. Furthermore, systematic

TABLE 6.17 Cost to National Parks and Wildlife Service per Pig Killed By Different Methods (N.S.W.)

District	Ground Shooting	Trapping	Poisoning	Helicopter shooting
Moree-Narrabri	\$0.70	-	\$2.00	-
Griffith	-	-	\$4.60	-
Kinchega	\$2.00	-	\$27.00	\$6.00
Glen Innes	-	\$50.00	-	-
Morton-Nowra	-	\$50.00	-	-
Lismore	-	\$50.00	-	-
Upper Hunter	\$0.93	\$53.50	-	-

shooting might have little effect in wooded country where pigs are present but rarely seen. Also cost effectiveness in terms of expenditure is not the only consideration. Side-effects must be taken into account, such as the possible side-effects of 1080 poisoning on native species - even though this is likely to be small when grain bait is used.

Expenditure on feral pig control by N.P.W.S. Districts. Table 6.18 sets out the total expenditure on pig control in each of the N.P.W.S. Districts. It can be used to calculate the expenditure on pig control per pig present in parks in each District. When this is done, it is found that there is a tendency for expenditure per pig present to decline as pig numbers rise in districts. Two of the three districts with the greatest number of pigs (Coonabarabran and Moree-Narrabri) have expenditures on pig control (per head of pigs estimated to be present) well below the average for the group. Expenditures on pig control in these districts amount respectively to 57¢ and 14¢ per head of feral pigs present in parks and reserves in the district, whereas the group average is $1.44. Possibly political consideration and the desire to improve public relations with adjoining landholders are the prime determinant of current expenditures rather than a rational calculation of social gains.

Economies of scale in control? The cost per pig taken or killed for the N.P.W.S. Districts providing sufficient information is shown in Table 6.19. When this cost is related to the number of pigs taken as in Fig. 6.13, it tends to fall

TABLE 6.18 Annual Expenditure on Pig Control by Districts Providing Details Related to Estimated Number of Pigs Present in Parks and Reserves (N.S.W.)

District	Estimated number of pigs in parks	Total expenditure $
Grafton	20	570
Glen Innes	70	700
Morton-Nowra	150	500
Lismore	200	500
Kinchega	300	2 540
Griffith	1 028	2 300
Tibooburra	1 030	3 000
Moree-Narrabri	1 050	147
Upper Hunter	2 101	4 364
Coonabarabran	7 050	4 000
TOTAL	12 999	18 621

TABLE 6.19 Cost per Pig Taken in Ten N.P.W.S. Districts Related to Number of Pigs Taken (N.S.W.)

District	Number of pigs taken in last year	Cost per pig taken $
Grafton	0	$\to \infty$
Glen Innes	8	87.50
Morton-Nowra	10	50.00
Lismore	10	50.00
Moree-Narrabri	80	1.84
Upper Hunter	170	25.67
Coonabarabran	187	21.40
Tibooburra	400	7.50
Kinchega	420	6.05
Griffith	500	4.60
TOTAL	1 785	AVERAGE COST 10.43

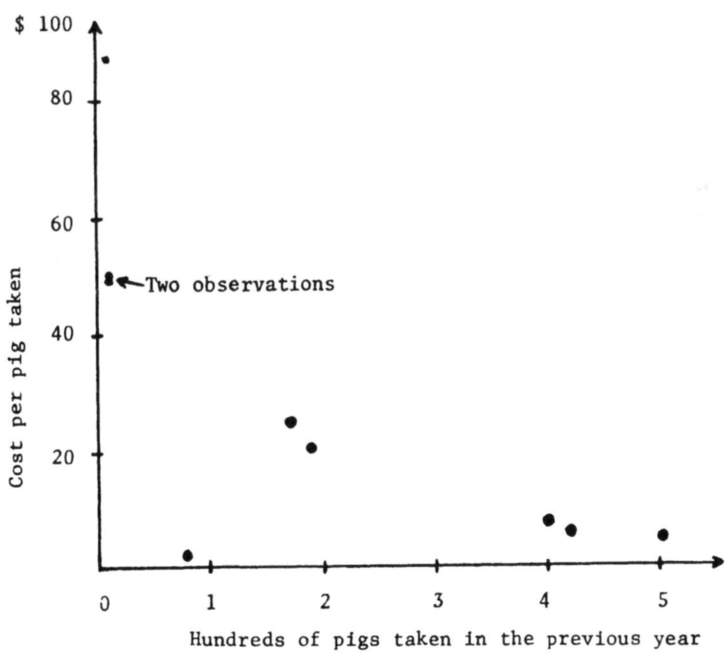

Fig. 6.13 Cost per pig killed related to the number of pigs killed in the last year for nine individual N.P.W.S. Districts. Control expenditure per pig killed tends to decline as the number of pigs taken increases.

with the number of pigs taken or killed. Whether or not this indicates economies of scale in control is uncertain, even though overhead costs are more widely spread as more pigs are taken.

Kill rates and effectiveness of controls. Another interesting piece of information are kill-ratios of pigs, (a) based on the present population, and (b) based on the (maximum) population without control in the previous year. These are set out in Table 6.20 for different Districts and vary greatly, although for the whole state they are respectively 13.8 per cent and 12.1 per cent for this sample. Kinchega, Griffith and Tibooburra have the highest kill rates (on their estimated raw figures). Their kill rates are over one-third of the population in their District. Kill rates in other Districts are substantially lower, being in most cases under 10 per cent. No dramatic decline in present pig populations can be expected in parks in these other Districts, given the rapid rate at which feral pig populations increase under favourable conditions.[31] This is not to suggest that the present programmes in these Districts are useless. Control is more difficult and costly to achieve in some of these Districts, though marginal control is being achieved, and park personnel are obtaining experience in control methods

TABLE 6.20 Rates of Kill of Pigs in N.P.W.S. Districts
(a) As a Proportion of Current Pig Populations and
(b) As a Proportion of Current Population plus Kill in the Previous Year (N.S.W.)

District	Estimated current number of pigs (1)	Number taken in previous year (2)	Current number plus those taken (1)+(2)	(a) Kill rate % (2)/(1)	(b) Kill rate % (2)/(1)+(2)
Coonabarabran	7 050	187	7 237	2.6	2.6
Upper Hunter	2 101	170	2 271	8.0	7.5
Moree-Narrabri	1 050	80	1 130	7.6	7.0
Tibooburra	1 030	400	1 430	38.8	28.0
Griffith	1 028	500	1 528	48.6	32.7
Kinchega	300	420	720	140.0	58.3
Lismore	200	10	210	5.0	4.8
Morton-Nowra	150	10	168	6.6	6.2
Glen Innes	70	8	78	11.4	10.3
Grafton	20	0	20	0	0
All of above	12 999	1 785	14 711	13.7	12.1

and demonstrating to farmers and graziers the seriousness with which the National Parks and Wildlife Service views the feral pig as a pest.

Under general comments on control of feral pigs (the last item of the questionnaire), a number of District Officers stressed the shortage of funding for control operations and the need for co-operation by *all* adjoining landholders in a control programme. However, one District Officer suggested that as far as feral pigs are concerned, there should be:

'1. immediate decommercialization;
2. eradication areas set up on natural boundaries;
3. high penalties for keeping feral pigs enclosed;
4. high penalties for allowing domestic pigs to range freely.'

Comparative Data on Feral Pigs and Other Introduced and Feral Animals in National Parks and Nature Reserves, N.S.W.

Table 6.21 gives an indication of how prevalent different feral and introduced mammals are in parks and reserves, by N.P.W.S. Districts. On a state-wide basis, cats and rabbits occur in most Districts, followed by pigs and foxes, dogs, goats, dingoes, cattle, horses and sheep, and deer in that order. Some regional differences exist in the extent to which species are spread. Pigs are present in all Districts on the Tablelands and in the Western Slopes and Plains but in only a little over half of those on the Coast. Goats occur more frequently as one goes west. Brumbies and deer are most common on the Tablelands, and sheep are most widespread in the Western Slopes and Plains. Dogs and dingoes have the greatest range of occurrence on the Tablelands and Coast. Rabbits and cats are common in all regions but a little more widely present away from the Coast. Feral cattle are most widespread on the Coast, and foxes are most prevalent on the Western Slopes and Plains. It can be concluded that although pigs are not present in as many Districts as cats and rabbits, they are widespread since they are reported present in 81 per cent of N.P.W.S. Districts in New South Wales.

Comparative population levels of feral or introduced mammals. Table 6.22 summarizes information reported about the sizes of populations of introduced and feral mammals present in parks and reserves in New South Wales. Especially as far as the smaller mammals are concerned (rabbits and cats), the estimates as reported are subject to wide margins of error; the estimates for pigs and goats, while subject to considerable error, might be expected to be more accurate. While the estimated number of pigs (13 450) exceeds that for goats (10 745), numbers of goats are high.[32] Populations of both pigs and goats are concentrated in the Western Slopes and Plains, with this concentration being greatest for goats. In fact more

TABLE 6.21 Frequency of Presence of Feral and Introduced Mammals Reported in N.S.W. National Parks and Nature Reserves by Districts

SPECIES	COAST Frequency	% of Respondents	TABLELANDS Frequency	% of Respondents	WESTERN SLOPES & PLAINS Frequency	% of Respondents	ALL REGIONS % of Respondents
Pigs	5	55.55	5	100	7	100	81.00
Goats	5	55.55	3	60	6	86	66.66
Horses	2	22.22	3	60	2	29	33.33
Deer	-	-	3	60	-	-	14.00
Sheep	1	11.11	1	20	5	71	33.33
Rabbits	8	88.8	5	100	7	100	95.00
Dogs	7	77.7	5	100	4	57	76.00
Dingoes	6	66.6	4	80	3	43	62.00
Cats	8	88.8	5	100	7	100	95.00
Cattle	7	77.7	1	20	2	29	48.00
Foxes	7	77.7	3	60	7	100	81.00

than 97 per cent of the reported goat population in parks and reserves is located in the Western Slopes and Plains.

Species sufficiently serious pest to warrant control in parks. Turning to the question of whether introduced or feral mammals are a sufficiently serious pest to warrant control in parks under District control, the greatest number of Districts (17) believed feral cats to be such a pest. In terms of the number of Districts considering species to be a sufficiently serious pest to warrant control, pigs (14 Districts) and rabbits (14) were most frequently ranked as the next pests to cats warranting control, followed by goats (13), foxes (11), dogs (8), cattle (6), horses (5), dingoes (5) and deer (1), in that order.[33]

To some extent, a better indication of the seriousness of an introduced or feral animal as a pest may be obtained by examining the number of Districts in which it is present and considered to be a pest warranting control (in parks) as a proportion of the total number of Districts in which the species is present. On

TABLE 6.22 Populations of Introduced and Feral Mammals Reported in Parks in New South Wales

Species*	Coast		Tablelands		Western Slopes & Plains		All regions
	No.	% of total	No.	% of total	No.	% of total	No.
Pigs *Sus scrofa*	580	4.3	2 371	17.7	10 499	78.0	13 450
Goats *Capra hircus*	125	1.18	120	1.12	10 500	97.7	10 745
Horses *Equus caballus*	30	5.0	350	59.0	210	36.0	590
Deer *Cervus sp.*	?	?	60	100.0	-	-	60
Sheep *Ovis aries*	20	18.0	-	-	90	82.0	110
Rabbits *Oryctolagus cuniculus*	51 100	64.0	1 100	1.4	27 500	34.6	79 700
Dogs *Canis familiaris*	215	66.6	100	30.9	8	2.5	323
Dingoes *Canis familiaris dingo*	107	41.5	110	42.6	41	15.9	258
Cats *Felis catus*	1 420	49.5	150	5.2	1 300	45.3	2 870
Cattle *Bos taurus*	459	92.0	-	-	40	8.0	499
Foxes *Vulpes vulpes*	595	35.0	10	6.0	1 100	59.0	1 705

*Eden also reported rats and mice.

this basis, pigs rank as the most serious introduced or feral mammal pest in parks and reserves in New South Wales and are very closely followed by goats, cats and brumbies, and then rabbits, foxes and cattle. Control is considered desirable in a high proportion of Districts in which these animals are present. N.P.W.S. Officers in 87.5 per cent of the Districts where pigs occur (excluding a non-response to this question) regard them as a sufficiently serious pest to warrant control. The comparable figures for other species (excluding non-responses) are goats (86.6 per cent), cats (85 per cent), horses (83.3 per cent), rabbits (77.7 per cent), foxes (64.7 per cent), cattle (60 per cent), dogs (50 per cent), deer (33 per cent), dingoes (31 per cent) and sheep (0). It is interesting to note how infrequently it is thought desirable to control dingoes.

Fig. 6.14 Goats Capra hircus *shot near the Darling River, western N.S.W., by a hunter for their winter skins. Feral goats are considered to be a pest and are abundant in some of the western national parks in N.S.W. - for example, in Kinchega National Park.*

Rankings of feral and introduced mammals as pests in parks. Respondents were asked to rank feral and introduced mammals by the seriousness of the present threat posed by these animals, the preservation of parks and nature reserves in a natural state and their native wildlife. Eighteen of the 21 responding Districts gave a rank. Each of the three non-rankers were in different regions of the state.

The rankings given by District Officers were used to calculate a mean rank score of the seriousness of the present threat. The number 10 was assigned to that species ranked most serious, and lower consecutive numbers for species ranked as less serious. The mean rank scores for New South Wales as a whole are set out in Table 6.23. Cats rank highest in terms of the mean score of their seriousness as a pest, followed by rabbits, pigs, goats, foxes and dogs. It must be remembered, however, that this is an averaging procedure involving *all* responding Districts. The detailed distribution of ranks for the state as a whole and by regions are given in my original report.[34]

If one considers how frequently the different species are ranked in the top

Fig. 6.15 A rusa deer Cervus timoriensis *roaring in Sydney Royal National Park. In most parks where deer (an introduced species) are present in New South Wales, National Parks and Wildlife Service Officers do not believe that they are a sufficient pest to warrant control.*

TABLE 6.23 New South Wales (All Regions) Ranking of Feral and Introduced Mammals by Seriousness of Present Threat to Preservation of Parks in Natural State and Their Native Wildlife

Species	Mean rank score
Cats *Felis catus*	8.16
Rabbits *Oryctolagus cuniculus*	7.59
Pigs *Sus scrofa*	7.06
Goats *Capra hircus*	5.82
Foxes *Vulpes vulpes*	5.53
Dogs *Canis familiaris*	5.29
Cattle *Bos taurus*	2.77
Dingoes *Canis familiaris dingo*	2.44
Deer *Cervus* sp.	2.00
Horses *Equus caballus*	1.77
Sheep *Ovis aries*	1.53

four places by regions and in the state as a whole, it is found that cats, rabbits, pigs and goats are most frequently included. All of these species are regarded as a serious pest in parks and reserves by District Officers.

Comparative straying by species and comparison of complaints about their straying. Straying of species from parks and reserves is sometimes a cause for complaint by adjoining farmers and graziers. Table 6.24 sets out the relative frequency with which N.P.W.S. District Officers reported that species strayed and the relative frequency of District Officers indicating that they receive complaints about such straying. A high proportion of Districts with pigs present reported that they stray from parks (70.6 per cent) and the proportions were also high for goats (64.3 per cent) and for dingoes (61.5 per cent). Many Districts received complaints about pigs straying from parks. Ordering species - by Districts reporting complaints about their straying as a proportion of Districts where they are present - pigs top the list. Complaints about pigs straying from parks were received by 58.8 per cent of Districts where pigs are present. Dingoes follow next on the basis of complaints (53.8 per cent), and then dogs (46.7 per cent). Despite the fact that a high proportion of Districts containing goats report that they stray, comparatively few Districts received complaints about their straying. Regional details of straying and complaints about straying of introduced and feral mammals from parks are set out in the original report.[35]

TABLE 6.24 Relative Frequency of Reports and Complaints of Animals Straying from National Parks (N.S.W.), by District

Species	Number of Districts reporting presence	% of Districts reporting straying	% of Districts reporting complaints of straying
Pigs	17	70.6	58.3
Goats	14	64.3	14.3
Horses	5	40.0	0
Deer	3	33.3	-
Sheep	-	-	-
Rabbits	20	40.0	20.0
Dogs	15	46.7	46.7
Dingoes	13	61.5	53.8
Cats	19	36.8	10.5
Cattle	10	50.0	20.0
Foxes	17	52.9	29.0

Pigs in National Parks and as a Threat to Wildlife

While the feral pig is not the only feral or introduced pest of concern in parks and reserves, District Officers of the National Parks and Wildlife Service regard it as a serious pest. It frequently strays from parks, and complaints about this straying are widespread. Straying is important, because social decisions about the control of the wild pig in parks must take into account its damage in parks as well as the straying and damages caused while straying. At the same time, it needs to be remembered that the movement of pigs is a two-way process. While pigs may move out from parks to adjoining properties, pigs also enter parks from adjoining properties and can repopulate them. Effective control of pigs requires co-operation by all adjoining landholders. Steps have been taken to foster such co-operation.[7]

Concluding Comments on the N.S.W. Survey

The survey verifies the widespread presence of feral pigs in parks administered by the National Parks and Wildlife Service in New South Wales. On the basis of estimates given by District Officers, the population of wild pigs in N.S.W. national parks and nature reserves exceeds that of any other large introduced mammal. While District Officers pointed to damages caused to fauna and flora in parks and reserves by pigs and their effects on erosion, most officers pointed out that their knowledge about these effects is limited. Nevertheless along with cats, rabbits and goats, feral pigs were considered to be a serious pest - a threat to the preservation of parks in their natural state and a danger to native wildlife. In-depth study is needed to identify and quantify these damages.

Feral pigs more frequently stray from parks than any other mammal, and complaints about their straying are relatively more frequent than for any other animal; damage is caused to farming and grazing property as a result of such straying. Straying also creates problems for the control of pigs in parks because surrounding properties can be a source of repopulation of parks.

In nearly all districts where feral pigs are present, the National Parks and Wildlife Service has instituted control programmes. Ten Districts providing details spent more than $18 000 on pig control in the year preceding the survey. While expenditures tend to be greater in Districts with more pigs, expenditure rises less than proportionately. Kill rates between Districts vary greatly. In most Districts they are below 10 per cent. In the West, however, three districts have achieved estimated kill rates of over 30 per cent.

One difficult question is how much is it worthwhile to spend on feral pig control in national parks? The decision should at least take into account the damage caused in the park by pigs, the straying of the pigs and the damage caused while straying. On average it cost $10.43 to take each pig in parks in the year preceding the survey. Was the overall savings of damages in monetary terms in excess of this? Even if it was not, the effort still might have been justifiable from a political point of view, given the frequency and widespread complaint from farmers and graziers of pigs straying from parks and complaints about their presence in parks. There might also be a spin-off benefit from N.P.W.S. control. Adjoining landholders might be more willing to take action to control pigs on their properties when the National Parks and Wildlife Service demonstrates that it is taking action to control pigs.

RESULTS FROM A SURVEY OF WILD PIGS IN NATIONAL PARKS IN WESTERN AUSTRALIA

While feral pigs are not present in Western Australia in the numbers and to the extent in New South Wales and Queensland, they are present and appear capable of causing considerable damage to fragile ecosystems, such as those occurring in Western Australia, even when their densities are low. With this in mind, I approached the National Parks Authority of Western Australia in May 1979 about the possibility of distributing a questionnaire to national parks in Western Australia designed to provide information on the presence of pigs in parks, damages caused by pigs, the straying of pigs, measures being taken to control pigs in parks and to provide some comparative data about other introduced or feral mammals. The head office of the Authority kindly distributed my draft questionnaire[37] to the rangers of seven parks in which it was thought possible that feral pigs might be present and returned the completed survey forms to me in August 1979. Results were therefore not provided for all parks but for *all* those in which the regional superintendent of the authority thought that pigs could be present. All parks were surveyed in which there seemed a possibility of pigs being present.

Presence and Population of Pigs in W.A. Parks

The parks in the survey are shown in Fig. 6.16. Four of the seven parks reported the presence of feral pigs. While no pigs were reported from Warren National Park, they are known to be present in the district. The responses indicate that feral pigs are present in national parks close to Perth and in the south-west corner of the state generally, in the Murchison River area and in the Kimberleys. The presence or absence of feral pigs in the parks surveyed is indicated in Table 6.25.

The total number of feral pigs estimated to be present in parks containing

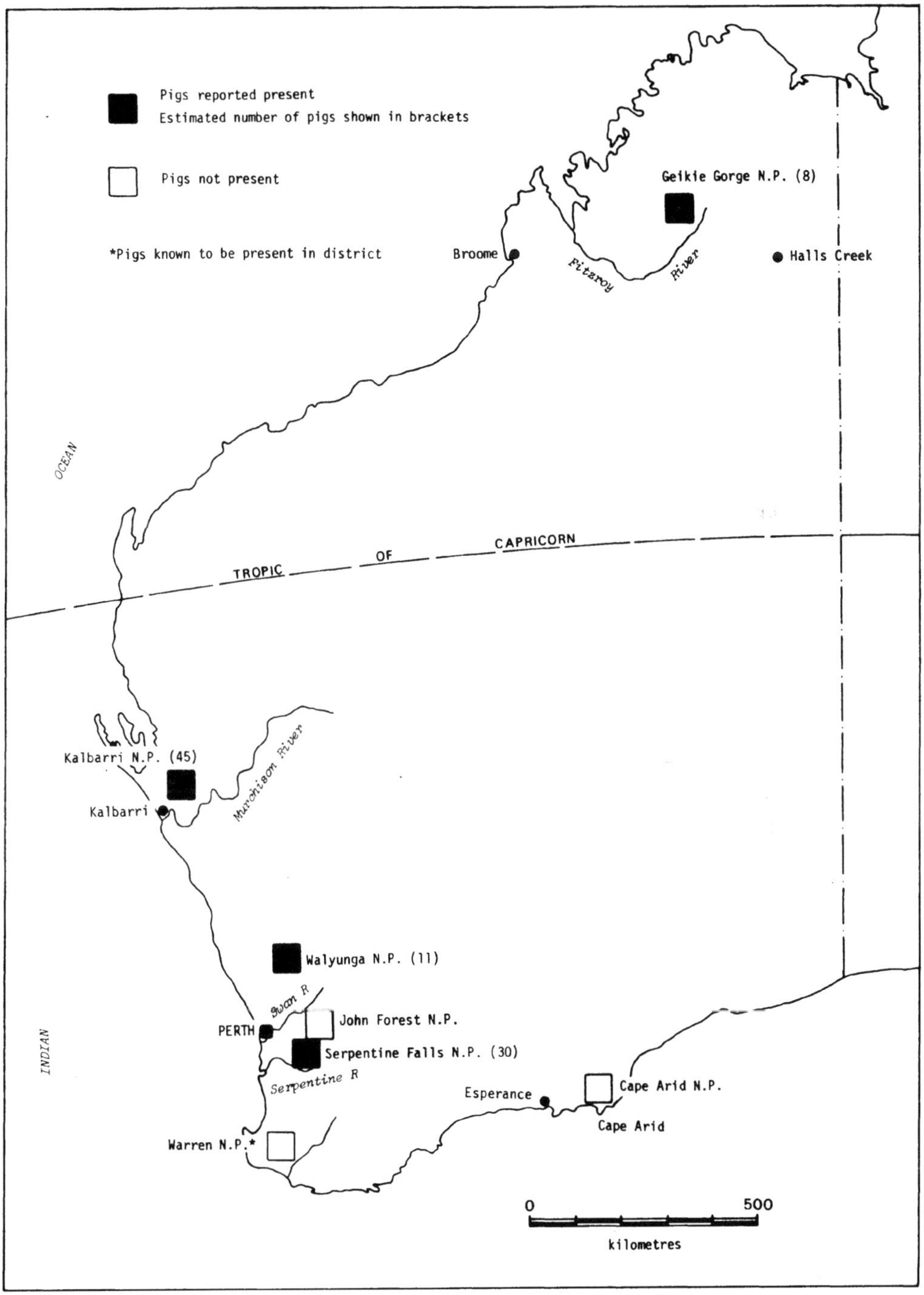

Fig. 6.16. Estimated number of feral pigs in seven national parks surveyed for feral pigs in Western Australia.

TABLE 6.25 Presence or Absence of Feral Pigs as Reported by National Parks in Western Australia

National park	Location	Present	Absent
Geikie Gorge	Fitzroy River, Kimberleys	1	
Kalbarri	Murchison River	1	
Walyunga	Swan River, Perth area	1	
John Forrest	On a tributary of Swan River, Perth area		1
Serpentine Falls	Serpentine River, South of Perth	1	
Warren	Warren River, near Pemberton		1*
Cape Arid	East of Esperance		1
	TOTAL	4	3

*But feral pigs are known to be present in the district.

them is small. The estimated total is 94, with the highest number being in Kalbarri National Park, followed by Serpentine Falls. Densities are low except at Serpentine Falls, where they are high even by comparison with New South Wales parks.[38] Numbers of pigs and density of pigs for those parks reporting their presence are set out in Table 6.26. In most cases the estimates are a 'guestimate' by rangers.

Damages to Flora and Fauna and Benefits of Wild Pigs in W.A. Parks

All four national parks having feral pigs reported that they damage flora. All complained that they cause damage to established plants by rooting around them for food. Two parks said that they destroy seedlings and young trees; and one said that they spread weeds, especially blackberry *Rubus vulgaris* and the arum lily *Arum italicum*. The responses are summarized in Table 6.27.

As for damage to fauna by pigs, the ranger in charge at Walyunga reported that the extent of this was unknown in his park. Rangers in the other three parks with pigs reported damages to ground nesting birds and small animals and gave the following comments about damage to fauna:

> Geikie Gorge: 'Minimal probably. Destruction of habitat of ground nesting birds and shelter of small mammals and marsupials.'
>
> Kalbarri: 'Mainly bird nests, bob-tailed goannas, etc., marsupials with young, in trouble and sick.'

Rangers in charge of parks with pigs were asked if any other animal posed as great

TABLE 6.26 Number of Pigs Estimated to be Present in National Parks in Western Australia, and Densities

National park	Area in hectares	Estimated number of pigs	Density hectare/ pig
Kalbarri	186 097	45	4 135.5
Serpentine Falls	635	30	21
Walyunga	1 790	11	163
Geikie	3 136	8	392
OVERALL	191 658	94	2 038.9

a threat to native flora and fauna as feral pigs. The ranger at Geikie Gorge said that he did not know, whereas the rangers at Kalbarri and Serpentine Falls suggested goats as posing a comparable threat and the ranger at Walyunga nominated feral cats. Their specific comments were:

Walyunga: 'Feral cats - damage to most forms of small fauna including birds.'

Kalbarri: 'Goats because of their damage to flora and role in increasing soil erosion.'

Serpentine Falls: 'Goats damage flora with browsing and also appear to assist in spreading exotic weeds such as wild oats *Avena fatua*, onion grass *Cuperus rotundus*, etc.' (see Fig. 6.17).

TABLE 6.27 Damage to Flora by Pigs: Frequency Reported by National Parks (W.A.)

Type of damage	Geikie	Kalbarri	Walyunga	Serpentine Falls	Total frequency
Caused by rooting around established plants for food such as suckers, bulbs and insects	1	1	1	1	4
Destruction of seedlings and young trees	1			1	2
Spread of weeds (especially blackberry *Rubus vulgaris* and arum lily *Arum italicum*)				1	1

Fig. 6.17 Onion weed Cuperus rotundus *was introduced to Australia and is widespread. Feral pigs eat the bulbs and have rooted out those above.*

Walyunga, Kalbarri and Serpentine Falls said that there were no benefits from wild pigs in their parks; the ranger at Geikie Gorge suggested that the uprooting of bare ground to provide a simple seed bed for new plants could be beneficial.

Damages Other than to Flora and Fauna Caused by Pigs in Parks and Disease Potential, W.A.

Park rangers were asked whether feral pigs caused damages to other than flora and fauna in their park and whether or not the potential of feral pigs for carrying disease and parasites posed a problem for their park.

Kalbarri, Walyunga and Serpentine Falls rangers said that wild pigs are responsible for erosion in their parks. The ranger at Geikie Gorge said that they are not responsible for erosion there. Responsibility for erosion was reported to be minor at Kalbarri and at Walyunga but more serious at Serpentine Falls. The rangers commented:

Kalbarri: 'Owing to the small number of pigs in parks the problem is not serious.'

Walyunga: 'Minor erosion near walking tracks and water soaks.'

Serpentine Falls: 'This is a problem which over extensive areas and in time could lead to vast erosion.'

As for types of damage other than damage to native flora and fauna or erosion caused by feral pigs, rangers at Walyunga and Serpentine Falls did not know of any but rangers at the other two national parks containing pigs reported as follows:

Kalbarri: 'Foul up water supplies.'

Geikie Gorge: 'Disturb tourists by scavenging in rubbish bins at night.'

The potential of wild pigs for carrying disease and parasites was said not to pose a problem at Kalbarri. The rangers at Walyunga and Serpentine Falls said that they did not know whether it posed a problem. The ranger at Geikie Gorge said there is a problem because pigs in the park scavenge in rubbish bins and at the rubbish tip.

Straying of Pigs from Parks in W.A.

Only one of the national parks containing pigs received complaints about pigs moving from the park and causing damage - the ranger at Walyunga reported a few complaints from landholders, about damage to pastures and dams caused by pigs.

While the ranger at Serpentine Falls had received no complaints about pigs straying from the park, he has observed damages caused by feral pigs on private property in his area. He reported,

> 'Damage seen by me on private property consisted of large areas of pasture badly ripped up and sections of muddy creek bed churned up enough to trap sheep.'

In all the national parks where pigs were reported present, rangers in charge reported that pigs stray from their parks. But feral pigs from private property also stray into parks and some groups of feral pigs appear to be migratory, visiting parks only for short periods.

Control Measures Against Feral Pigs in Parks, W.A.

As for *policy* on the control of pigs within parks, rangers in charge of parks containing pigs responded as follows:

Geikie Gorge: 'Destroy if possible.'

Kalbarri: 'Eradication.'

Walyunga: 'Shoot on sight.'

Serpentine Falls: No answer.

All four national parks reporting the presence of feral pigs were taking action to control them. Rangers in each park shot pigs and a trap for pigs had been installed in addition at Geikie Gorge. The control measures specified were:

Geikie Gorge: 'Trap constructed near rubbish tip. Shooting where possible.'

Kalbarri: 'Shooting by national parks rangers.'

Walyunga: 'Destruction by shooting. Vermin patrol on foot in suspected areas during cool months when pigs are likely to be present.'

Serpentine Falls: 'Control by me (ranger in charge) shooting any pigs sighted.'

In all parks, shooting was considered satisfactory as a control measure. The comments of the respondents on the effectiveness of their control measures are set out in Table 6.28.

Rangers in each of the national parks containing pigs reported some special difficulties in controlling feral pigs in their park. Special difficulties mentioned are:

Geikie Gorge: 'Proximity of open pastoral land and stock.' (Neighbouring pastoral land appears to be a source of invasion).

Kalbarri: 'Break-away country is difficult to negotiate. Movements of tourists limit shooting.'

Walyunga: 'Pigs are not permanently resident and only pay brief visits. Access to their preferred area is difficult.'

Serpentine Falls: 'Lack of time to keep more rigid control measures. Also pigs are supplied with excellent cover by large thickets of blackberry making searches of areas most difficult.'

The Selective Opening of Parks to Amateur Hunters

On the question of whether national parks should be open at certain times of the year to permit selective shooting or hunting of wild pigs, three of the parks containing pigs opposed this, and one favoured it under special conditions and

TABLE 6.28 Effectiveness of Control Measures for Feral Pigs taken in W.A. National Parks

Park	Method	Comment by ranger on effectiveness
Geikie Gorge	Trapping Shooting	Trapping appears to be useless. Heavy calibre weapon required.
Kalbarri	Shooting	Very effective.
Walyunga	Shooting	Satisfactory results.
Serpentine Falls	Shooting	(Apparently of some effect). Pigs not entering park in as great a number as previously. Total reduction is impossible.

restrictions - but not in his Park. The reasons given by those opposed to the suggestions were:

A: 'Irresponsibility of some members of the public.'

B: 'Previous personal experience in allowing selective shooting has shown it to have undesirable repercussions because the word soon gets around and unauthorized shooting does commence and is extremely difficult to stop.'

C: 'Most shooting club members cannot follow instructions. Also have a tendency to think that once permission is given for one shoot, it then authorizes to shoot whenever they want.'

The ranger suggesting that selective shooting might be allowed in some circumstances said:

D: 'Depends on size of park and number of feral pigs. Not in this park, but could be of use in parks with larger area and number of pigs. Would have to be organized very carefully.'

All of the rangers in charge of national parks containing pigs gave suggestions about what should be done to control feral pigs in national parks. Most favoured shooting by national park staff. Specific suggestions were:

Geikie Gorge: 'Not in favour of shooting unless done by professionals. Poisoning of carcasses is a possibility, but how many native animals would be affected?'

Kalbarri: 'Shooting by national park staff.'

Walyunga: 'Full time hunter for control of all vermin, not just pigs.'

Serpentine Falls: 'Selective shooting by park rangers where applicable.'

Comparative Data on Feral Pigs and Other Introduced and Feral Animals in Western Australian Parks

Rangers in the seven parks surveyed were asked to rank feral and introduced mammals by the seriousness of the threat they posed to the preservation of the park (in its natural state) and its native wildlife. This would provide a basis of comparison with pigs.

In *all* parks in which pigs, goats, cats and donkeys occurred, they were regarded by rangers as a sufficiently serious pest to warrant control and rabbits were so regarded in most. Cats occurred in all the parks and rabbits in all but one. In parks containing pigs, pigs were most frequently ranked in the first four places, followed by goats and cats, and then rabbits and foxes. However, goats and cats were more frequently placed in first place than were pigs (see Table 6.29). It is clear that pigs, goats and cats are all considered a serious pest in parks where pigs occur. Taking *all* parks together, cats were most frequently ranked in

TABLE 6.29 Frequency with which Feral and Introduced Animals were Reported Present in Seven W.A. National Parks and Frequency with which They were Regarded by Rangers as a Sufficiently Serious Pest to Warrant Control

Species	Number of parks in which species present	Number of parks in which species serious enough for control	% of parks (where species present) advocating control
Pigs *Sus scrofa*	4	4	100
Goats *Capra hircus*	4	4	100
Horses *Equus caballus*	3	0	0
Donkeys *Equus asinus*	1	1	100
Sheep *Ovis aries*	2	1	50
Rabbits *Oryctolagus cuniculus*	6	4	66.6
Cats *Felis catus*	7	7	100
Foxes *Vulpes vulpes*	6	2	33.3
Cattle *Bos taurus*	2	1	50
Camels *Camelus dromedarius*	1	0	0

the first four places as a threat and, like goats, were high in the ranking of seriousness. Rabbits and foxes were next to cats most frequently ranked in the top four places, but they tended to be placed towards the lower end of the seriousness scale. Then followed pigs and goats.

As for the straying of feral and introduced animals from national parks, pigs, goats and sheep were reported to stray from national parks whenever they are present. No rangers received complaints about introduced or feral animals straying, except in one park for pigs, another for foxes and another for cattle. The point was made in a number of cases that feral animals stray *into* parks. The following comments were received from rangers:

Walyunga: 'Pigs, goats and sheep all stray into the park from outside and are not resident within the Park.'

Geikie Gorge: '"Straying from Park" questions should be rephrased as "straying into Park".'

Serpentine Falls: 'Complaints about cattle wandering out of the park onto roads. We complain about cattle wandering into the Park too.'

Estimates of the number of feral animals present in the parks are of limited reliability, except possibly for the larger mammals. However, the estimates for goats are interesting: a total of approximately 1250 goats were estimated to be present. Most of these were in Kalbarri National Park (1000) and the next greatest number were at Serpentine Falls National Park (25 to 200). It is clear that overall goat numbers are well in excess of pig numbers in national parks in Western Australia - whereas the reverse is the case in New South Wales.

Concluding Comments on W.A. Survey

Only two general policy suggestions were received from rangers about the control of feral pigs. The head ranger at Kalbarri said,

'In cases of heavy infestations of pigs, the Agriculture Protection Board of W.A. could be requested to carry out control.'

The ranger in charge at Serpentine Falls said,

'Feral pigs have been in this and surrounding areas for years and only systematic shooting or trapping will effectively reduce numbers. I do not consider existing poisons a good control because there is no control over dead pigs or missed baits.'

In summary, the estimated number of feral pigs in Western Australia is not large, but they are still regarded by rangers as serious pests in national parks in Western Australia whenever they occur. Rangers said that feral pigs were a sufficiently serious pest to warrant control in all parks in which they were present, and all of these parks were taking control measures against pigs. Other serious pests in National Parks as mentioned by rangers were feral cats, goats, rabbits and foxes.[39]

RESULTS FROM A SURVEY OF WILD PIGS IN NATIONAL PARKS IN QUEENSLAND

In the diversity of wildlife Queensland is perhaps Australia's most interesting state; important remnants of tropical rainforest occur along its eastern coast, and in the north it experiences a 'monsoonal' wet season and a dry season. Much of the state is located in the Torresian fauna region[40] and the state contains fauna either not present or not significantly present in New South Wales or Western Australia. Thus the survey of Queensland National Parks and Wildlife officers covering feral pigs in their national parks and nature reserves, undertaken in 1979 and 1980, provides an additional perspective on the problems posed by feral pigs in the national parks of Australia.[41] As with the other states, and in accordance with the model of assessment outlined earlier in this chapter, I sought information on the presence and population levels of feral pigs in state national parks and nature reserves, damages caused in and to national parks, and any benefits, the straying of pigs and complaints about this, measures taken to control pig populations in national parks, and the comparative threat of various species of feral and introduced mammals to indigeneous wildlife as seen by the regional officers of the National Parks and Wildlife Service.

Population Levels of Pigs in Queensland National Parks

All of the N.P.W.S. Regions reported the presence of feral pigs. The greatest number of pigs was reported to be present in the Cairns Region (more than 25 000), the next largest number in the Rockhampton Region (2000), followed by the Brisbane Region (660), and the Gympie Region (10 or more). The largest population is in the north of the state: Lakefield National Park, a newly established park on Cape York Peninsula, (in the Cairns Region) was estimated to have the greatest number of pigs - around 20 000. Fig. 6.18 shows the administrative Regions of the National Parks and Wildlife Service in Queensland and the population of pigs (by Regions) reported to be present in national parks and in wildlife reserves. For the state as a whole, almost 28 000 feral pigs were reported to be present in national parks.

The *density* of feral pigs in national parks is also greatest in the north of the state. As indicated in Fig. 6.19, the density is highest in the Cairns Region (less than 25 hectares per pig), next highest in the Rockhampton Region (80 hectares per pig), followed by the Brisbane Region (171.5 hectares per pig), and the lowest density in the Gympie Region (greater than 1161 hectares per pig).[42]

Damage to Native Flora and Fauna

All Regions reported that feral pigs cause damage to native flora, and all except Maryborough recorded damage to native fauna.

Damage to flora. Pigs were said to damage plants growing in swampy areas, especially tuberous and bulbous plants and those with extensive root systems. From Lakefield (Cairns Region) it was reported that 'Entire swamps, miles long have been

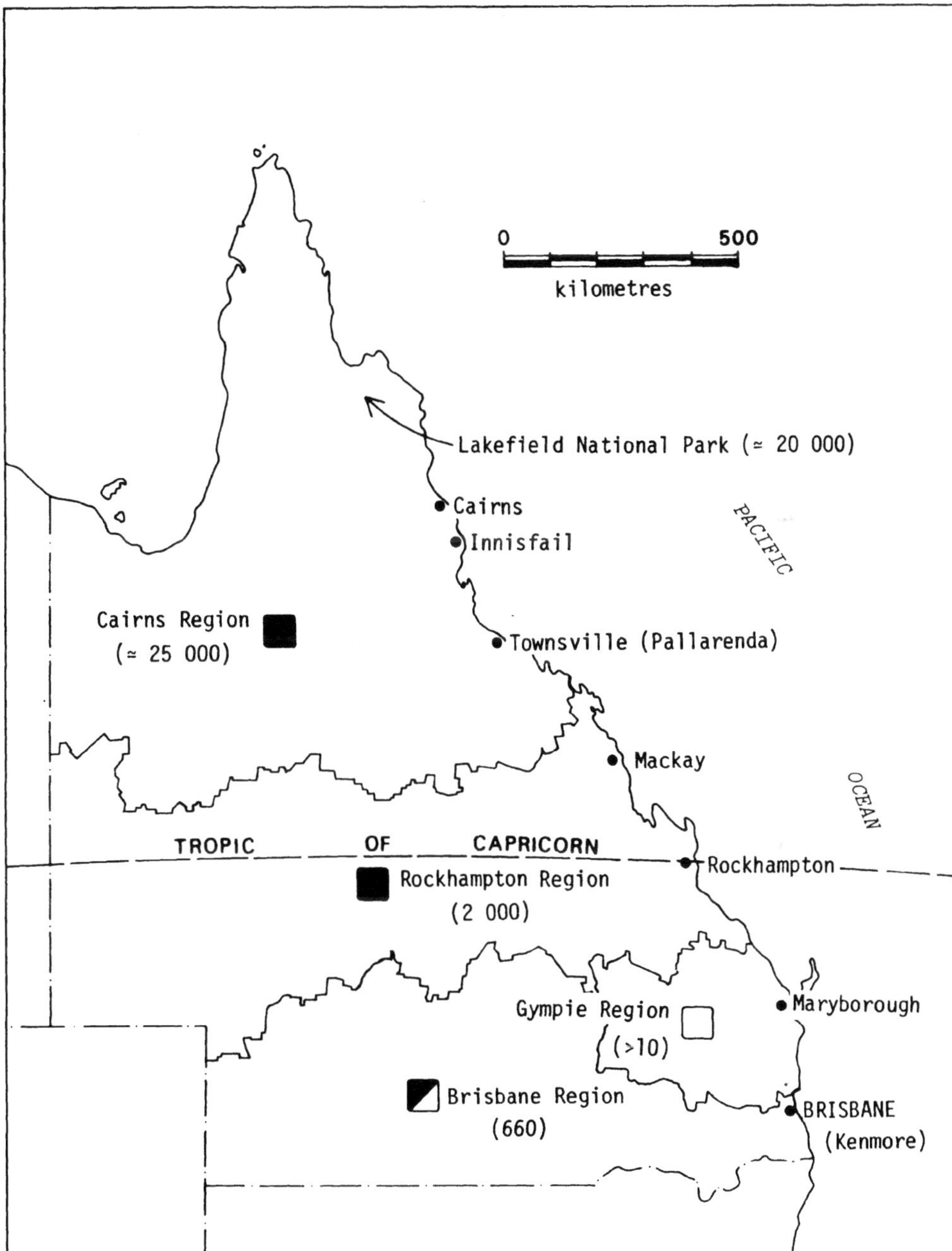

Fig. 6.18 National Parks and Wildlife Service Regions in Queensland, and the number of feral pigs reported by N.P.W.S. officers to be present in national parks and nature reserves.

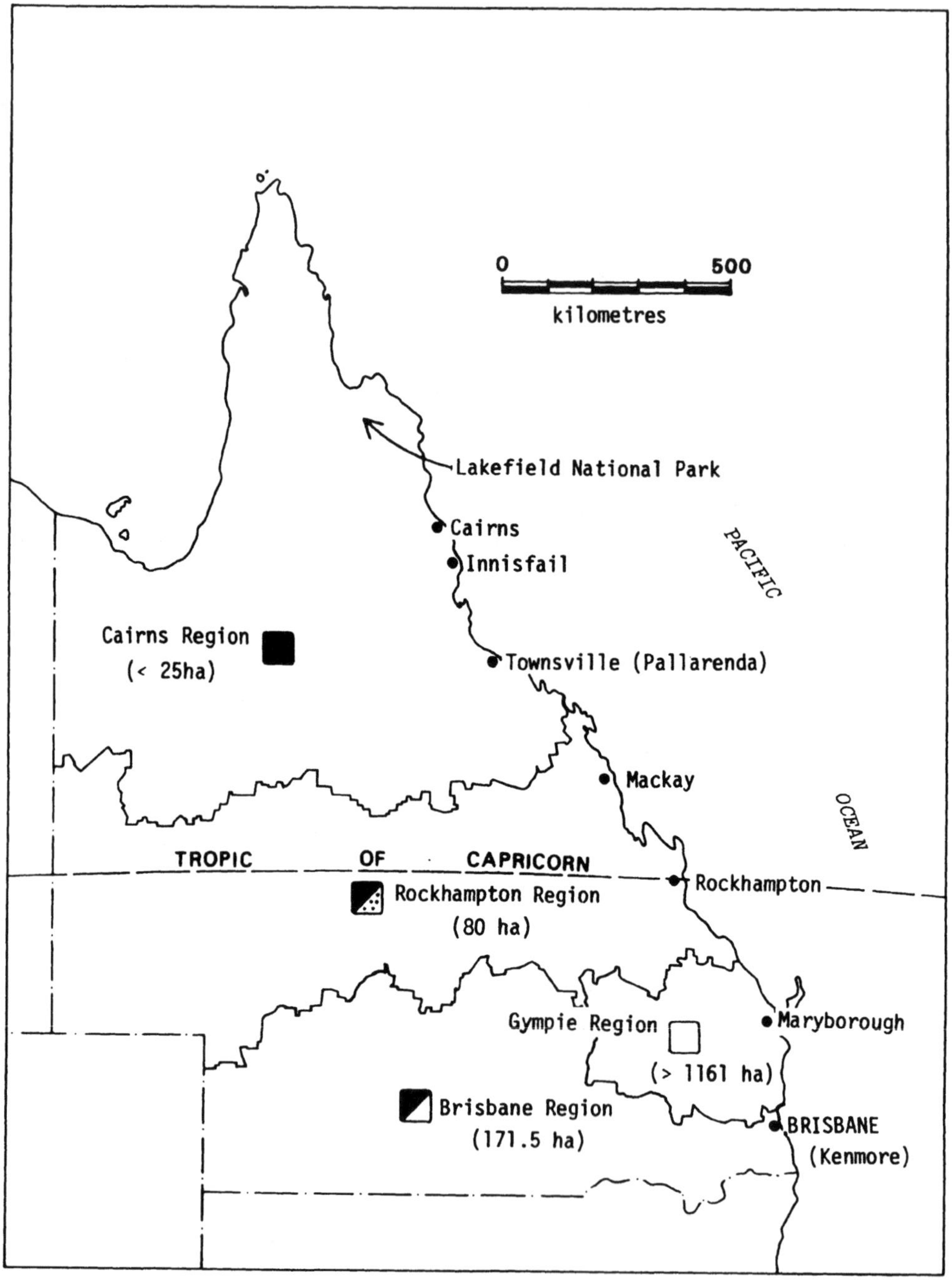

Fig. 6.19 Estimated densities of feral pigs in national parks and nature reserves, Queensland, by Regions. Hectares per pig are given in parentheses.

entirely ploughed up.' The Pallarenda office in Townsville said 'Swamp grasses and water lilies especially are rooted and destroyed.' The grasses that grow along the edge of swampy areas are a major target for pigs. As the Lakefield officer points out, 'Ploughing up of mud prevents the growth of grasses along the edge of swamps and unfortunately these grasses also provide habitat for native animals.' Furthermore, after the destruction of the swamp grasses, 'the ground is then open to erosion when the next 'wet' arrives.' In the Brisbane region feral pigs were said to destroy terrestrial orchids and bulbous plants unique to swamp areas - for example, in the Girraween National Park.

In some regions the uprooting of seedlings and young trees was attributed to the presence of feral pigs. The Mackay office in the Rockhampton Region suggested that 'small mobs of pigs (which may average six) in the rainforest area show a preference for young palms.' The comments of the N.P.W.S. officers on flora effects of pigs are summarized in Table 6.30.

Impact on native fauna. Most of the respondents felt that feral pigs had a serious impact on native fauna. The Cairns region appeared to be most affected. Habitat destruction, competition for food and predation were all mentioned as possibilities. The general responses are set out in Table 6.31.

The most important single effect cited of feral pigs on native fauna was habitat destruction, mentioned by every N.P.W.S. office who thought the impact of pigs on native wildlife to be serious. 'The destruction of habitat has a serious effect, especially on waterbirds. Some areas become devoid of waterbirds as the dry progresses and the water levels recede. The pigs dirty up the hole in the shallow water.' (comment from Pallarenda office, Cairns Region) From the Brisbane Region it was reported that 'habitat destruction in swamp areas and granite country, particularly along streams' is serious.

Competition for food with native fauna was regarded as the second most important impact of the feral pig presence: in Lakefield National Park it is claimed that pigs create 'severe competition for food, for example, with brolgas *Grus rubicundus* and magpie geese *Anseranas semipalmata* which also consume tubers and bulbs.'

Predation on native species by feral pigs was considered to be serious in the Cairns Region. It was reported that 'pigs may consume lizards and snakes along with young chicks and the eggs of waterfowl. Pigs have also been recorded digging up sand around the nests of freshwater crocodiles *Crocodylus johnstoni*.' The eggs and the young of these crocodiles may be eaten. Pigs adversely affect brush turkeys and scrub hens. The Innisfail N.P.W.S. office reports that 'pigs dig out the nests (mounds) of the brush turkey *Alectura lathami* and of the scrub hen *Megapodius freycinet*.' The Mackay office also reported that pigs damage the nests of brush

TABLE 6.30 Impact of Feral Pigs on Flora in Queensland National Parks

Region	Damage to flora	Comment received
Cairns Region		
Lakefield	Yes	Pigs root up and consume plants, particularly those plants with extensive roots/tubers found in swamps and lagoons. Pigs consume water lilies, particularly those with tubers or bulbs and other aquatic vegetation. Entire swamps miles long have been completely ploughed up. Ploughing up of mud also prevents the growth of grasses at the edge of swamps, as the water level drops during the wet season.
Innisfail	Yes	Uprooting of young trees.
Pallarenda	Yes	Heavy damage to low areas coming out of the mountains, especially near water. Large areas of swamp lands and grass lands are uprooted. In particular, swamp grasses and water lilies are rooted and destroyed. The ground is then open to destruction when the next 'wet' arrives.
Rockhampton Region		
Mackay	Yes	A preference for young plams.
Rockhampton	Yes	Uprooting of seedlings and creek bank vegetation.
Gympie Region		
Maryborough	Yes	Damage to bulbous roots.
Brisbane Region		
Kenmore (Brisbane)	Yes	Destruction of habitat especially in swamp areas - terrestrial orchids and bulbs unique to these areas. Very serious damage in granite country such as at Girraween National Park (near Stanthorpe).

turkeys. The detailed comments received are set out in Table 6.32.

Threat Posed by Other Feral and Introduced Animals to Native Flora and Fauna

In the Cairns Region, pigs were considered of all introduced animals to pose the greatest threat to native flora and fauna, specifically in Lakefield National Park. The Pallarenda office also thought pigs were the greatest threat, though the Innisfail office suggested that cats were as great a threat overall. Table 6.33 summarizes the responses of the various offices to the question of whether any other feral animal posed as great a threat to native flora and fauna as feral pigs.

The pigs were considered to have no benefits at all in any of the national parks.

Erosion and Other Damage, Parasites and Diseases

In all N.P.W.S. regions in Queensland it was agreed that pigs were responsible for increased erosion in both swamps and creeks on slopes. The effect was said to be very serious in the area covered by the Pallarenda office and in the Brisbane region. The comments of the district officers are summarized in Table 6.34.

As for other damages caused by pigs, most officers did not know of damages other than those already mentioned, but the Kenmore office pointed out that public relations, especially with adjoining landholders, suffers as a result of damage caused by feral pigs using margins of national parks for harbour.

Officers were asked whether the potential of wild pigs for carrying disease and parasites posed a problem. In the case of Lakefield National Park the answer

TABLE 6.31 Impact of Feral Pigs on Native Animals as reported by N.P.W.S. Officers* in Queensland

Region	Habitat destruction	Competition for food	Predation	Overall impact
Cairns Region				
Lakefield	Serious	Serious	Serious	Serious
Innisfail	Serious		Serious	Serious
Pallarenda	Serious			
Gympie Region				
Maryborough				Not significant
Rockhampton Region				
Mackay				Not significant
Rockhampton		Serious		Not significant
Brisbane Region				
Kenmore	Serious			Serious

*Classified by reporting officers

TABLE 6.32 Impact of Feral Pigs on Native Fauna in Queensland National Parks

Region	Damage to fauna	Comment received
Cairns Region		
Lakefield	Yes	The main damage is through habitat destruction and severe competition for food - for example, with brolgas *Grus rubicundus* and magpie geese *Anseranas semipalmata* which also consume tubers and bulbs. Pigs may consume lizards and snakes along with young chicks and eggs of waterfowls. Pigs have also been recorded digging up sand around the nest of freshwater crocodiles *Crocodylus johnstoni*.
Innisfail	Yes	Digging out of nests of brush turkey *Alectura lathami* and scrub hens *Megapodius freycinet* in the top section of Palmerston Park.
Pallarenda	Yes	Probably little direct damage to fauna but the destruction of habitat has a serious effect, especially on waterbirds. Some areas become devoid of waterbirds as the 'dry' progresses and the water recedes because the pigs dirty up the holes in the shallow water.
Rockhampton Region		
Mackay	Yes	Reputed damage to brush turkey nests resulting in depletion of the population on the eastern side of Eungella Range. Population reduction may have been caused by shooting as pigs were not sighted in this area.
Rockhampton	Yes	Minimal, but some competition for food.
Gympie Region		
Maryborough	Not recorded	
Brisbane Region		
Kenmore	Yes	Habitat destruction in swamp areas and granite country, particularly along streams.

was yes, 'although this possibility has yet to be looked at.' Presumably the northern position of the park, close to Asian countries having exotic animal diseases, could be a consideration here. The Innisfail office did not feel that the potential of pigs for carrying parasites and diseases was a problem. All other respondents did not know if it was a problem and Pallarenda commented 'In the experience of N.P.W.S. staff, so far there is no evidence that this problem exists.'

TABLE 6.33 Feral Animals Posing a Serious Threat to Native Flora and Fauna (Qld)

Region	Other feral animals as great a threat as pigs	Other feral animals causing a serious threat				
		Cattle	Goats	Cats	Horses	Dogs
Cairns Region						
Lakefield	No	√		√		
Innisfail	?			√		
Pallarenda	No					
Rockhampton Region						
Mackay	Yes		√			
Rockhampton	Yes	√			√	
Gympie Region						
Maryborough*	Yes	√		√	√	√
Brisbane Region						
Kenmore	Yes		√	√		

*Also mentioned were cane toads because of their impact on insect population and fouling of water. The cane toad *Bufo marinus* was introduced into Queensland to control insect pests in sugar cane.

Straying of Feral Pigs from Parks

Feral pigs are relatively mobile and given their destructive habits this can be a source of friction between adjoining landholders. As in other states, complaints are made by farmers about feral pigs straying from national parks on to their properties and causing damage. Table 6.35 provides information about the frequency of complaints to N.P.W.S. officers in Queensland about damages done by feral pigs migrating from their parks.

In the Cairns region, the Innisfail office reported that bird watchers complain of pigs destroying terrestrial and arboreal bird life and farmers complain of damage to cane crops. The Pallarenda office said that cane farmers complain about pigs sheltering in the national park by day, coming out at night to destroy their young cane. Some vegetable (small) crops are also affected. The Mackay office also reported damage to small crops and cane. In the Brisbane region, there are complaints from landholders adjoining parks about damage caused by pigs to crops and rabbit-proof fences, and local authorities, particularly councils, express concern.

TABLE 6.34 Erosion Caused by Feral Pigs in Queensland National Parks

Region	Degree of problem	Comment received
Cairns Region		
Lakefield	Restricted	Erosion has been found to occur on tracks and on slopes where pigs have dug.
Innisfail	Restricted	Erosion of tracks and on creek edges.
Pallarenda	Very serious	Pigs loosen up creek banks during the dry period. During the next 'wet' there may be 20 to 30 feet of water over the banks of these creeks.
Rockhampton Region		
Rockhampton	Restricted	Pigs cause limited erosion to creek beds.
Mackay	Restricted	Erosion of walking tracks at times a nuisance.
Gympie Region		
Maryborough	Minor	Pigs cause a minor amount of erosion.
Brisbane Region		
Kenmore	Very serious	Erosion is very serious in some areas with swamps and drainage lines being heavily disturbed by pigs.

TABLE 6.35 Complaints to N.P.W.S. Officers About Damage by Pigs Straying from Parks (Qld)

Region	Frequency of complaints
Cairns Region	
Lakefield	(Not applicable; park not operational at time of survey)
Innisfail	Many complaints from landholders
Pallarenda	A few complaints
Rockhampton Region	
Mackay	A few complaints
Rockhampton	No complaints in the past 2-3 years
Gympie Region	
Maryborough	No complaints (though there are few pigs in the park)
Brisbane Region	
Kenmore	Many complaints, particularly to N.P.W.S. officers in parks that adjoin freehold agricultural land

Control of Feral Pigs in Queensland National Parks

In all but the areas under the control of the Pallarenda and Mackay offices, some systematic effort was being made by the N.P.W.S. to control pig numbers. Shooting and trapping were used - shooting being the most common method employed.

The N.P.W.S. officers were asked to comment on the effectiveness of control measures taken. (Their answers are summarized in Table 6.36.) None of the measures were without problems and all seem to have a small impact on the overall pig population.

On the question of whether there should be more control of pigs in parks under their control, most officers said that this was impossible given their existing staffing, and one (Mackay) felt that greater effort would not be effective in reducing pig numbers in his area - a rather sobering but not unrealistic thought

TABLE 6.36 Effectiveness of Measures Taken to Control Feral Pigs in National Parks (Qld)

Region	Control	Effectiveness	Problems
Cairns Region			
Lakefield	Shooting (trapping to commence shortly)	Partially effective, more so when a number of shooters are involved	Pigs migrate out of shooting areas and become extremely wary
Innisfail	Shooting	Not very effective	Expense, and activity of pigs confined to marginal areas of parks
Pallarenda	None		Staff shortages, and inaccessibility of pig areas
Rockhampton Region			
Rockhampton	Trapping	Effective in small areas	Staff availability
Mackay	None		Control would not be effective in extensive rainforest
Gympie Region			
Maryborough	Shooting	Unknown	
Brisbane Region	Shooting and trapping	Shooting not very effective. Trapping reasonably effective in small areas	Considerable time and effort needed for a thorough job

given the fecundity and mobility of pigs. Table 6.37 sets out the recommendations of the officers for greater control of pigs in the parks under *their* control.

N.P.W.S. officers were asked to state any special difficulties that they encounter in controlling feral pigs in parks under their control. Most frequently they mentioned lack of manpower and thick forest cover as problems, then lack of access to areas containing pigs and difficult topography. The large areas to be policed and the mobility of pigs were also considered problems. A typical comment was the one from the Pallarenda office:

> 'Access to areas where pigs are doing damage and the deficiency in staff numbers to look after traps. Other commitments are of a higher priority at this time. The country the pigs are using is impossible to shoot over because of the density of cover, and dogs are not used because of the damage to native fauna.'

Table 6.38 summarizes the responses.

On the general question of what should be done to control wild pigs in national parks, the following comments were received:

> Lakefield National Park: 'Every possible measure should be used with due consideration to the effects upon other wildlife (e.g. no poisoning).

TABLE 6.37 Recommendations by N.P.W.S. Officers for Greater Control of Feral Pigs in Parks under Their Control (Qld)

Region	Present control	Recommendations to reduce problem
Cairns Region		
Lakefield	Shooting	Trapping programme planned for 1980 as more manpower and funds are provided
Innisfail	Shooting	Trapping programme
Pallarenda	None	Control methods linked to staff availability; at present 8 men control 5 parks (with a total area of 37 289 hectares)
Rockhampton Region		
Rockhampton	Trapping	No availability of staff at present
Mackay	None	Not practical
Gympie Region		
Maryborough	Shooting	(No comment)
Brisbane Region		
Kenmore	Trapping and shooting	Implementation of an intensive trapping and shooting programme. At present limitations of staff and funds are a problem.

TABLE 6.38 Special Difficulties in Controlling Feral Pigs in National Parks (Qld)

Region	Large area	Mobility of pigs	Topography	Access	Forest cover	Limited Manpower
Cairns Region						
Lakefield	✓	✓	✓	✓		
Innisfail					✓	
Pallarenda				✓	✓	✓
Gympie Region						
Maryborough			✓			✓
Rockhampton Region						
Mackay	✓				✓	✓
Rockhampton			✓			✓
Brisbane Region						
Kenmore		✓		✓	✓	

Primarily trapping in small parks and shooting in larger parks such as Lakefield.'

Innisfail: 'A regular shoot aimed at eradication of all pigs carried out by national park employees, together with the setting of traps.'

Pallarenda: 'In this area more staff are needed to carry out the higher priority work so that time is available to do pig trapping.'

Mackay: 'Rangers to shoot pigs when they make their periodic forays.'

Rockhampton: 'Extensive trapping and shooting programme to be carried out by N.P.W.S. staff when they are available.'

Kenmore: 'Trapping by N.P.W.S.'

The Opening of National Parks for Hunting

Officers were asked whether they felt that national parks should be open in certain times of the year to permit shooting or hunting of wild pigs. All but one were opposed to this. Specific comments were:

(*Cairns Region*) Lakefield National Park: 'Possibly, if national park procedures are not sufficiently effective.'

Innisfail: 'No. People on such shooting expeditions bring dogs with them and some people will shoot everything that moves. Dogs bail up cassowarries

Casuaris casuaris and these are a special feature of the Park.'

Pallarenda: 'No. The only effective way to control pigs in this area is by trapping. Shooting or hunting is of minimum benefit and may cause more disturbance to the wildlife than the pigs, especially if dogs are used.'

(*Rockhampton Region*) Mackay: 'No. Men and dogs have been hunting pigs in rainforests for generations and it does not seem to have had much effect. I think more damage would be done to the flora and fauna by men and dogs trampling all over the place than would be done by the pigs.'

Rockhampton: 'No. Shooters would create far too many management problems.'

(*Gympie Region*) Maryborough: 'No. Difficulties in control of shooters.'

(*Brisbane Region*) Kenmore: 'No. Shooting is not very effective especially in areas of dense vegetation. Native fauna would be disturbed especially if dogs were used. Trapping by N.P.W.S. would be the preferable method of control.'

Indiscriminate shooting by hunters and damage to wildlife caused by their dogs is one of the chief concerns of the N.P.W.S. officers about opening parks for the hunting of wild pigs.

The Cost and Effectiveness of Measures Taken by N.P.W.S. to Control Pigs

Officers were asked to list the control measures that they had used during the previous twelve months and the number of pigs killed by each method. The responses are summarized in Table 6.39. The cost per pig taken varies considerably, and the sample is unfortunately so small that is doubtful whether any conclusion can be drawn about the relative effectiveness of the different methods. But it is clear that as in New South Wales only a small proportion of pigs is being destroyed in the national parks. The proportion is certainly below 5 per cent of the population and is probably around 2-3 per cent (640/28 000). At such a low kill rate and given the high fecundity of pigs, very little reduction in their numbers can be expected from current control measures. The recorded expenditure on pig control by Queensland N.P.W.S. officers in the 12 months before the survey was $1460.

The question of how much control of pigs in national parks is worthwhile is a difficult one. It is important to take account not only of the damage that pigs do within national parks but also the damage that occurs when they roam to agricultural land. Then there is the matter of the control method to use. The N.P.W.S. officers are clearly correct in emphasizing that the spillover effects of any control measures on native wildlife must be taken into account.

TABLE 6.39 Control of Pigs in National Parks (Qld) Over a Twelve-month Period

Region	Control method	Cost	Pigs killed	Cost per pig killed
Cairns Region				
Lakefield*	Shooting	$800	450	$1.78
Innisfail	Shooting	$10	50	$0.20
Rockhampton Region				
Rockhampton	Trapping	$200	100	$2.00
Gympie Region				
Maryborough	Shooting	Not recorded	5	-
Brisbane Region				
Kenmore	Shooting	$300	15	$20.00
	Trapping	$150	10	$15.00
TOTAL		$1460	630	

*Four months only

Comparative Data on Feral Pigs and Other Introduced and Feral Animals in Queensland National Parks

N.P.W.S. officers were asked to rank feral and introduced animals in their parks as pests. Their rankings are set out in Table 6.40.

Pigs were ranked as the most serious pest throughout the Cairns Region. On a simple weighted index, they ranked as the most serious pest in Queensland national parks, closely followed by cats and cattle. From the worst pest downwards the rankings are: (1) feral pig, (2) feral cat, (3) cattle, (4) feral horses or goats, (5) feral dog, (6) fox, (7) cane toad and rat, (8) rabbit and sheep. Feral pigs, cats and cattle occurred in all Regions, but the distribution of other species of feral and introduced animals mentioned appeared more restricted. However, the distribution of these is greater than indicated - for example, the distribution of the cane toad.

Estimates of the species and number of feral or introduced animals in national parks, whether they stray, and the complaints about their straying were given in my original report of the survey.[43]

TABLE 6.40 Queensland N.P.W.S. Officers' Ranking of Feral and Introduced Animals as Pests - 1 indicates the pest ranked as most serious

Region	Pig	Cat	Cattle	Horse	Goat	Dog	Fox	Cane toad	Rat	Rabbit	Sheep
Cairns Region											
Lakefield	1	3	2	4							
Innisfail	1	3	2			3			3		
Pallarenda	1	3	3	3	2		3				
Rockhampton Region											
Mackay	2	3	7		1	4	5				6
Rockhampton	3	4	2	1							
Gympie Region											
Maryborough	2	1	1	1	1	2	2				
Brisbane Region											
Kenmore	2	1	7		5	4	8	3		6	
PEST INDEX*	9.3	8.4	7.5	5.0	5.0	4.4	3.7	1.1	1.1	0.7	0.7

*Respondents were asked to rank feral or introduced animals by the seriousness of present threat posed by these animals. These rankings given by district officers were used to calculate a mean rank score. The number 10 was assigned to that species ranked most serious and lower consecutive numbers for species ranked as less serious. The total of these rankings were divided by the regional office observations; for example, feral pigs scored 65 rank points divided by the 7 regional office observations, giving an Index of seriousness of 9.3.

Concluding Comments on the Queensland Survey

Feral pigs are present in national parks in all N.P.W.S. administrative Regions of Queensland. Their numbers and densities are highest in the north of the state and least in the Gympie Region. They have a serious impact on flora in all regions, but especially so in the Cairns and Brisbane Regions. The Cairns and Brisbane Regions also report serious effects on native fauna caused by pigs as a result of habitat destruction, food competition and predation. Some water birds - for example, brolgas - seem to be at risk. Birdwatchers have complained about pigs to one N.P.W.S. officer, and adjoining landholders not infrequently complain to N.P.W.S. staff about pigs straying from parks and destroying their crops.

Shooting and trapping is being used by N.P.W.S. staff in a number of parks to control pig numbers, but present measures are of limited effectiveness because the number of pigs taken is relative small. However, several factors limit greater control; these include shortages of N.P.W.S. manpower, density of forest cover, difficulties in gaining access to areas where pigs are located, rugged topography of some of the parks and their large areas, and the mobility of pigs. But N.P.W.S. staff are opposed to opening parks to public hunting at selected times of the year. This is because some hunters pose a threat to native wildlife, may damage flora and are difficult to manage; their dogs, in particular, are considered to be a problem for native fauna.

Pigs were said to contribute to erosion in parks in all areas. The problem was considered to be serious in the Pallarenda area (Townsville) and the Brisbane Region.

For Queensland as a whole, feral pigs were rated as the most serious pest in parks, followed by cats, cattle, feral goats and horses. However, only in the Cairns Region were pigs ranked as the worst pest; they were placed second or third in other Regions. Nevertheless the feral pig is clearly a very serious pest in Queensland national parks. No simple solution to the problem appears to be in sight.

CONCLUSIONS ABOUT FERAL PIGS IN NATIONAL PARKS

Feral pigs are most numerous in the national parks and nature reserves in New South Wales and Queensland. They are also present in small but widely spread populations in national parks in Western Australia, and although the Northern Territory did not participate in my survey, the pigs are known to be present in parks there.

It is clear that the feral pig - an animal introduced to Australia - can and does cause considerable damage to native flora and fauna in national parks in each of the states surveyed. The damage occurs in a wide range of environments - from sub-alpine regions to tropical rainforests. The pigs contribute to erosion, seriously so in some areas such as northern Queensland; they help spread some noxious, introduced weeds[44]; and they cause other damage within parks - for example, to footpaths and roadways. Many N.P.W.S. officers regard feral pigs as a potentially serious carrier of exotic diseases.

Feral pigs commonly stray from national parks and reserves, and complaints about their straying, mostly from agriculturalists, are more frequent than for any other species. On their forays they destroy crops and do other damage, such as that discussed in Chapter 2.

N.P.W.S. officers regard the feral pig as one of the more serious introduced animal pests in national parks. The most serious pest varied, as one would expect, with the locality of N.P.W.S. districts but overall the feral cat seems to be looked on with most disfavour. Rabbits and the goats were also regarded as prominent pests. In some localities, feral horses, feral cattle, foxes and feral dogs were looked on with particular dislike by N.P.W.S. officers from a conservation viewpoint.

The National Parks and Wildlife Service of New South Wales seems to be spending more on the control of feral pigs in parks than similar services in other states. Shooting, trapping and poisoning are employed in New South Wales, and shooting and trapping in the other two states surveyed. It is doubtful, given the fecundity and mobility of pigs, that current kill rates are sufficient to reduce noticeably the population of feral pigs in parks. (A cynic might say that, at least, current programmes help to appease local landholders who complain of national parks harbouring pigs.)

Control of feral pigs in national parks is clearly difficult to achieve. Often pigs are located in inaccessible areas, they have considerable cover, the terrain is difficult, the area patrolled is large with many watering points, and so on, and of course the funds made available to local N.P.W.S. regions are limited. However, almost without exception N.P.W.S. officers were opposed to opening the national parks, even at specified times, to (amateur) hunters out to kill feral pigs. As pointed out in Chapter 3, some hunters have suggested that national parks should be opened to enable feral pigs to be controlled. The main argument put forward by the N.P.W.S. officers against this is the possible adverse side-effects of hunters and their dogs on native fauna and flora and on other park-users.

As pointed out, in theory it is possible to do a cost-benefit analysis of feral pig control in any park. If, for example, the annual damage by one extra pig in a park is valued at \$15 and the spillover loss (for example, damage caused by straying) is \$10, it would be worthwhile spending up to \$25 annually to reduce the pig population by one. Certainly in those parks where pigs not only cause considerable damage in the park but also cause great damage when they stray, more attention should be given to the economics of their control. For example, are there any parks in Australia where from a social viewpoint it would pay to eradicate feral pigs and exclude the entry of new colonies of pigs (for instance, by fencing)?[45] The economic assessment depends considerably, but not entirely, on the value placed on the conservation of native wildlife. Despite the difficulties, such a valuation cannot be avoided. To ignore the problem is to place by one's action of control or lack of control an *implicit* value on the conservation of wildlife. The valuation problem can be approached in several ways, as Sinden and Worrel point out in their recent book.[46]

However, whatever values are placed on the conservation of wildlife and natural environments, it is necessary to know the ecological impact of feral pigs on these. The surveys and my research indicate that in Australia there is considerable scientific ignorance about the impact of feral pigs upon native fauna and environments. Rational economic decisions cannot be made about the control of feral pigs until more is known about these effects, since they form *part* of the economic decision. Present control programmes (including poisoning, with adverse side-effects) are being carried out in a state of ignorance. Overseas studies, while of value, are only a guide to conditions in Australia since Australian wildlife and environmental conditions are unique - they evolved along a different path to that in the rest of the world, when Australia was relatively isolated from the rest of the world.[47]

CHAPTER 7

Pigs in Forests and as a Threat to Wildlife

Pigs are frequently present in forest environments, and this raises many questions. To what extent does their presence aid or hinder silviculture? What other benefits and drawbacks do pigs have in forests? What spillover effects are generated by pigs in forests? Can we expect these to be of the same order in countries (such as Australia, New Zealand and U.S.A.) where the pig's introduction is relatively recent as in countries where pigs have been present for a considerable amount of time? How should pigs be managed or controlled in forest environments? These are some of the questions touched on in this chapter.

In this chapter, observations on the impact of pigs on forest environments overseas are summarized and, in the light of these, the Australian situation is considered, at first generally and then on the basis of my surveys of foresters in New South Wales and Queensland. This chapter also extends the discussion of the previous chapter of the consequences of the presence of pigs for flora and fauna, erosion, straying, and so on - and particular attention is given to their impact on flora. Given that the objectives of forest management *are* multiple purpose, these aspects are relevant even though less weight may be given to them than in the management of national parks.

OBJECTIVES OF FOREST MANAGEMENT AND WILD PIGS

One's approach to the management of a forest and therefore to controlling wild pigs in the forest depends upon the ends, or objectives, set for the forest. Is it to be managed to maximize profit (or the net present value of returns) from timber production? Or are multiple means, not just timber production, to be used to maximize profit? Is profit maximization, in fact, to be pursued, or are multiple objectives to be aimed for? If the latter, what emphasis is to be placed on the different objectives? To what extent are spillover or neighbourhood effects to be taken into account in forest management?

Thus policies towards controlling pigs in forests depend on:

(a) whether the aim is to maximize profit (net present value) through timber production only;

(b) whether the aim is to maximize profit using all available means;

(c) whether the aim is to pursue other objectives, or multiple objectives, such as those combining a profit objective with conservation of wildlife, scenic values in the forest and its use for recreation; and

(d) the extent to which spillovers are to be taken into account; spillovers include consequences of the condition of the forest for water quality, erosion and siltation in catchment areas for dams, the damages caused when animals such as pigs stray from their refuge in the forest, and so on.

Ideally, and from a social viewpoint, a forest should be managed with multiple objectives in mind, and account should be taken of spillovers. But in profit-seeking societies, it may be idealistic to expect forests under private ownership to be managed in this way. Even the objectives of forests under state ownership may be uncertain. Strong pressures may be placed upon state forest authorities by commercial timber users, which result in timber production being stressed at the expense of other objectives; forest authorities may be satisfied to follow this path - thus avoiding the difficult problem of socially evaluating their policies. Nevertheless in state forests it is not unreasonable to expect foresters to give attention not only to timber production but also to wildlife conservation, recreational opportunities, and spillovers and other considerations of a social nature. But, of course, at the same time less emphasis is *expected* to be put on wildlife conservation in forests than in nature reserves or national parks. My discussion in this chapter proceeds on the assumption that it is appropriate to take a social perspective in forest management. Those who wish to take a narrower view may ignore the wider issues.

The general principles that apply to determining the optimal level of control and techniques to use to control wild pigs in forests are the same as those discussed in the previous chapter in relation to pigs in national parks. However, the net social benefit function or relevant objective function to be maximized in a forest may differ from a national park - for example, because less emphasis is given to the wildlife conservation objectives in forests.[1] Table 7.1 summarizes some of the costs and benefits that need to be considered in deciding on the optimal level of control of wild pigs in forests. These are elaborated on in the next section.

GENERAL IMPACT OF PIGS ON FORESTS

Opinions are divided to some extent about the impact of wild pigs in forests. While wild pigs can have an adverse impact on forests, in some circumstances they appear to have a beneficial effect. Let us consider the possible effects of wild pigs on timber production in a forest, the damages the pigs do to forest installations,

TABLE 7.1 General Benefits and Costs of Control of Wild Pigs in Forests

Benefits:

Benefits internal to the forest

Reduced losses of commercial timber (increased profit) because of less pig activity.

Less damage to forest facilities such as roads, fire trails, dams etc., from pig rooting, wallowing, and so on.

Increased income from grazing rights.

Possibly increased diversity of wildlife (fauna and flora) and improved scenic quality.

Benefits external to the forest

Less straying of pigs from forest and attendant damage, e.g. to agricultural crops.

Decreased erosion in the park with less siltation downstream, e.g. of dams.

Possibly reduced risk of transmitting diseases to domestic livestock.

Possibly increased control of pigs by neighbours via the demonstration effect, which in turn may give internal benefits.

Direct Costs of Control Program to Forest Authorities:

(a) Explicit costs of poison, bait, ammunition, of additional travelling costs, of labour, etc.

(b) Implicit costs - for instance, labour used in pig control should be costed at its next most valuable use, so that if this exceeds the wage paid an addition would have to be made to the actual wage paid.

Foregone benefits from pigs in forests

Some authorities indicate that pigs in moderate numbers may have a beneficial effect in some forests, e.g. by eating insects (pests of trees) and improving vegetation.

Reduced income from hunting for pigs where hunting is permitted for a fee.

Side-effects of control measures

Unwanted side-effects of control measures on other wildlife, farm dogs, recreational visitors to the forest, etc.

their impact on forest fauna and especially flora, and some of their other possible effects in a forest.

Implications for Timber Production of the Presence of Wild Pigs

Trees may be affected directly and indirectly by the presence of wild pigs in forests. The eating, rooting or grubbing, tusking, wallowing and rubbing activities of pigs can all have an impact.

Direct effects on trees. Wild pigs eat the seeds of many trees. This may reduce regeneration of a forest; and to the extent that pigs have definite preferences for the seeds of one species over another, this can alter the composition of the forest. Rooting by pigs may also favour regeneration of one species (in some case, weed-pests) in comparison to another. Fradrich has pointed out that in Europe 'observations show that where wild boars occur, oak forests turn into birch woods or fir trees replace oak.'[2] Fradrich attributes this to digging and rooting by wild pigs but it is also possible that the liking of pigs for acorns has an effect. No information seems to have been collected in Australia on this aspect, although it is clear that pigs do eat the seeds, nuts, etc., of native trees - for, as mentioned in the previous chapter, Giles and colleagues found acacia seeds in the stomachs of pigs in the Macquarie Marshes, N.S.W.

Rooting or grubbing by pigs, especially if pigs are present in large numbers, can destroy the seedlings of trees. This type of damage not only occurs in forests but sometimes happens in forest nurseries. Established trees can be undermined by pigs grubbing around their roots and may topple in wet or windy weather, especially in wet tropical areas. The growth of some trees may be checked, by damage to their roots as a result of digging by pigs, and lateral roots of some species may be eaten.

Tusking (pigs ripping the bark of tree trunks with their tusks), rubbing (such as may be done repeatedly by pigs against the same tree to remove mud and vermin after using a wallow) and eating of the bark of trees (which may occur when pigs are very hungry) can result in the ringbarking (girdling) of trees and consequently death or a much reduced growth rate. Pigs may also break down and root out small trees and shrubs to make nests or beds.[4]

On the other hand, some trees may benefit from the rooting of pigs. Such rooting may help to aerate the ground and speed up the decomposition cycle. The seeds of some trees may also be placed into an ideal seed bed by the pigs' ploughing of the soil.

Indirect effects on trees. It has been claimed that pigs may have a beneficial impact on forests by eating the larvae of insects that damage trees. On the other hand they may also consume insect predators of such pests, beneficial insects or invertebrates in the decomposition cycle, and so their effect on balance might vary from region to region. Heptner *et al.* say:

> 'The significance of the wild pig for the destruction of forest pests has been known for a long time in Western Europe. The wild pig is of particular significance since it destroys the larvae of cockchafers and leaf wasps. These have on occasions been reduced by 20-30% by the wild pig ... In cases where the incidence of forest pests have risen to levels threatening catastrophe, wild pigs have been known to reduce the pest population sufficiently to avoid the catastrophe.'[5]

But no concrete published information is available in the Soviet Union on the overall significance of wild pigs as destroyers of forest pests.

In the United States, Henry and Conley suggest that 'beneficial aspects of hogs should be studied, such as increasing the quality of seed beds for certain economically important tree species and decreasing the abundance of forest insect pests.'[6] These aspects do not appear to have been studied in Australia yet.

Furthermore, pigs are known to consume rats and mice and to dig rabbits from their burrows. Each of these species can have serious adverse impacts on forests: rats and mice by consuming the seeds of trees, and rabbits by browsing some species of trees and/or ringbarking when food is scarce. On the other hand, by consuming snakes - predators of these animals - pigs have a reverse consequence. Without scientific research, the net impact is uncertain.

Wild pigs might assist in the spread of soil fungi - some beneficial to trees for timber, others not. The detrimental fungus *Phytophthora cinnamomi* which causes some trees to die back because it attacks their root system may be assisted in its spread and occurrence by the activities of pigs.

Activities of pigs that disturb the soil may promote erosion. In some areas, this could reduce the speed of rejuvenation of a forest; in others, it may be beneficial by providing a mineral soil seedbed.

Notes on overseas experience. In the United States, Wood and Lynn undertook a survey of 733 foresters, wildlife biologists and land managers in 11 south-eastern states to determine the impact of pigs on forests and on wildlife in forests.[7] After processing the replies, Wood and Lynn concluded that in 8 per cent of the total forest area inhabited by wild hogs timber regeneration damage was frequent, in 56 per cent of the area it was infrequent, and in 29 per cent of the area it was nonexistent; in 7 per cent of the area it was not indicated. Respondents were asked to rank the various effects of pigs on timber: as (1) important, (2) minor, (3) nonexistent, or (4) beneficial. The frequency with which they were rated important or beneficial is shown in Table 7.2. Hardwood seed consumption, soil disturbance, seedling damage, and sapling girdling (ringbarking) were considered to be significantly adverse by several respondents. Nevertheless, Wood and Lynn felt after the survey that damage to timber regeneration was not important except in certain local areas.

In Europe and the Soviet Union, wild pigs are considered on the whole to make a positive contribution to silviculture. Heptner *et al.*, as mentioned earlier, indicate that pigs benefit trees by consuming tree pests. Furthermore while they eat the seeds of trees, in an average year their consumption is insufficient to affect the regeneration of forests.[8] But they also claim other positive benefits.

'Of great significance for the life cycle of the forest is the digging habit of

TABLE 7.2 Effects on Timber Regeneration of Wild Hogs in South-eastern Forests of the U.S.A.

Hog effect	Important damage* %	Beneficial* %
Hardwood seed consumption	50.3	3.4
General soil disturbance	28.6	23.1
Hardwood seedling damage	18.9	0.7
Longleaf seedling damage	16.2	0.0
Pine seed consumption	13.0	0.0
Loblolly seedling damage	10.9	0.0
Sapling girdling	7.9	0.0
Slash seedling damage	6.6	0.0
Root damage to mature trees	5.6	0.6
Pond Pine seedling damage	3.0	1.0

*Percentage of respondents reporting these hog effects. These percentages do not add across to 100%. The difference is accounted for by respondents who said the hog effects were minor or non-existent.

Source: Based on Table 4, in G. W. Wood and T. E. Lynn, Jr, 'Wild hogs in southern forests', *Southern Journal of Applied Forestry* 1 (1977): 2.

the wild pig. By loosening wide areas of ground it helps to deposit the seeds properly into the soil and therefore contributes to the rejunevation of tree growth. In this connection the wild pig is significant in the pine forests with moss covered ground in Tienshan, in the cembria pine forests in East Siberia and in the oak forests of Europe.'[9]

But the positive contribution of the pig depends upon its density not being too high. If densities become too high, regeneration is adversely affected because seeds that have already germinated are rooted out. Heptner *et al.* suggest that the composition of a forest may be altered over a long period of time by the presence of pigs. They say that '... In central Europe oak forests changed to birch forests because of the activities of the wild pig. In the Belowesh primeval forest the wild pig is contributing to the replacement of oak forests by spruce forests.'[10] Whether or not this is a satisfactory state of affairs depends, of course, on the value placed on the climax tree species in the presence of pigs.

One of the more enthusiastic reports about the beneficial consequences of wild pigs for silviculture was recently given by S. Erl, on the basis of experience in a 3000-hectare conifer forest in Germany:

'According to our experiences the wood growth has increased because of the sows: wild pigs encourage the breaking down of the raw humus cover through constant turning over of the soil and the aerated soil improves the moisture content of the forest floor. The wild pig fertilizes the soil and destroys forest pests such as insects and mice. The wild pigs even destroy Seegrass (a tall forest grass) and clear the young planted forests of competitive weeds. For this reason alone if not for hunting purposes, wild pigs should be kept in forests.'[11]

But the impact may depend on the type of forest, and without systematic study we cannot be sure that Erl's conclusions will apply to Australia, New Zealand or other countries.[12]

Impact on Fauna and Flora

Pigs can have a significant impact on fauna and flora in forests, especially in countries and areas where they have been recently introduced. Their impact on fauna was discussed in some depth in the previous chapter, so here I shall concentrate more on damage to flora.

Fauna damage. It is worthwhile considering again the study by Wood and Lynn in the United States. They questioned their respondents about the impact of wild pigs on wildlife in south-eastern forests of the United States. Respondents were asked to rate wildlife effects of wild hogs on the same scale as that previously mentioned for their impact on timber. The percentage of respondents rating various effects on wildlife as an important damage or as beneficial, is set out in Table 7.3.

In Europe and the U.S.S.R. there is some predation by wild boar on other wildlife. They eat large quantities of mouse-like rodents, carrion which they are able to smell out over long distances, and sometimes rabbits. Under food stress they may become temporary predators attacking the young of wild ruminants and pursuing weakened or straying deer and hinds of all kinds. Sludskii found that

'The wild boar everywhere feeds on mouse-like rodents, the carcasses of many species of animals - both wild and domesticated - and sometimes attacks large wild animals which are in a weakened state. Because of this feature of the diet of the wild boar, it may easily pick up various infections, and does, in fact, pick them up.'[13]

The diet of the wild pig varies from area to area but in the Ila Delta of Kazakhstan, Sludskii observed that wild pigs incidentally eat musk-rats, pheasant and duck eggs; perch is a secondary food source; and basic food sources are carp, grasshoppers and their larvae, larvae of beetles and earthworms.

Flora damage. Considerable damage can be done to the understorey of forested

TABLE 7.3 Effects on Wildlife of Wild Hogs in South-eastern Forests of the U.S.A.

Hog effect	Important damage* %	Beneficial* %
Competition for food with turkeys	48.9	0.0
Competition for food with squirrels	47.2	0.0
Competition for food with deer	44.5	0.7
Turkey nest predation	35.0	0.0
Predation on snakes	20.5	17.1
Quail nest predation	20.4	0.0
Rabbit nesting predation	12.2	0.0
Songbird nest predation	10.4	0.0
Turkey poult predation	9.8	0.0
Quail poult predation	9.1	0.0
Predation on mice and moles	8.2	13.3

*Percentage of respondents reporting these hog effects. These percentages do not add across to 100%. The difference is accounted for by respondents who said the hog effects were minor or non-existent.

Source: Based on Table 4, in G. W. Wood and T. E. Lynn, Jr, 'Wild hogs in southern forests', *Southern Journal of Applied Forestry* 1 (1977): 2.

areas and to the ground flora. Bratton's research in the United States indicates extensive damage by pigs to the shrub and herb components of the forest of the Great Smoky Mountains National Park. Bratton found that in Gray Beech Forest, understorey cover was severely damaged by exotic wild pigs.

> 'Quadrat sampling of both disturbed and undisturbed sites in Gray Beech Forest indicates that hog rooting can reduce the cover of the herbaceous understorey to less than 5% of its expected value. Over 50 non-woody species are known to be eaten, uprooted or trampled by hogs. The damaged wild flowers include such species as *Claytonia virginica, Dicentra cucullaria, Lilium superbum, Phacelia fimbriata, Stellavia pubera* and *Trillium erectum*. Disturbed species exhibit changes in population structure including a reduction in the percentage of mature flowering individuals and reduction in clump size. Changes in species composition favour plants with deep or poisonous roots. Local extinction may be expected as rooting continues.'[14]

As a result of further study in the Great Smoky National Park, Howe and Bratton concluded: 'The overall environmental impact of wild hogs includes the following: (1) rooting activities expose and loosen the soil surface and lead to considerable erosion and probably compaction and stream siltation; this erosion was particularly evident in areas of *Dioscorea* and Mesic herb rooting, which were often on stream banks and steep slopes, respectively; (2) selective feeding directly reduces the populations of a number of species with starchy bulbs, tubers, rhizomes and corms; certain families are much more strongly affected than others; and (3) the hogs are modifying normal successional processes, and may alter the future floristic composition of both low and high elevation plant communities.'[15]

In Hawaii Smith and Diong are undertaking a study in Kipahulu Valley, Haleakala National Park, Maui Island, of the impact of wild pigs on plant populations in rainforest. The number of exotic plant species present appears to be increased by pig digging (as also appears to be the case in Australia) and by selective predation especially on starchy plants; and damage to the floor canopy appears to be altering plant communities in their study area. The diversity of plant species may well be declining, and their sparsity is increasing as a result of pig activity. Smith and Diong suggest that the 'reduction of individual plant populations in Kipahulu Valley is probably a function of time and density of pigs per unit area.'[16]

In New Zealand, there are reports of significant damage to native flora by feral pigs. Wodzicki[17] quotes Kirk[18] as claiming that the formerly plentiful orchid *Gastrodia cunninghamii* has been to a large extent destroyed by pigs and rats. Also Thomson[19] reports that feral pigs have adversely affected the abundance of the ground orchid *Gastrodia sesamoides*. Their damage to flora on the Auckland Islands and the Chatham Islands is well documented[20] - some species of plants, such as the Chatham Island lily *Myosotidium nobile*, have been able to continue to survive only in spots inaccessible to pigs.

Very little scientific information is available about the impact of feral pigs on flora in Australia[21] but intrinsically it seems to be a subject of worldwide interest. The CSIRO has reported that

> 'The Australian flora has fascinated science for two centuries. An estimated 80% of its species and more than 30% of its genera are found nowhere else in the world, easily the highest figures for any continent.'[22]

Dr Len Webb and Mr Geoff Tracey, of the Rainforest Ecology Section of the CSIRO Division of Plant Industry, claim that

> 'Far from being a receptacle for 'second-hand' immigrant floras of other continents, Australia was probably a centre, or part of a centre, for the evolution and radiation of primitive flowering plants.'[23]

Feral pigs are present in a wide range of Australian forest areas including rainforests. For instance, in the Barrington Tops area (mentioned in the previous chapter) they are present in rainforests as well as in other flora associations, including sub-alpine communities. (See Fig. 7.1). In this area there are at least four species of plants found nowhere else, and parts of the area are rich in ground orchids - which, judging from New Zealand experience, are liable to severe damage

Fig. 7.1 'Pol Blue' sphagnum swamp, Barrington Tops State Forest, N.S.W. Mountain cord rush Restio australis *dominates this part of the swamp, which is bordered by* Eucalyptus pauciflora *woodland. Feral pigs do considerable damage to flora in and on the margins of these swamps.*

from pigs. Fairley points out that on the plateau of the Barrington Tops itself,

> 'The ground cover consists mainly of a carpet of *Poa* (snow grass) with a large number of ground orchids and other herbaceous plants. There are many swampy areas on the plateau with a distinctive swamp flora and a definite community of plants. Other distinctive communities are found along the creek banks where there are often very dense and tall shrubs and areas of sub-alpine grassland bordering the swamps. Each of these communities has specialised plants restricted to that area.'[24]

Feral pigs have invaded these areas (as was observed in Chapter 6), they like to root along creek beds and in bogs and their margins, and they are causing considerable damage to the flora[25] - a point also substantiated in this chapter from reports of foresters in the Barrington Tops area. This pattern is repeated in several other unique floral areas of Australia. Yet ecological information about the subject is scanty, to say the least. Nevertheless, all circumstantial evidence points towards a serious adverse impact on Australian plant communities in some forests from the presence of feral pigs.

Other Effects of Pigs in Forests

Wild pigs in forests can affect forest installations, erosion, hunting, grazing rights and relationships with neighbouring properties - given the penchant of pigs for straying.

Forest installations and erosion. As a result of pigs rooting or wallowing on or at the side of roads, tracks and fire trails, these may wash out and need more maintenance. Dams for the supply of water for fire-fighting or livestock may be eroded, silted or fouled by the activity of pigs. Erosion especially along gullies may be accelerated by pig activity leading to siltation of dams where the forest is in a catchment area.

Hunting. In some forests hunting is permitted and even a fee may be charged for this.[26] The presence of wild pigs can add to opportunities for hunting. On the other hand, its net impact depends on its consequences for other huntable species, such as deer. In some forests, however, hunters are not welcome because of the possible damage they and their dogs may do to non-target species, the possibility that they may start fires accidentally, disturb other recreational forest users, and so on. In some forests in Australia pigs are considered as a pest because they attract hunters, often illegally, into the forest. In the United States, where controlled hunting in the forests is permitted, wildlife food plots are often damaged by pigs - an adverse impact on other wildlife. Wood and Lynn report that some national forest rangers in their survey in south-eastern U.S.A. expressed the opinion that

> 'The losses in other resources due to destruction by hogs outweighed the hog's potential as a game animal. This was not a unanimous opinion, however, and several respondents commented that the feral hog should be considered a game animal which provided hunters with "off-season sport" ... there were some comments on the adverse effects on the forest by European wild hogs in Tennessee and North Carolina. Several consultant foresters from North Carolina did comment that there was a strong need to keep the population in check.'[27]

Grazing rights. In some forests it is possible to graze domestic livestock for a fee. For example, in Australia domestic cattle are not infrequently allowed to graze at a fee. The presence of feral pigs in a forest could alter the availability of food for such animals, especially because pigs appear to consume a considerable amount of green grass when it is available.

Straying. It is common for pigs to stray from forests to neighbouring farms in search of food, and frequent complaints are made about this by landholders. Occasionally pigs may also migrate.

Heptner *et al*. believe that, at least as far as the U.S.S.R. is concerned, damage caused to crops by wild pigs straying from forests is often exaggerated.[28] Nevertheless, they report that when pigs live in the vicinity of vegetable gardens, orchards and fields they cause damage by digging up potatoes and by eating sugar melons, water melons, sweet corn cobs and grain; they start eating these items as soon as they are weaned. The main damage to young grain is the consequence of such a large area being trampled upon and stamped into the ground. But major damage to crops occurs only when the fields are small and are surrounded by forests. When food is scarce in the forest wild pigs are likely to frequent fields and do considerable damage. Heptner *et al*. quote Lebedeva's finding that in the Belowesh primeval forest 34-47 per cent of the pigs normally resident in the forest stray into the fields in years of average food supply. Straying to this extent appears normal but it increases in years when food is in short supply. Lebedeva also found that most damage to crops occurs near the forest margin.[29] At the *edge* of the forest wild pigs destroyed up to 30 per cent of the potato and buckwheat crop, whereas the average damage in fields adjoining the forest was only 0.08 to 4.7 per cent.

Whether or not feral pigs in Australia are more prone to stray from forests than wild boar in the Soviet Union is not known. If food supply in forests in Australia happened to be more variable then greater straying could be the case. Circumstantial evidence seems to suggest that, as in the U.S.S.R., damage to crops is greatest near the forest margins when pigs are using the forest as their refuge. Let us look more closely at the Australian situation.

BACKGROUND ON AUSTRALIAN FORESTS

Australian forests are 98.5 per cent natural forest, and of these the bulk (slightly more than 80 per cent of the productive or potentially productive area) is dominated by *Eucalyptus* species. The productive or potentially productive forest areas in New South Wales and Queensland are the largest of those of the Australian states, and between them these two states account for almost two-thirds of the forest areas of Australia. The forest types of New South Wales, Queensland and Australia are indicated in Table 7.4. From the table it can be seen that Queensland is relatively well endowed with rainforest and contains tropical eucalypt and paperbark (*Melaleuca* spp.) areas absent in New South Wales.

TABLE 7.4 Forest Areas Classified by Forest Type:
New South Wales, Queensland, and Australia, 30 June 1977

Forest type	N.S.W. ('000 ha)	Qld ('000 ha)	Aust. ('000 ha)
Eucalypt	13 176	4 940	29 627
Tropical eucalypt and paperbark	-	4 078	6 535
Cypress pine *Cypress hugellii*	1 908	1 683	4 373
Rainforest	300	1 068	1 872
Plantations	166	134	660
	15 550	11 903	43 067

Source: Based on statistics in *Year Book Australia 1979*, Vol. 63, Australian Bureau of Statistics, Canberra, p. 315.

Plantations in Australia consist mainly of coniferous species, and two-thirds of Australian total plantings consist of *Pinus radiata*. However, in New South Wales the proportionate planting of *Pinus radiata* is much higher than two-thirds, whereas (as Table 7.5 indicates) in Queensland it is much lower. In Queensland, plantings are dominated by *Pinus elliotti* and significant plantings of native *Araucaria* species are made.

Private ownership of Australian forests is relatively small but is increasing. More than three-quarters of the area of Australia's forest is publicly owned. Publicly owned land reserved or used primarily for timber production is usually under the control of a forest commission or department of the state in which the timber is located. Forests in the states are the responsibility of state governments rather than the Commonwealth Government of Australia. In New South Wales, the Forestry Commission of New South Wales administers state forests, and in Queensland the Department of Forestry.

The *Year Book* suggests that

'The main object of [state] forestry authorities is to manage the forests of the country in a manner that will provide maximum benefits, both direct and indirect for the community. The authorities aim to promote the multiple use concept in management under which forests remain in perpetuity as sources of valuable raw material, areas of natural beauty, sanctuaries for fauna and flora, and areas for scientific investigation and watershed protection. The provision of special protected areas such as forest parks for recreational use and for the conservation of plants and animals is an objective. Forestry also aims at improving existing forests and woodlands by properly controlled harvesting, by protection from such destructive agencies as fire, insects and

TABLE 7.5 Forest Plantation Areas Classified by Species: New South Wales, Queensland, Australia, 31 March 1977.

Species	N.S.W. (ha)	Qld (ha)	Aust.
Coniferous:			
Pinus radiata	131 568	2 487	431 776
Pinus elliotti	3 874	73 917	77 791
Araucaria species	1 459	38 434	39 893
Pinus pinaster	-	-	28 088
Pinus caribaea	9 013	-	10 213
Other coniferous species	13 489	6 774	30 133
	150 390	130 625	617 894
Broadleaved:			
All species	15 499	3 243	42 108
Total plantation area	165 889	133 868	660 002

Source: Based on statistics in *Year Book Australia 1979*, Vol. 63, Australian Bureau of Statistics, Canberra, p. 315.

diseases, and by inducing regeneration.'[30]

My surveys of foresters in New South Wales and Queensland in 1979 concerning the impact of feral pigs on forests took into account all of these multiple objectives.

It should be pointed out that forests are of special environmental importance in Australia. Native forests cover only 6 per cent of the land area of Australia. More than one-third of Australia's species of birds and more than 50 per cent of its species of mammals are found only in forests. McIlroy indicates that in Australia 16 per cent of birds and 17 per cent of mammals are confined to rainforest, and 16 per cent of birds and 35 per cent of mammals occur only in sclerophyll forest.[31]

It was with this background in mind that I surveyed foresters in New South Wales and Queensland in 1979 to determine the impact of feral pigs on forests. The questionnaire sent was the same as that given to National Parks and Wildlife officers except that foresters were also asked what they estimated the commercial loss from pig activity in their forests to be and whether or not they thought that feral pigs helped to spread dieback. Pigs were present in the forests in most districts in the two states. Approximately 35 000 pigs were estimated to be present in New South Wales state forests, and more than 22 500 in Queensland state forests and timber reserves.

Fig. 7.2 Young Eucalyptus grandis *trees in a private plantation on the north coast of New South Wales. Most Australian plantations are of conifers and are government owned. Pigs were using the fire and access trail illustrated to visit a nearby property.*

RESULTS FROM A SURVEY OF WILD PIGS IN FORESTS IN NEW SOUTH WALES

In June 1979, survey forms were sent to foresters at the sub-district level in the state forests of New South Wales, to collect information about feral pigs in forests.[32] It was intended to obtain information on the presence and population levels of feral pigs in forests, the damages caused by wild pigs in forests and any benefits, the straying of pigs and complaints about this, the measures taken in forest sub-districts to control the numbers of feral pigs, and the comparative threat of the various species of feral and introduced mammals to indigenous wildlife as seen by sub-district offices of the Forestry Commission. As well, views were sought about the desirability of allowing hunting of wild pigs in the forests. Co-operation was excellent, Responses were received from all of the 57 sub-districts.

Statistical Results in Summary[33]

Approximately 35 000 feral pigs were reported to be present in New South Wales forests at the time of the survey. While all Forestry Commission districts reported the presence of feral pigs, Baradine (covering an area known generally to have a high population of feral pigs) accounted for more than two-thirds of the total. The distribution of numbers by forest districts is shown in Fig. 7.3.[34] The largest feral pig populations are on the whole away from the coast. Casino in the north is exceptional. Even in the Newcastle district, most of the pigs are situated in the Gloucester sub-district near the Barrington Tops. Densities of feral pigs as indicated in Fig. 7.4 follow a similar geographical pattern to that of numbers.

Damage to flora and fauna. Forty-one per cent of sub-districts containing pigs said that they caused damage in their forest areas to native flora and fauna; an equal number said that they did not; and 18 per cent said that they were uncertain about this.

Relative threat to wildlife. In sixty-one per cent of sub-districts where pigs are present, some feral animals other than pigs were considered to pose as great or greater threat to native flora and fauna as feral pigs. Feral cats were mentioned as a great or greater threat in 15 sub-districts,

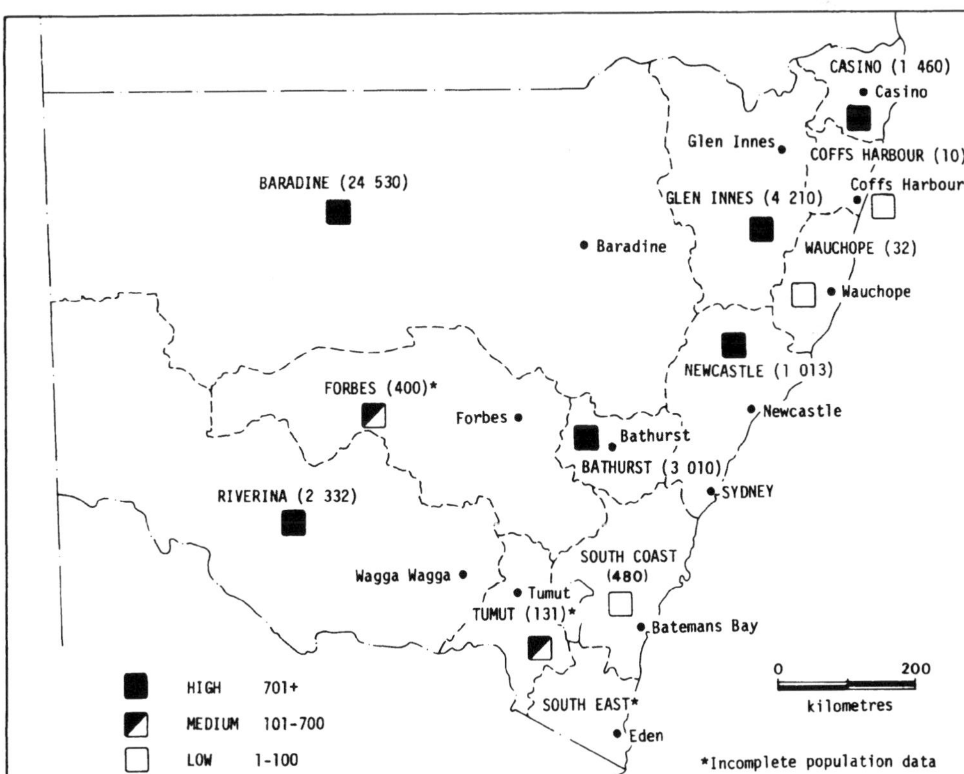

Fig. 7.3 Estimated populations of wild pigs in the forests of New South Wales, by Forestry Commission district.

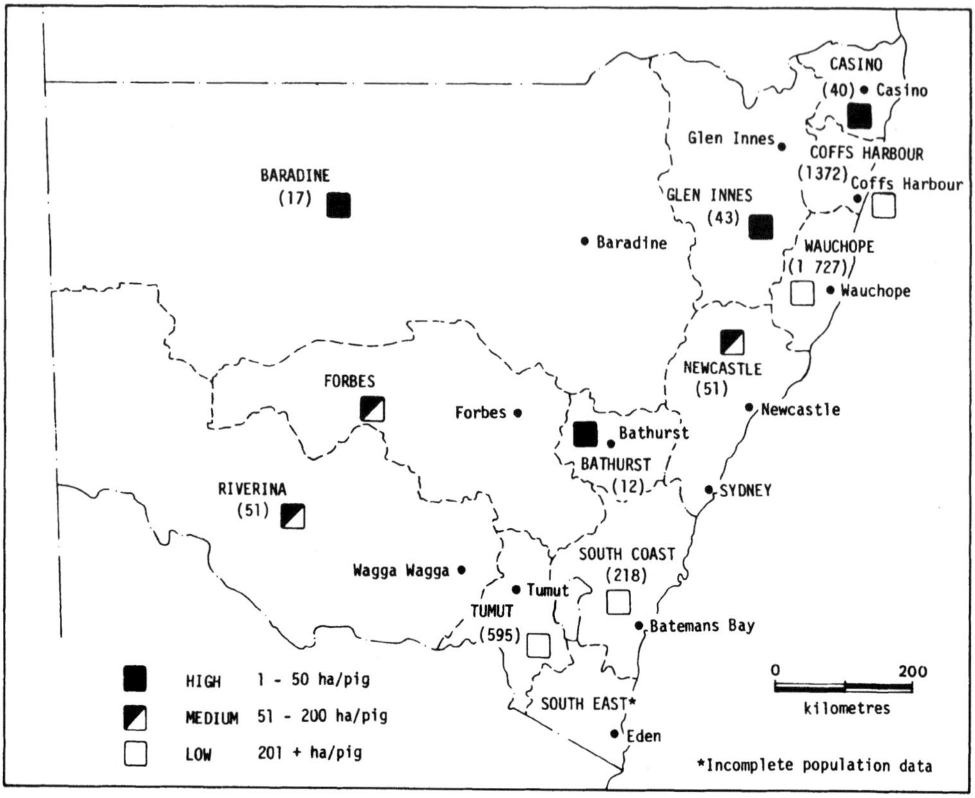

Fig. 7.4 Estimated densities of wild pigs in the forests of New South Wales, by Forestry Commission district.

rabbits in 11, foxes in 10, goats in 7 and dogs in 5. Pigs were considered to be the greatest threat in 11 sub-districts; three sub-districts were uncertain; and no reply was received from one sub-district.

Impact of pigs on commercial value of forests. In only five of the sub-districts where pigs were present (12.8 per cent of the cases) were they believed to reduce the commercial value of forests; and in only four sub-districts was it not known whether pigs reduced the commercial value of forests. In the majority of cases (almost 80 per cent) pigs were believed not to reduce the commercial value of forests.

Complaints about the straying of pigs from forests. In about half of the forest sub-districts with pigs, complaints were received by forest officers about the straying of pigs from forests. In all but one case these complaints were said to be few; but in one sub-district in the Casino district, there were many complaints.

Control measures taken in the forests. A little over one-third of those forest sub-districts containing pigs reported that they were taking measures to control their populations. Methods used include shooting, trapping and poisoning with 1080 (sodium monofluoroacetate). In the 12 months prior to the survey, sub-districts spent over $16 000 in controlling wild pigs. As was found in my survey of national parks in New South Wales,[35] trapping was on average more expensive per pig taken than poisoning with 1080. But poisoning is not suitable for all areas, because 1080 is rapidly leached from the bait by rain and in closely settled areas it poses a risk to domestic dogs.

The number of pigs estimated to have been taken by foresters during the previous year was about 1300, so the overall average cost per pig taken was about $12. While in one sub-district (Inverell) approximately one-third of the feral pig population appears to have been captured in 1978-79, in most cases the actual percentage taken in sub-districts is much lower. Given that a pig population can double in about six months, the forest kill rates may not have a significant impact on the population of pigs in forests. Similar results were observed in the control programme of the National Parks and Wildlife Service in New South Wales. Because of its high fecundity, mobility and adaptability to varied habitats the feral pig is a difficult animal pest to control economically. The motivation for the Forestry Commission to control wild pigs appears to stem not so much from their impact on the commercial value of forests (which appears to be negligible) but from their adverse effect on the economic operations of neighbouring agriculturalists and pastoralists. Other considerations may be the threat posed to native wildlife by wild pigs, their potential as vectors of disease, and their role in erosion.

Ranking of feral and introduced mammals as pests in forests. Officers in forest sub-districts were asked to rank all feral and introduced mammals by the seriousness of the threat they pose to the preservation of forests in a natural state and their native wildlife. The mammals most frequently ranked as the most serious pest were rabbits (17 cases) followed by feral cats (14 cases) and pigs (12 cases) using the averaging procedure of mean rank scores.[36] The most serious pests are ranked as (1) rabbits *Oryctolagus cuniculus* (2) cats *Felis catus* (3) foxes *Vulpes vulpes* (4) pigs *Sus scrofa* and (5) feral dogs *Canis familiaris*, in that order. While the mean rank is probably not a very useful indicator in this instance, there is little doubt that the first four mentioned are considered to be serious threat to the native environment, and in some localities others (such as goats) are also a serious pest.

From the comments made by New South Wales foresters responding to the survey, the following selection pertains to the impact of pigs in state forests.

Comments on flora and fauna damage. Comments received from foresters about damage caused to native flora and fauna by feral pigs are set out in Table 7.6. As far as flora is concerned, damages mentioned include: destruction of the forest understorey in some areas, damage to ground flora in wet or swampy sites; damage to tree and shrub seedlings from rooting, damage to saplings from rubbing and tusking, spread of exotic weeds by providing these with suitable seedbeds as a result of rooting, disturbance of swamps by wallowing, and destruction of small pine by pigs to build their 'nests'. Damaged flora included: plants with bulbs and fleshy roots; fleshy herbaceous plants all of which are eaten (as Bratton's U.S. study[37] indicates), swampland vegetation, ferns, sedges and heaths *Macrozamia* spp., sphagnum plant associations, native grasses, and snow grass *Poa* spp. (see Fig. 7.5). Tree seedlings said to have been damaged by pig activity include small pines (probably *Pinus radiata* and *Pinus elliotti*), the Riverina red gum *Eucalyptus camaldulensis* and blackbutt *Eucalyptus pilularis*. Introduced plant pests suggested to be advantaged in their spread by pig rooting were saffron thistle *Carthamus lanatus*, Bathurst burr *Xanthium spinosum*, noogoora burr *Xanthium chinense* and boxthorn *Lycium ferocissimum*. Independent information was received to the effect that pigs also aid in the spread of Spanish broom *Teline monspessulana* by rooting (see Fig. 6.10).

Fauna damage mentioned included: destruction of the habitat of ground dwelling animals as a result of rooting, predation on reptiles and small native fauna, eating of eggs of birds, and competition with native animals for food and water. Reptiles, ground nesting birds, waterbirds, marsupial mice and other small marsupials and mammals were specifically mentioned as being disadvantaged by the presence of pigs in some forests.[38]

Threat posed by other feral animals to forests. Cats and foxes were mentioned as causing a particular threat to birds, reptiles and small native animals by predation. Rabbits and goats were mentioned in areas where they are present in noticeable numbers as being extremely damaging to seedling regeneration in both native forests and plantations. Their damage to silviculture in western areas of New South Wales appears to exceed that caused by pigs. In several western subdistricts it was emphasized that rabbits have a significant impact on natural cypress pine *Callitris hugelii* regeneration and in pine plantations because they eat young pine seedlings.

Benefits from pigs in forests. Australian foresters do not seem to be as enthusiastic about wild pigs in forests as some of their European counterparts, but some did suggest advantages that have also been reported in Europe. Foresters in three sub-districts suggested that the rooting of pigs could provide a suitable seedbed for beneficial tree species and thus aid forest regeneration. One mentioned in particular that this could be advantageous around swampy areas. The eating

TABLE 7.6 Damage to Native Flora and Fauna
Reported by Foresters in Forest Sub-districts (N.S.W.)

District	Sub-district	Damage reported
Baradine	Gilgandra	Slight damage to small pine, which is used in nests. No damage to fauna observed.
Bathurst	Lithgow	Damage to ground flora only in wet or swampy sites. No damage to trees, problem not considered serious.
	Oberon	Ground-dwelling marsupials (habitat destruction).
	Orange	Pigs root around digging for worms (habitat destruction).
Casino	Urbenville	Rooting native grasses and herbs.
	Casino	Damage to young seedlings - several years ago 5 hectares of newly planted blackbutt *Eucalyptus pilularis* rooted up by pigs. Presently concerned that damage to new pine seedlings may eventuate. Damage to nesting waterbirds and to reptiles.
Coffs Harbour	Woolgoolga	Evidence of rooting confined to reedy or poorly drained areas (not good tree regeneration areas).
Forbes	Condobolin	Pigs root for bulbs and fleshy roots in damp areas. Do not know what effect they have on preferred species. Disturbed ground provides habitat for weeds: saffron thistle *Carthumus lanatus*, Bathurst galvanised burr *Xanthium spinosum* and noogoora burr *Xanthium chinense* and boxthorn *Lycium ferocissimum*.
Glen Innes	Tenterfield	Not serious to tree species; cause death of understorey species; magnitude of effect on understorey species not known. Adverse impact on populations of birds, small mammals and reptiles.
	Walcha	Rooting around swamps and trail edges.
	Glen Innes	Damage to swampland vegetation, ferns, sedges, heaths which may affect swamp function.
Newcastle	Gloucester	Pigs root up shrubs and fleshy herbaceous plants; disturbance to snow grass. Rub off bark on sapling eucalypts. Can cause breakdown of spaghnum moss swamps by wallowing. Destroy habitat of ground-nesting birds, marsupial mice, lizards. Will eat mice, eggs, etc.
Riverina	Mildura	Some damage to native grasses. Possible damage to mammals and reptiles.
	Mathoura	Pigs damaging red gum seedlings *Eucalyptus camaldulensis* when rooting. Possibly eat reptiles.
	Griffith	Pigs root up seedlings, magnitude not known.
South Coast	Queanbeyan	Ground disturbance near gullies.

Cont.

Table 7.6 *cont.*

District	Sub-district	Damage reported
South Coast *cont.*	Batemans Bay	Dig up burrawangs *Macrozamia* spp., but not seriously.
Tumut	Tumut	Light and scattered uprooting of herbage cover.
	Tumbarumba	Minor damage to wetter areas and from digging for roots.
	Batlow	Pigs grubbing in grassy areas - possible damage to tree and shrub seedlings. Competition with native animals for food and water.
Wauchope	Wauchope	Pigs being omnivorous, any fauna that was caught would be eaten.

habits of pigs was also claimed to control some exotic weed pests. The two species mentioned were cumbungi *Typha latifolia* and Paterson's curse *Echium lycopsis*.

The presence of pigs in the forests was also claimed to be beneficial in providing a resource for hunting for sport. Furthermore it was suggested in one area that their presence attracted unauthorised hunters away from hunting protected fauna. Carrion removal by pigs was also mentioned as a minor benefit.

Fig. 7.5 Feral pigs have rooted out snow grass Poa sieberana *and ferns in this area, where they might have been consuming (amongst other things) the roots of the black wattle* Acacia decurrens.

The possibility of pigs consuming forest pests was not directly mentioned. However, one forester suggested that the larvae of the scarab beetle, popularly also called Christmas beetle, *Anoplognathus* of the family Scarabaeidae might be eaten by pigs. These beetles can cause considerable damage by eating the leaves of *Eucalyptus* species (see Fig. 7.6). But on the whole, it seems New South Wales foresters do not see feral pigs as important in the control of forest pests.

Commercial forest loss from pigs. Only one forester gave an estimate of the commercial loss from feral pigs in his forest sub-district. He suggested an annual timber loss of around $500 p.a. from feral pig activity in the Gloucester sub-district of the Newcastle forest district. Commercial losses caused to forest by pigs in New South Wales *appear* to be small. These include: (a) possible losses in revenue from grazing leases (rights to graze domestic stock in forests); (b) minor uprooting of tree seedlings; (c) rooting of roads and of firebreaks, and rooting in dams which tend to silt and are a source of water for firefighting, and this consequently means more maintenance work; (d) damage to saplings (see Fig. 7.7) by boars sharpening their tusks and pigs' rubbing against them - for example, to remove lice or mud after wallowing. In the survey it was also mentioned that pigs removed small pines to build their nests; this behaviour has also been observed in other countries.

Acceleration of erosion by pigs. The foresters who suggested that pigs in their forests accelerate erosion made comments of the following type:

Fig. 7.6 In Europe, wild pigs sometimes have a beneficial impact in forests by eating the larvae of forest pests. In Australia, little research has been done into this aspect, but pigs do consume insect larvae. In this case, pigs have shifted a decaying tree trunk at Pampoolah, N.S.W., to forage for larvae.

Fig. 7.7 Fresh mud left by a pig rubbing on a Melaleuca *tree on north coast of New South Wales. The rubbing has removed some of the bark.*

the main damage occurs along creeks, in billabongs and swamps;

increase in turbidity of streams as a result of pig activity;

minor damage to and near roadways (see Fig. 7.8);

confined to specific areas such as water holes, wet areas, fire trails and roadside batters;

possible erosion if pig numbers were to build up, since forest is in granite area.

As mentioned earlier, it is possible that the breeding of native fish and frogs could be affected by rooting in billabongs and wet areas.[39] The increased turbidity of streams may adversely effect trout spawning.[40]

Other forest damages caused by pigs. When forest sub-district officers were asked whether feral pigs caused damage in the forest other than that to native flora and fauna and in accelerating erosion, the answers (by sub-district) were: Yes (pigs cause other damage), 11; No (pigs don't cause damage), 33; Do not know (if pigs cause damage), 9.

Other forest damages mentioned were:

Fig. 7.8 Roadside pig wallow in a New South Wales state forest. This is likely to wash out after heavy rain and extend erosion to the road itself.

digging up roads and roadside embankments (mentioned 3 times); respondents did not equate digging up roadside embankments with erosion;

damage to fences (mentioned 3 times); in Tumut district this leads to a reduction in rabbit control, and rabbits eat pine seedlings;

spreading of blackberries during the fruiting season and spreading of other noxious weeds;

fouling of dams or waterholes used by domestic stock grazing in forests, under grazing leases;

damage to pasture grasses;

attracting illegal and indiscriminate shooters;

predation on lambs and fly-struck sheep;

serious damage to research work related to soil fauna, understorey vegetation, forest wildlife and forest fuel weights.

Problems posed by pigs as carriers of disease and parasites in forests. Nine sub-districts reported that the potential of wild pigs for carrying disease and parasites does pose a problem for their forests. Responses to the question 'Does the potential of wild pigs for carrying disease and parasites pose a problem to your forests?' were as follows: Yes, 9; No, 17; Do not know, 13 (total respondents, 39).

Comments from foresters suggesting that the potential of feral pigs for carrying diseases and parasites was a problem for forest management included the following:

the proximity of grazing lands to forests; the fact that most forests are already grazed by leaseholders; and the mobility of the pigs could aid in the spread of disease and parasites injurious to livestock;

if an exotic livestock disease outbreak occurs, there would be serious economic consequences on the local and Australia-wide grazing industry;

such an outbreak of disease would be difficult to control because of the large areas of state forests and the time available to forestry commission personnel to initiate control measures;

feral pigs may spread disease to native fauna.

Mention was made of the potential of pigs to spread exotic diseases, such as foot-and-mouth disease, and endemic diseases such as leptospirosis, anthrax and various skin diseases. It was also indicated that pigs have a role in the transmission of the parasites: hydatids, sparganosis and tapeworms.

Comments from foresters suggesting that the potential of pigs for carrying parasites and diseases was not a problem for forest management included:

no pig diseases known to me that are likely to affect either trees or native fauna (mentioned twice);

pigs in this area are more domestic than wild; in past years domestic boars have been let free to keep the quality of pig populations high;

pigs are regularly trapped and sold (see Fig. 7.9); they have always been disease free;

no problem unless pigs carried *Phytophthora* (fungus dieback);

the Forestry Commission is involved as a responsible landholder, in order to control potential disease problem to domestic pig industry;

pigs may carry disease that affect pine plantations (the respondent did not elaborate).

The following views were expressed by some of the foresters who were uncertain whether the potential of pigs for carrying diseases and parasites poses a problem for management of their forests:

the mobility of the pig coupled with their known ability to spread blackberries and other noxious weeds would suggest they have the ability to spread disease (mentioned twice);

insufficient is known - no studies having been undertaken; most information comes from casual observation of pigs and their effects on the physical environment;

the outbreak of diseases spread by pigs would reduce the relative value of the forests to grazing leasees;

whilst there are no known cases of pigs spreading leptospirosis, problems of hydatids, sparganosis, tape-worms, various skin diseases and anthrax could potentially occur;

Fig. 7.9 Healthy feral porkers trapped in a state forest near Nundle, N.S.W.

local farmers raise pigs on state forest boundaries;

if disease outbreak occurs it may be possible for people to contact the disease by drinking polluted water.

Feral pigs and the spread and occurrence of dieback in trees. There seems to be at least two types of dieback of trees;

(a) the type induced by eucalyptus leaves being eaten by leaf-devouring insects such as Christmas beetles *Anoplognathus* spp.

(b) the type caused by fungal infection of the roots of trees particularly by the fungus *Phytophthora cinnamomi*.

Presumably pigs might reduce the incidence of the first form of dieback by eating larvae of the insects involved, but pigs may foster the spread and occurrence of the fungal form. Some foresters in New South Wales did indicate this as a possibility. However, 23 respondents felt that pigs did not spread dieback, and 33 respondents said they did not know. It is useful to divide the comments of respondents according to whether they are referring to type (a) or type (b) dieback mentioned above.

The forester in the Walcha sub-district said that pigs did not spread dieback in his area of the northern tablelands, where it is a serious problem and is of type (a) mentioned above. He pointed out that pig populations were low in his dieback areas and he explained the occurrence of dieback in his areas in the following detailed way:

'It is felt that dieback is caused by a combination of pasture improvement and leaf-eating insects and grazing. Dieback in northern tablelands of New South

Wales occurs over very large areas. Causes: (1) clearing of forests (for grazing land) leads to habitat destruction - removal of ground-dwelling carnivores and insect-eating birds; (2) pasture improvement - affects soil structure by heavy and persistent grazing; soil micro habitat changed through introduction of exotic pastures which are fertilized. These fertilizers, e.g. superphosphate, kill soil fauna and allow large populations of insect nymphs to develop, e.g. Christmas beetle; (3) insects hatch and as there are no birds to eat insects and no ground-dwelling carnivores, e.g. *Rattus*, or *Antechinus* spp., then large insect populations hatch which then feed on leaves of eucalypts; (4) over time, energy reserves of eucalypts are used up and trees slowly die - no replacement of trees because of continued grazing; (5) positive feedback situation, as there are fewer trees - more insects - greater damage over a bigger area.'

Other foresters suggested that pigs may reduce the incidence of this form of dieback because:

pigs may eat larvae as they develop in the soil; however, impact on populations would be insignificant;

in sparsely timbered areas, pigs may eat scarab beetle larvae.

As for pigs and the occurrence of *Phytophthora cinnamomi* attacks and its spread, foresters pointed out that the fungus is endemic to coastal New South Wales where it survives in wet spots. Views were divided about the possible role of pigs in the occurrence of attacks. Some representative comments were:

Eden: '*Phytophthora cinnamomi* is endemic to this area, and transfers by pigs would only be possible in wet areas where the fungi already exist.'

Oberon: 'Do not know, but quite possible in areas of dieback species.'

Gloucester: 'No evidence at this stage, but a real possibility exists.'

Kyogle: 'No. Fungi that cause dieback widespread but passive in east-coast forests. No evidence of dispersal of fungus by pigs.'

Woolgoolga: 'Do not know. May help to spread *Phytophthora*, but dieback from scarab attack a different proposition.'

Coffs Harbour: '*Phytophthora cinnamomi* is not a problem in east-coast forests but is in Western Australia where it is spread by machinery and man.'

Even though evidence from New South Wales forests suggests that pig activity does not increase the incidence of *Phytophthora* attacks, the possibility cannot be dismissed that wild pigs assist the spread and occurrence of *Phytophthora cinnamomi* attacks on Australian trees. Reports from Queensland forests strongly suggest that they assist the occurrence and spread of *Phytophthora cinnamomi* in rainforests.[41] Ways in which wild pigs could assist in the spread and more frequent occurrence of the dieback fungus could be:

(a) By disturbing soil communities by rooting and feeding and so reducing animals that feed on soil fungi and compete for the same resources as the dieback fungus. Springett reports that 'Dieback fungus *Phytophthora cinnamomi*, which attacks eucalypts such as jarrah is present in most Australian soils, but normally kills few plants. In a healthy soil, it is probably controlled by many animals such as mites that feed on soil fungi, and by other microflora that compete for the same resources as the dieback fungus. When the soil is disturbed or the diversity of the soil community is reduced, dieback fungus becomes more abundant and trees are killed.'[42]

(b) By increasing the probability, through their rooting and movement of soil, of dieback spores, Zoospores and fungi coming into contact with roots of a susceptible tree species.[43]

(c) By creating conditions for more rapid parasitic penetration of susceptible tree species by the fungi, by damaging tree roots when rooting[44] and by weakening some trees (see Figs. 7.10 and 7.11).

To the extent that this phenomenon occurs it could help to change the composition of some forest areas (reduce the occurrence of species not resistant to dieback) and result in some commercial loss of timber.[45]

Damages caused by pigs straying from forests. Damage caused by pigs straying from forests needs to be taken into account in managing forests so as to maximize *social* benefits. Pigs are prone to stray and to be the cause of complaint. New South Wales foresters reported that complaints about damages caused by pigs straying

Fig. 7.10 Pigs have wallowed and dug around this tree and rubbed themselves on the bark. A wallow and the exposed and damaged roots are shown.

Fig. 7.11 Pig rooting at the base of a tree in a Eucalyptus *plantation. Such rooting could help to increase the incidence of fungal dieback.*

from forests came from landholders adjoining forests and engaged in agricultural or pastoral pursuits, plus complaints of lawn and garden damage in a small coastal village. The most frequent types of damage complained about from straying are specified in Table 7.7.

A number of foresters pointed out the Forestry Commission is sometimes unjustly criticized for harbouring pigs. One said that despite 'the co-ordinated pig control effort by the Forestry Commission and the Pastures Protection Board, neighbours still blame the Commission for harbouring pigs, but the populations concerned exist on neighbouring properties.' Another pointedly remarked 'Pigs are not confined to forests. Pigs also inhabit unimproved private property.'

Policies and Methods of Control of Wild Pigs in N.S.W. Forests

In most forest sub-districts where feral pigs occur, the policy is for foresters to try to control the number of pigs in forests, usually in co-operation with the

TABLE 7.7 Complaints about Damage Caused by Pigs Straying from Forests. Number of Sub-districts Receiving Same Type of Complaint

Order	Type of damage	Number of sub-districts
1.	Damage to crops (mostly cereal crops mentioned)	12
2.	Damage to fencing	6
3.	Damage to pasture (see Fig. 7.12)	4
4.	Attacking and or killing lambs and sheep (usually only poorer individuals, probably genetically weak)	3
5.	Damage to sugar cane	1
6.	Damage to moist swampy areas not associated with crop or pasture production	1
7.	Silting of dams and waterways	1
8.	Damage to lawns/gardens in small coastal village	1

Fig. 7.12 Pigs harbour in the margins of the forest in the background and visit the property in the foreground to damage pasture especially along the creek in the centre. The property 'Tippaburgh' near Nundle, N.S.W., is used by Mr Ed Vanwarderen for grazing livestock.

local Pastures Protection Board and a group of landholders adjoining forests. External pressure, mostly by landholders, seems to be a significant determinant of the amount of action taken against pigs in forests. The policies of the sub-districts in which foresters indicated a policy are set out in Table 7.8.

TABLE 7.8 Policies Adopted by Forest Sub-districts for the Control of Feral Pigs, as Stated by Sub-district Foresters (N.S.W.)

District	Sub-district	Policy
Baradine	Baradine	Co-operation with local Pastures Protection Board (P.P.B.).
	Gunnedah	Assistance [not specified] to landholders in control programs.
	Narrabri	Poisoning with 1080 when numbers warrant.
Bathurst	Lithgow	Trapping and poisoning programs will be introduced in conjunction with P.P.B.
	Oberon	Eradication of feral pigs if possible, but in any case keep population as low as possible.
	Orange	Trapping pigs, but needs co-ordinated action with local landowners. Trapping discontinued until increased effort by neighbours assured.
Casino	Casino	Co-ordinated co-operative action with local P.P.B. and groups of landholders. Trapping program instituted in pine plantation where a large pig population is present.
	Casino West	Co-ordinated co-operative effort most effective.
Forbes	Condobolin	Co-ordinated effort by Forestry Commission in conjunction with grazing leaseholders. Control methods are: intermittent shooting; poisoning; and trapping, using pig pellets as bait in weld-mesh circular traps.
	Forbes	Unrestricted shooting.
Glen Innes	Tenterfield	General Forestry Commission policy is to co-operate with other bodies when control measures are requested.
	Walcha	Persistent trapping and poisoning, mostly in winter months when pigs return from dormant pastures for harbour.
	Armidale	Small numbers do not warrant expensive control measures. If sightings increase, control measures will be instigated.
	Tamworth	Pigs trapped by Forestry Commission employees. Trapped pigs destroyed. Co-operation with local P.P.B. in 1080 poisoning.

Cont.

Table 7.8 *cont.*

District	Sub-district	Policy
Glen Innes *cont.*	Inverell	Co-operation with local P.P.B. and groups of landholders in control.
	Glen Innes	Co-operation with local P.P.B.
Newcastle	Gloucester	Trapping using panel traps and single traps with corn and pellets as bait.
Riverina	Mildura	No Forestry Commission control but most forests are leased for grazing. Lessees are required under the lease agreement to destroy noxious animals, including feral pigs.
South Coast	Batemans Bay	Allow shooting with game permit away from tourist areas; trapping program near coastal towns; 1080 poisoning if sufficient damage and complaints, but not a common practice.
South East	Bombala	Trapping where evidence of occurrence of pigs.
	Bega	If encountered would be controlled.
Tumut	Tumut	Co-ordinated action by Pastures Protection Board, National Parks and Wildlife Service and Forestry Commission. Attempts made to shoot pigs sighted (but see Table 7.9).

Control methods used. As for control methods used by foresters, 11 used trapping (see Fig. 7.13), six used poisoning by means of 1080 (sodium monofluoroacetate) in non-meat bait, and four used shooting. Another forester pointed out that in his sub-district the Forestry Commission required grazing leaseholders to take action to control pigs in their leases; and in two sub-districts amateur shooting of feral pigs was actively encouraged by foresters as a control measure. Nevertheless the proportionate kill of pigs appears to be relatively small and is well below the replacement rate of the population in most sub-districts.

Effectiveness of control measures. Foresters were asked to comment on the effectiveness of the various measures employed by forest personnel to control feral pig numbers. The responses are set out in Table 7.9. Trapping appears to be relatively expensive per pig taken and is not sufficient in itself to eradicate most populations. Poisoning is less expensive in terms of direct cost per pig taken but unlike trapping can have adverse spillovers on wildlife and other animals and cannot be used successfully in all areas, and all pigs may not take the bait. Shooting in many areas is less expensive per pig taken than poisoning but is clearly of limited value in scrubby or wooded country even if carried out in conjunction with the use of dogs. No available method appears to be 100 per cent effective in

Fig. 7.13 Mr Kevin Ribbons, forestry foreman at Barrington Tops State Forest, demonstrating a forest weldmesh trap for feral pigs. Pigs are enticed into the trap to eat corn. The trap mechanism consists of a swinging gate which is pushed in as pigs enter the trap but cannot be pushed out from inside, the gate being automatically closed by rubber straps.

forests for a reasonable outlay. A number of foresters pointed to the mobility and fecundity of pigs as factors reducing the effectiveness of their control programs.

Special difficulties in controlling pigs in forests. Foresters were asked what special difficulties they faced in controlling pig populations in their forests. The main difficulties mentioned were:

spread of pigs over large areas in forests, sometimes in inaccessible places and isolated pockets;

high mobility and fecundity of pigs;

dense habitat - for example, lignum *Muehlenbeckia cunninghami* scrub, dense pine forests;

migratory nature of pigs;

difficulty of pinpointing pigs (see Fig. 7.14);

interference with control measures by visitors to forests;

remoteness of populations from some forest bases - infrequency of visits and cost of travel;

fickleness of eating habits - difficulty to ensure that they take bait;

the partiality of some native species (for example, emus *Dromaius novaehollandiae*) to bait set for pigs;

lack of control or unco-ordinated control by neighbours.

TABLE 7.9 Effectiveness of Methods used to Control Pigs in Forests: Comments Received from New South Wales Foresters

District	Sub-district	Methods*	Effectiveness
Baradine	Baradine	Poisoning in co-operation with P.P.B.	No comment.
	Narrabri	Poisoning	Gives good results.
Bathurst	Lithgow	Trapping	Very effective on limited areas although reinfestation occurs when trapping ceases.
	Oberon	Shooting Trapping	Control methods are effective. Number of pigs limited to manageable levels.
	Orange	Trapping	Trapping is reasonably effective but is a high-cost operation as it is labour-intensive. It is impossible to totally control pig populations using trapping only.
Casino	Casino	Trapping	Appears to be successful but dogging program may be better.
Forbes	Condobolin	Trapping	Relatively effective.
	Forbes	Nil	Shooting (by amateur hunters) is an effective control.
Glen Innes	Walcha	Trapping Poisoning	Control at the beginning of summer appeared very effective, but there was an influx of pigs in the autumn as pigs returned to the forests for harbour. The influx consisted mostly of young pigs from the previous season. High fecundity reduces effectiveness of controls.
	Armidale	Nil	Migratory nature of pig populations are such that control is not warranted.
	Tamworth	Trapping	Trapping and poisoning most effective in reducing local pig numbers.
	Inverell	Poisoning	Very effective.
Newcastle	Gloucester	Trapping	Effort has considerably reduced sightings, but a slow process.
Riverina	Mildura	Shooting by lessees	Not known.
South Coast	Queanbeyan	Trapping Shooting Poisoning	No comment.
	Batemans Bay	Trapping Shooting Poisoning	Trapping gives good results near villages. Effect of shooting not known.

Cont.

Table 7.9 *cont.*

District	Sub-district	Methods*	Effectiveness
South East	Bombala	Trapping	No success.
Tumut	Tumut	None	Dingoes are thought to keep population under control.
	Tumbarumba	Shooting on routine duties	Little effect without co-ordinated effort.

*Poisoning in all cases refers to the use of 1080 (sodium monofluoroacetate) in non-meat baits.

Opening of Forests for Hunting

In New South Wales, hunting in forests (of non-protected species) is allowed under permit from the Forestry Commission. As mentioned earlier, at least two forest sub-districts considered amateur hunters to have a role in the forest program to control feral pigs.

Foresters in sub-districts were asked whether they had issued game permits for the hunting of wild pigs in the 12 months preceding the survey. Only 20 per cent of sub-districts containing wild pigs said that they had done so. In seven

Fig. 7.14 Wild pigs have been shot in this Eucalyptus *forest, but it is difficult to pinpoint pigs in such areas.*

sub-districts, permits were not issued because of danger of hunting to other fauna and to other recreational users of the forest, damage to forest property - for example, water tanks and signs - and the shooting by hunters of livestock on leaseholds. In two sub-districts, permits were not issued because hunters were believed to disperse pigs, making them more difficult to trap; and another two said that no permits had been requested. Several foresters pointed out that illegal shooting in forests was common.

Foresters were asked whether forests should be open at certain times of the year to permit amateur hunters to hunt wild pigs. Foresters in 54 sub-districts responded. Twenty-two sub-districts thought that forests should be open in certain times of the year to permit hunting of wild pigs; 25 sub-districts thought that they should not be; and 7 were undecided. Views varied considerably. Those in favour suggested in general that *selected* shooters should be allowed because they could be effective in keeping pig numbers down, would be cheaper than control by the Forestry Commission, and would provide an opportunity for shooting recreation, a 'legitimate' sport. Those against argued that hunters could be a bigger problem than pigs. The risks from hunters were:

shooting of other fauna - for example, brush turkeys *Alectura lathami*;
other recreational users at risk;
domestic stock in forests could be killed;
danger of fires being started in some seasons;
damage to signs, other forest property, regeneration areas, etc.

Some argued that shooters were ineffective because pigs had too much cover (from cypress, lignum, etc.) and hunters may also scatter pigs and so make control measures by the Forestry Commission less effective. While several officers saw a conflict between shooting and other recreational uses of forests, some believed that these could be resolved with forethought.

Recommendations for the Control of Wild Pigs in Forests

Foresters in the sub-districts were asked what should be done to control wild pigs in forests. The views varied, but several foresters stressed the need for a co-ordinated co-operative effort by landholders in a region to control feral pigs. The desirability of poisoning seemed to be the most divisive issue: some foresters wished to see poisoning extended, even to the extent of being on a regional basis, using meat baits; whereas others said that it should not be allowed because it was not sufficiently species-specific. Some of the comments about what should be done were:

trapping and shooting; no poisoning should be allowed;

shooting by selected hunters;

continue present practice of forest boundary control of pigs in co-operation with Pastures Protection Boards;

trap and poison where effects can be limited to pigs;

leave problem to grazing lessees;

organized shooting to keep numbers down;

poisoning with 1080 using meat as bait;

co-ordinated co-operative effort with Pastures Protection Boards, etc. (mentioned several times);

trapping, or alternatively dogging, by Forestry Commission personnel;

expand trapping program, and 1080 poisoning under close supervision;

co-ordinated poisoning program;

organized poisoning on a regional basis;

continued hunting; poisoning is indiscriminate;

continue trapping; allow some supervised shooting and dogging;

persistent trapping and poisoning in co-ordination with neighbours;

poisoning and trapping, followed up by shooting;

set up a monitoring program and assess need; professional control, not *ad hoc* action;

selective poisoning, trapping and hunting;

1080 poisoning by professionals away from dingo population.

In general, foresters near the coast showed less support for poisoning as a control measure than those farther west. The adverse side-effects of poisoning on other animals appears to be a matter for concern by some foresters. A number of foresters were in favour of greater use of controlled dogging to reduce pig numbers in forests.

Conclusions from Survey of N.S.W. Forests

Feral pigs are widespread in New South Wales forests. In almost half of the sub-districts where feral pigs are present, foresters said that they caused damage to native flora and fauna. Damage to flora in wet areas is often serious, the under-storey of some forests is removed and bulbous and rhizomous herbs and plants are rooted out and eaten by pigs. Commercial damage to timber and forest installations, however, is reported to be slight. Fauna and egg predation, habitat destruction and food competition with other animals occurs in some forests. But foresters consider rabbits, goats or cats to be worse pests in most New South Wales forests, even though feral pigs are regarded as a sufficient pest to warrant control measures in one-third of the forest sub-districts where they occur. Foresters in some areas said that feral pigs accelerate erosion in their forests and that their capacity for carrying diseases and parasites infectious to domestic livestock, human beings and native fauna was another disadvantage of their presence in forests. Complaints about the straying of pigs from forests, and damages caused while straying, occurred and were often an important reason for instituting or expanding forest control programs. As for dieback in trees, some foresters thought that activity by pigs by eating the larvae insects might actually marginally reduce the occurrence of the type of dieback caused by insects' devouring the leaves of trees; and others thought that the occurrence of *Phytophthora cinnamomi* attacks might be increased by feral pig activity. Most foresters did not know what role pigs would play in the spread and occurrence of dieback in trees. Few benefits were seen from feral pigs in forests.

The main methods used by foresters to control feral pigs in New South Wales forests are trapping, poisoning and shooting. None of these measures are 100 per cent effective in eliminating pigs from an area. While current kill rates of pigs in forests help to reduce criticism of the Forestry Commission by farmers and others affected by straying pigs, they possibly have little impact on feral pig populations because of the fecundity of pigs. The mobility of pigs is an additional problem. Opinions of foresters are almost equally divided about whether forests should be open for the hunting of wild pigs by amateurs. This method would involve little or no direct cost to the Forestry Commission but, as pointed out by some foresters, can have undesirable side-effects, and these indirect costs would need to be taken into account. In this regard, the policy in Queensland, to be discussed in the section below, is interesting.

RESULTS FROM A SURVEY OF WILD PIGS IN FORESTS AND TIMBER RESERVES IN QUEENSLAND

In mid-1979, survey forms were sent to foresters in Queensland state forests and timber reserves to obtain information about the presence and population levels of feral pigs in forests, the damages caused in and to forests and any benefits, the straying of pigs and complaints about this, the measures taken to control pig populations in forests, and the comparative threat of various species of feral and introduced mammals to indigeneous wildlife - as seen by the sub-district (forest management group) officers of the Forestry Department. A questionnaire form[46] was forwarded to each forest sub-district (forest management group, in this context).[47] Responses were received from all manned sub-districts: 54 out of 55.

Distribution of pigs in Queensland forests. Eight of the nine forest districts of Queensland reported the presence of pigs. They are present in the forest districts of North Queensland, Rockhampton, Monto, Maryborough, Murgon, Brisbane, Warwick and Dalby, but absent in Yarraman. They are present in 30 of the 54 sub-districts (approximately 55 per cent of sub-districts). The total reported number of pigs in state forests and timber reserves was 22 500, but three North Queensland sub-districts with pigs did not provide estimates so the actual number of pigs could well be almost as high as in New South Wales forests (35 000). The distribution of the pig population in Queensland forests is shown in Fig. 7.15 by forest districts.[48] The largest pig populations are in the districts of North Queensland, Rockhampton, Warwick and Dalby, but Maryborough and Gympie districts also have a considerable population. The highest (average) densities of pigs in forests occur in Rockhampton district, followed by Gympie, North Queensland, Maryborough and Warwick, which have approximately equal densities. Densities for forest districts are indicated in Fig. 7.16. The estimated average overall density of pigs in Queensland forests is lower than that in New South Wales if only forest areas are considered where pigs occur. In Queensland the average is 68 hectares to a pig, whereas in New South Wales it is 30 hectares per pig.

An interesting feature of the geographical distribution of pigs in Queensland forests is the relatively high populations and high densities in coastal areas, many of which specialize in sugar cane production. This distribution of pigs contrasts with New South Wales, where pig populations (except at Casino) are light along the coast.

Threat of pigs to flora and fauna. For 43 per cent of sub-districts containing pigs, it was reported that wild pigs cause damage in forest areas to native flora and fauna; in 40 per cent this was said not to be the case; and in 17 per cent of sub-districts with pigs it was not known whether they damaged native flora and fauna. These percentages are similar to the corresponding ones for New South Wales.

In 14 (or 46.66 per cent) of sub-districts containing pigs, feral or introduced mammals other than pigs were considered to be as great (or greater) a threat to native flora and fauna as pigs. Pigs are believed to be the *greatest* threat to native flora and fauna in 11 (36.66 per cent) of sub-districts with pigs. Five sub-districts with pigs were uncertain whether any other feral or introduced mammal posed as great a threat to native wildlife as pigs. A higher percentage of forest sub-districts in Queensland than in New South Wales considered the pig to be the major threat to native flora and fauna in forest areas. As in New South Wales, cats were most frequently mentioned as posing as great (or greater) a threat to native wildlife as pigs; then came foxes - whereas in New South Wales, the order was rabbits and then foxes.

Impact of wild pigs on the commercial value of forests. In seven (23.34 per cent) of sub-districts with pigs, pigs were said to reduce the commercial value of

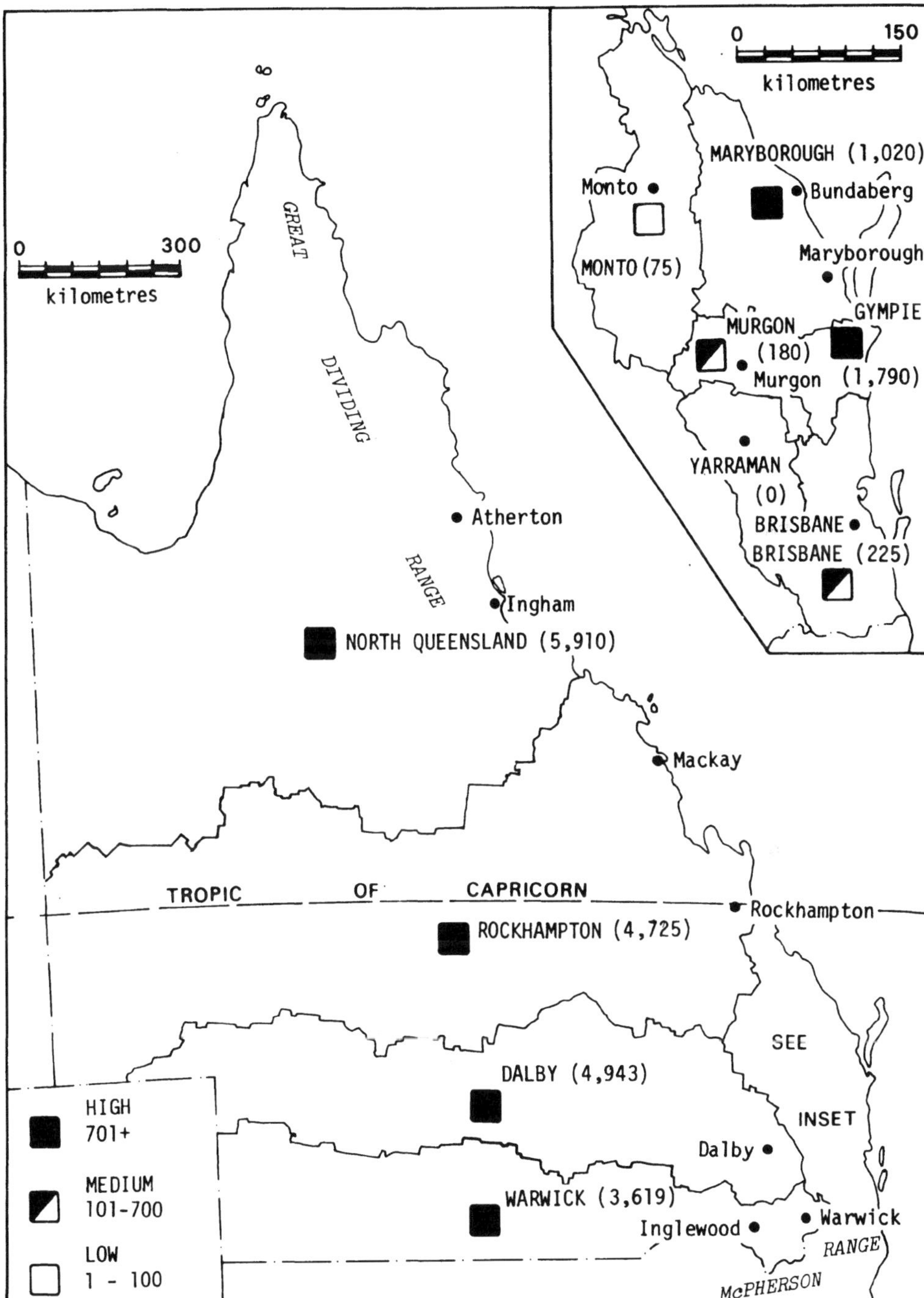

Fig. 7.15 Estimated populations of wild pigs in the forests of Queensland, by official forest district.

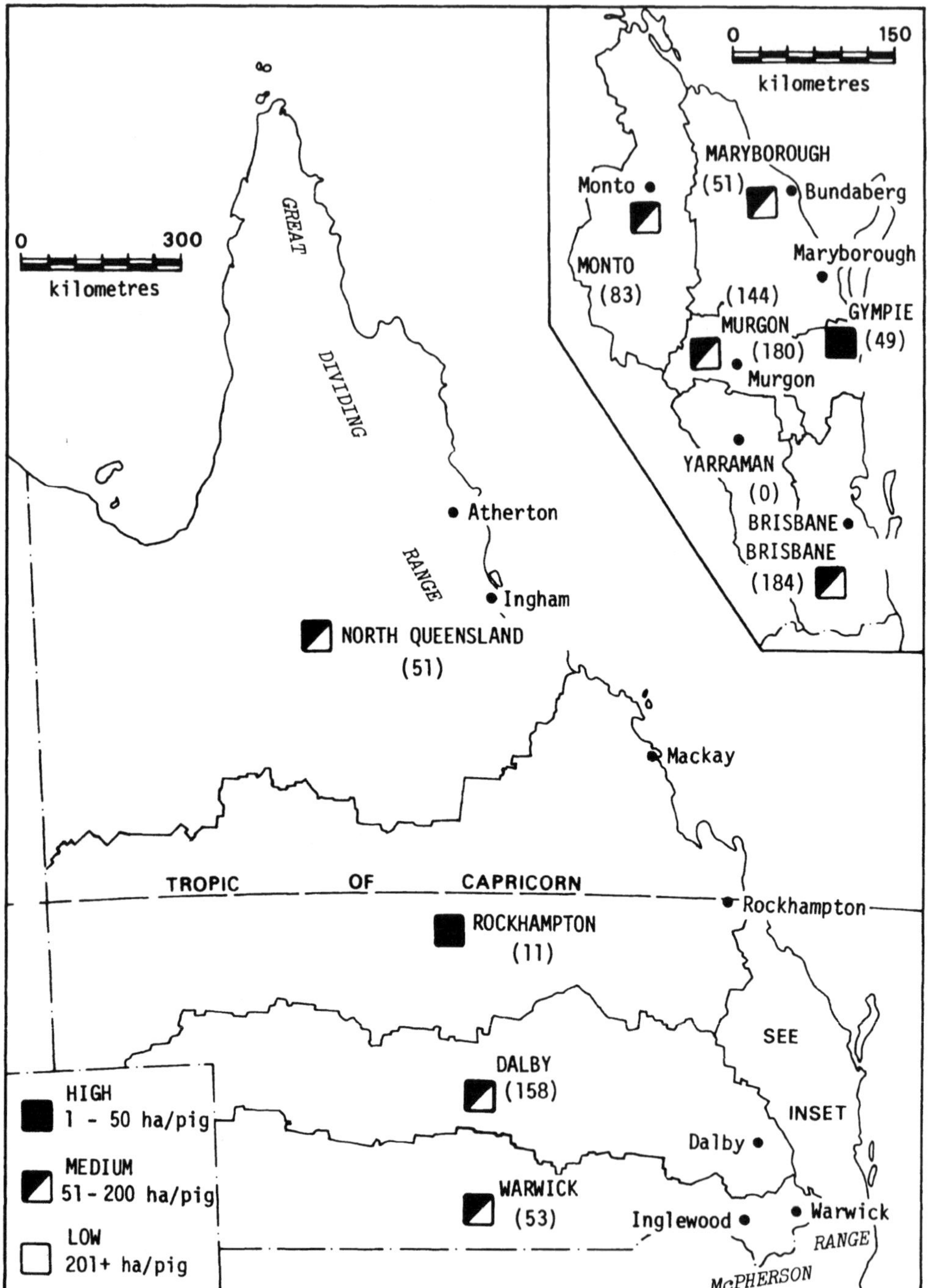

Fig. 7.16 Estimated densities of wild pigs in the forests of Queensland, by official forest district.

forests; in 17 (56.66 per cent) they were said not to; and in six (20 per cent) the commercial effect of pigs on forests was not known. A greater proportion of sub-districts with pigs in Queensland than those in New South Wales believed that pigs reduced the commercial value of forests.

Complaints about straying. Complaints were received by forest officers about the straying of pigs from forests in 40 per cent of the forest sub-districts with pigs. In all but two cases, these complaints were said to be few. Many complaints were received in a North Queensland sub-district and in a Dalby sub-district.

Control measures. The policy on control of pigs in Queensland state forests differs from that in New South Wales. In general, the policy appears to be: no recreational shooting or hunting permitted, but cane growers, lessees of forest grazing rights and landholders adjacent to forest areas can obtain hunting permits subject to several restrictions (in New South Wales, recreational shooting is possible under a game permit). There is little or no control of pigs in Queensland forests at the Forestry Department's expense (whereas in New South Wales active control programs financed by the Forestry Commission exist in one-third of forest sub-districts with pigs). From the Department's own point of view, control may not be economical. However, given the mobility of pigs and other actual and potential spillover effects on other landholders, the social case for control becomes stronger. Nevertheless because of its high fecundity, mobility and adaptability to varied habitats the feral pig is a difficult pest to control economically. The economic, as opposed to the political, case for controlling pigs in all forests is not apparent, but in some forests where commercial damage to forests and damage to native wildlife and to neighbouring properties from pigs is high, co-operative control by all landholders could be justified on economic grounds.

Ranking of introduced or feral animals as pests. Each sub-district was asked to rank feral or introduced mammals by the seriousness of the present threat posed by each to preservation of its forest(s) in a natural state and its native wildlife. On the basis of the mean rank score, cats are considered to be the most serious threat, followed by dingoes, feral dogs, foxes and pigs. If dingoes are ignored on the basis that they were introduced so long ago by Aborigines, cats are the greatest threat followed on about equal scores by dogs, foxes and pigs. (By comparison the mean rank scores of threat for New South Wales forests were rabbits, cats, foxes, pigs and dogs, in that order.) The threat posed by rabbits in forests seems to be much greater in New South Wales than in Queensland but the situation for other species is similar, except that feral dogs are considered to be less of a problem in New South Wales. But of course, mean rank scores disguise differences between forest sub-districts. For instance, in one Queensland sub-district horses were considered to pose the most serious threat of all introduced mammals, and in another sub-district rats. Unlike in New South Wales, feral goats did not feature as a major pest in any of the Queensland forest sub-districts.

Selected Details about the Impact of Feral Pigs in Queensland Forests[49]

As pointed out earlier, in at least 43 per cent of sub-districts where pigs are present in Queensland forests, pigs are reported to cause damage to native flora and fauna.

Flora damage. As far as flora is concerned, seven sub-districts mentioned damage to tree root systems, four mentioned damage to damp areas and creek beds, four the uprooting of seedling trees, one the eating of seeds before germination, one the destruction of grassy bulbous plants, and one the uprooting of byfield fern *Bowenia serralata*. It was also mentioned that the roots of cycads or burrawangs *Macrozamia* spp. are eaten by pigs. Similar damage to burrawangs were reported in my New South Wales national parks survey[50] and in my New South Wales forest survey. In a number of areas it was claimed that pigs spread fungal dieback *Phytophthora cinnamomi*, for instance - by their rooting around and chewing the roots of hoop pine *Araucaria cunninghamii*, a valuable softwood timber tree with a wide range of uses and extensively used in Queensland for afforestation. In one sub-district (Barakula) it was pointed out that pigs interfere with the regeneration of cypress pine by rooting seedlings out. The northern cypress pine *Callitris columellaris* is a valuable timber tree, resistant to termite attacks and widely used in the building industry.

Fauna damage. As for specific comments on damages to native fauna caused by pigs, damages to the eggs and young of ground-laying birds (for example, brush turkeys, scrub hens and emus), including water-dependent birds such as ducks, were mentioned most frequently. Apart from this, destruction of lizards and goannas and small mammals (for example, ground-nesting rats and mice) by pigs was mentioned. In Northern Queensland, both Ingham and Palmerston sub-districts mentioned that pigs dig in the nests of brush turkeys and eat their eggs. Ingham sub-district reported this as a serious problem and Palmerston pointed out that scrub hens suffer the same fate. The brush turkey *Alectura lathami* and the scrub hen (or scrub fowl) *Megapodius freycinet* both build large incubator mound nests of organic matter and litter. Even if pigs only rooted in these nests (without deliberately seeking the eggs), they would be likely to damage the eggs or upset the stable temperatures required for the eggs to hatch. Given the unique nature of these bird species (along with the mallee fowl), the damages caused by pigs could pose a serious conservation problem. In the Toolara sub-district of Gympie, pigs are suspected of reducing the number of young emus *Dromaius novaehollandiae*, and in the Barakula sub-district of Warwick district of taking ducklings along watercourses.

Relative threat to wildlife of introduced animals. As pointed out, the cat and the fox are believed to pose a substantial threat to native flora and fauna because of their predation on birds, eating of eggs of birds and their taking of small mammals. But it was mentioned by some districts that feral dogs, sometimes breeding with dingoes, are a problem. In one sub-district, it was pointed out that deer browsing around hoop pine can be a nuisance, and occasionally wallabies ring-bark hoop pine *Araucaria cunninghamii* and Bunya pine *Araucaria bidwillii* under drought conditions. It was also mentioned by one sub-district that grazing requirement (lease of grazing rights) in state forests posed the greatest threat to flora and fauna. (A number of sub-districts in New South Wales also made this point about their forests.)

Beneficial effects of pigs in forests. Practically all forest groups were of the opinion that pigs in their forests have no or no known beneficial effects, although some minor benefits were not impossible. Palmerston thought that their only benefit would be as possible food for dingoes. Mapleton-Cooloolobin mentioned that they could encourage regeneration by their rooting. And Western Creek (Dalby district) observed that pigs appear to turn over soil, leaf matter and seeds and promote vigorous regeneration of cypress pine. Injune pointed out that pigs might have limited benefit as scavengers on dead animals. Overall, pigs have no significant beneficial effects in forests. As already observed, they have a deleterious impact on native flora and wildlife and, as will be seen, cause other damages.

Commercial consequences of pigs in forests. The activities of wild pigs may significantly reduce the commercial value of forests in some parts of Queensland. The main damage appears to occur in northern Queensland in the North Queensland and Rockhampton forest districts. In these areas they are considered to be a significant agent in the spread of dieback *Phytophthora cinnamomi* which is destroying a considerable amount of commercial timber. They also root around trees, so causing them to fall in wet and windy weather, they eat seeds and root out seedlings. It is very difficult to put a figure on the extent of the annual commercial loss, but in the Ingham sub-district it was estimated to be of the order of \$5000. For the north of the state it could run to more than \$10 000 annually. Minor damages to seedlings were reported from some sub-districts.

Erosion and other damages caused by pigs. There are other damages in forests resulting from the activities of pigs. Twenty-seven per cent of forest sub-districts with pigs said that pigs caused *erosion* in forests. Most said that it was not serious, but in some cases it was significant enough not to be ignored. Some comments of particular interest are:

Eungella (Rockhampton district): 'Yes, responsible for erosion but damage does not appear to be serious; but because of forest death from *Phytophthora cinnamomi* and heavy rainfall (2400 mm in eight days this year), it could become a serious problem.'

Cathu (Rockhampton): creating erosion of roads by blocking road drains through their rooting.

Western (Warwick): add to erosion but no more serious than cattle tracks.

Marmadua (Dalby): erosion in gullies is a fairly sizeable problem in most areas and pigs inevitably assist in the process.

As for other damages caused in forests by pigs, the following were mentioned:

Ingham (Rockhampton district): damage to road surfaces in many rainforest areas.

Byfield (Rockhampton): possible spread of fungal soil disease.

Jimina (Murgon): slight damage to roads as a result of wallowing.

Western (Warwick): damage to fences, including dog and rabbit fences.

Inglewood (Warwick): damage to fences.

Barakula (Dalby): reduction in grasses for cattle grazing (affects lease-holders of forest grazing rights).

Potential of wild pigs to spread diseases and parasites. As for the potential of wild pigs for carrying disease and parasites in forests and the problem posed for forests, 30 per cent of sub-districts with pigs said this is a problem. Seven sub-districts mentioned the spread of *Phytophthora cinnamomi* by pigs as a problem. Other possibilities mentioned were:

Byfield (Rockhampton district): help spread worms (parasites) through cycles, involving other animals.

Cathu (Rockhampton): pigs are known carriers of tuberculosis. In the event of the outbreak of an exotic disease such as rabies or foot-and-mouth disease, pigs as carriers would make it uncontrollable.

Bania (Maryborough): potential problem, since forests are grazed by domestic stock.

Yarraman (Yarraman): pigs could carry ticks into tick-free areas and so reduce their value for grazing.

Incidence of dieback and pigs. The occurrence of fungal dieback in Queensland's rainforests appears to be a serious problem, so the question of whether or not pigs spread it and promote its attacks on trees is of special importance. Of the 30 sub-districts with pigs, seven reported that pigs spread dieback, four that they did not, 13 did not know and six did not answer the question. Thus 29.16 per cent of respondents believed that pigs spread dieback, whereas 16.6 per cent believed it not to be so. Nearly all sub-districts in northern Queensland believe that pigs help to spread dieback and assist outbreaks. Specific replies suggesting this were:

Ingham (North Queensland district): 'Yes. Consistently *Phytophthora* is found in pig wallows throughout rainforests.' (Also elaborated on this elsewhere.)

Palmerston (North Queensland): 'Yes. Generally held that pigs help facilitate the occurrence of dieback. Has some research basis, but many pig wallows and dieback areas do *not* appear obviously associated.'

Eungella (Rockhampton): 'Yes. A necessary element for *Phytophthora cinnamomi* to become active is warm, wet soil. Pigs favour such soil and are active in such conditions. As it takes only 5 grams of infected soil in the right condition to spread *Phytophthora cinnamomi*, it is a fair assumption that pigs are a big factor in its spread.'

Cathu (Rockhampton): 'Yes. Pigs wallow in mud holes and then deposit this soil as they rub themselves against trees.'

Littabella (Maryborough): 'Strong circumstantial evidence in the Mackay area suggests that feral pigs help spread *Phytophthora cinnamomi*.'

A more cautious answer came from Kuranda, and Byfield doubted whether pigs were significant in causing dieback:

Kuranda (North Queensland district): 'Do not know. It is possible that pigs spread dieback, but there has not been enough sampling and research to link the two together.'

Byfield (Rockhampton): 'No. Dieback is more likely to be caused by excessive moisture on root systems.'

Thus it emerges that pigs appear to have practically no benefits in Queensland forests, they cause noticeable commercial loss of timber in some areas, damage native flora and fauna, carry some diseases and parasites that are a problem in forests, add to erosion problems, and appear in some areas to contribute to the occurrence of fungal dieback. The rainforest areas of northern Queensland appear to be most affected.

Complaints about pigs straying from forests. Of course, pigs are mobile animals and may stray out of (and into) forests causing damage on neighbouring agricultural and other properties, and this must be regarded as an additional cost of their presence. Forty per cent of Queensland forest sub-districts with pigs had received complaints about damages done by pigs straying from their forests. The types of damage complained of varied by district. In northern Queensland (North Queensland plus Rockhampton districts) the main complaints were about damage to sugar cane, bananas, melons, and fencing. In the Gympie district, there were complaints about straying pigs damaging grain crops, improved pasture and other crops. In the Warwick and Dalby districts (both inland) there were several complaints of damage to growing crops by trampling and eating and to temporary grain silos by rooting. Damage to orchards, killing of sheep and destruction of fences including dog fences were all reported.

Policies and Methods of Control of Wild Pigs in Queensland Forests

Looking at broad costs and benefits, it is clear that pigs are a pest problem in some Queensland forests. However, whether or not it is economic to control their numbers in these forests is an open question. While the Department of Forestry can undertake control measures against pigs at its own expense, there appears to have been small outlay to date in this regard (certainly a much smaller outlay and program than in New South Wales forests). Shooting (subject to certain conditions) of pigs in forests by cane growers and by lessees of grazing rights is permitted and trapping may be allowed. Lessees and permit holders and landholders adjoining forestry lands may be permitted to lay baits for pigs. Finally, as part of a Simultaneous Destruction Scheme, the District Forester may have baits laid and use Forestry Department funds for this purpose. *Recreational shooters are not permitted in state forests*, on the grounds of safety to the public and of other forest users. In relation to the general matter of control of pigs on forest lands, the Forestry Department has set out the following points:

> 'State Forests and Timber Reserves are used increasingly by members of the public for a wide range of recreational and educational activities and so safety of these people and other forest users is paramount. While it is appreciated that the control of feral pigs is desirable, the methods employed must be compatible with safety for forest users and the conservation of native fauna.'[51]

Control measures used. Actual policies and measures taken to control pigs in forests were stated to be as follows:

Palmerston (North Queensland district): dogging and shooting carried out by local cane farmers; pig hunting limited to ½ mile penetration into forest from adjoining landholder's property.

Byfield (Rockhampton): in past years, bait impregnated with 1080 has been laid but seems to have been ineffective.

Bulburin (Monto): permits to shoot pigs issued if pig damage warrants.

Toolara (Gympie): permits issued to hunt with dogs but not actively encouraged.

Inglewood (Warwick): one permit issued in 1975 for 1080 poisoning on a small area.

Marmadua (Dalby): aerial baiting using meat and 1080 carried out in 1976-77 in conjunction with neighbours.

Barakula (Dalby): no shooting of pigs but selective baiting carried out by lessees.

Effectiveness of control measures. It is apparent that very little is spent from the Forestry Department's own funds on pig control. Some interesting comments were received on the effectiveness of current control measures; in general, answers suggest that current measures are ineffective in controlling pigs:

Kuranda (North Queensland district): costs of pig control are prohibitive, and there has been no assessment of pig numbers and value of damages done.

Palmerston (North Queensland): nomadic pigs soon fill the gaps created by hunting pigs on the forest fringes.

Cathu (North Queensland): baiting has been unsuccessful and shooting has had no significant impact because pigs merely moved out of the area subject to shooting.

Toolara (Gympie): hunting and *dingoes* are effective in preventing pig population from increasing (see Fig. 7.17).

Passchendaele (Warwick): piecemeal control methods are ineffective as populations fluctuate widely.

Fig. 7.17 The dingo Canis familiaris dingo *is believed by some foresters and wildlife rangers to be an effective predator on wild pigs. Some have recommended that it should not be poisoned.*

Inglewood (Warwick): erratic nature of control measures make these ineffective.

Marmadua (Dalby): effectiveness of aerial baiting not known.

These comments highlight the main problems in pig control arising from their mobility and high fecundity.

Special difficulties in controlling pigs in forests. As for special difficulties in controlling pigs in forest areas, various sub-districts (groups) mentioned: size of forests, rugged or mountainous terrain, dense forest or scrub, lack of road access, scattered waterholes in large forest area, and lack of information on size and location of areas infested by pigs. Other difficulties mentioned were: baiting with 1080 is undesirable because it is dangerous to native fauna, pigs are migratory with fluctuating population levels, pigs are difficult to control because they are partially nocturnal, and shooting can lead to indiscriminate acts. It was also pointed out that in one sub-district dingoes appear to control pig numbers adequately unless baited.

Opening of Forests for Hunting

As pointed out earlier, game permits for recreational shooting are not issued for Queensland forests although permits may be issued to cane growers, lessees and landholders adjoining forests. But few such permits appear to have been issued because few have been requested.

On the question of whether forests should be open in certain times of the year to permit hunting of wild pigs, the 28 respondents to this question were divided equally between those in favour and those against. Many of the sub-districts in favour qualified their answer. Many of the sub-districts in favour said that hunting should be allowed only when it did not interfere with other forest users or native fauna; and some believed that it should be restricted to plantations. Many of those against hunting mentioned that recreational hunters were not selective in their prey species, created bigger problems than existed, were not effective in controlling pig numbers, and may help spread *Phytophthora* by the use of vehicles, etc., and they felt that hunting should be left to personnel of the Forestry Department. Some illuminating comments from those against hunting were:

Palmerston (North Queensland district): Recreational shooters are not selective in prey species and would exert major control pressures on forestry personnel.

Eungella (Rockhampton): Hunting damage may be greater than pig damage. Hunters and their vehicles would only add to the number of agents that could spread *Phytophthora cinnamomi*.

Bellthorpe (Brisbane): There would be a danger to the public, to forestry personnel and to native flora and fauna.

Injune (Dalby): Shooters could create problems for management and fire control.

Comments from those in favour of hunting included:

Kuranda and Danbulla (North Queensland district): Permits to be restricted to plantations and only those where hunting does not interfere with other forest users.

Cathu (Rockhampton): Shooting would help keep pig numbers down; however, care should be taken to guard against indiscriminate shooting.

Elliot (Maryborough): Although shooting is of limited value in controlling feral pig populations, it does provide an avenue for sporting pursuits and takes the pressure off native fauna.

Barakula (Dalby): Shooting to be under licence and strict policing measures should apply.

Recommendations for the Control of Wild Pigs in Forests

On the question of what *should* be done to control wild pigs in forests, shooting was the method most frequently suggested (mentioned 12 times), followed by selective baiting (nine times), 'do nothing' (five times), trapping (four times), hunting with dogs (twice), one sub-district suggested that the introduction of a disease such as a venereal disease might be effective, and another suggested that dingoes should not be baited as they checked the pig population. A sample of the specific recommendations are as follows:

On selective baiting of target species.

Passchendaele (Warwick district): if pig numbers become unacceptable then carefully controlled baiting programs during dry periods should be sufficient to lower pig populations.

On Shooting.

Marmadua (Dalby): selected professional shooters should be allowed to reduce number of pigs under very tight conditions.

Gallangowan (Murgon): game permits should be issued to public under strict controls.

Do Nothing.

Palmerston (North Queensland): do nothing. [No reason given, but could be economics of control or nomadic nature of pigs.]

Bellthorpe (Brisbane): do nothing as no damage is being done. [Note: Pig numbers low here.]

Trapping.

Western (Warwick): strictly controlled contract shooting or trapping. A bounty should be raised as an inducement. No poisoning.

Western Creek (Dalby): trapping by means of weldmesh traps. Poisoning causes losses to native carnivorous animals.

Hunting with dogs.

Byfield (Rockhampton): hunting with trained dogs with no firearms allowed.

Do not bait dingoes.

Kenilworth (Gympie): [suggests that dingo is a predator of pigs, especially young pigs.]

Disease introduction.

Cathu (Rockhampton): introduce a disease such as venereal disease or similar to make pigs barren or infertile.

In passing it might be noted that the last recommendation would not be workable in Australia if the disease would affect domestic pigs or stock.[52]

Conclusion from Survey of Queensland Forests

A considerable number of pigs are present in Queensland forests and are spread over a wide area and varied habitats. They appear to have few benefits in forests and can impose significant costs, especially in the north of the state. Pigs may have an adverse impact on the commercial value of forests by spreading fungal dieback[53] and to a minor extent, by rooting seedlings and larger trees and eating seeds, they may pose a threat to native flora and ground fauna especially to birds such as the brush turkey and scrub hen, add to erosion in forests and damage some forest installations such as roads and dams by rooting, may carry parasites, and may spread diseases infectious to domestic stock, could reduce the value of forest leases for grazing, and may stray from forests to the properties of neighbouring agriculturalists and pastoralists, who complain to foresters in many districts about the damage which can involve significant economic loss. Control of the species is difficult to achieve because of its adaptability, fecundity and mobility. The pig control program of the Forestry Department is a modest one and mostly relies on the initiative of affected cane growers, lessees and adjoining landholders. Whether or not greater control would be economic remains an open question; but considering the multiple uses of forests, further consideration may need to be given to this question, especially in the north of Queensland.

CONCLUSIONS ABOUT PIGS IN FORESTS

Very little scientific investigation has been done in Australia into the benefits and damages caused by feral pigs in forests. Reports from foresters suggest that tree regeneration and growth of trees in *Eucalyptus* and native cypress forests located mostly in New South Wales are little affected by the presence of pigs, but the rainforests of northern Queensland seem to be adversely affected. Reports from the United States indicate minor losses to timber production from wild pigs, but in Europe a number of writers suggest that pigs in moderate numbers are a positive net benefit in forests for timber production. However, European experience does not necessarily apply to Australia because the Australian ecological situation and range of native species differs from the European. Furthermore, hunting pressures are sufficient in Europe to keep wild pig populations within acceptable limits. In all European countries hunting is restricted so as to not deplete wild boar numbers excessively. In Australia, even if hunting were to be permitted in all forests hunting pressures, given Australia's relatively low human population hunting would probably be insufficient to exert control on pigs on the European scale.

Perceptible damage to flora and fauna is caused by introduced pigs in some Australian forests, especially in northern Queensland rainforests, in wet and boggy areas and in subalpine environments. Considerable damage to flora has been noted in other countries, such as the United States and New Zealand, where wild pigs have been introduced. As in the case of national parks in Australia, there is a need to assess those areas most at risk from wild pigs and consider whether attempts should be made to eliminate pigs from those areas and build exclosures.

There is concern amongst some Australian foresters about the ability of pigs in forests to carry exotic diseases and parasites. The Australian concern, like the New Zealand one, arises because Australia is relatively free of serious livestock diseases endemic to Europe and Africa. This matter, of widespread concern in Australia and New Zealand, is to be taken up in the next chapter.

CHAPTER 8

Health and Disease Risks from Wild Pigs

Cultural attitudes to pigs (often interwoven with religion) vary from society to society. To Jews and Moslems, the pig is an unclean animal and pork is not eaten as a rule.

The view that the pig is an unclean animal is stated in the Old Testament of the Bible. Leviticus, for example, lists the pig as an animal that is not to be eaten:

> 'Do not eat pigs. They must be considered unclean; they have divided hoofs but do not chew their cud. Do not eat these animals or even touch their dead bodies.' (Lev. 11. 7-8)[1]

This admonition is also repeated in Deuteronomy, 14.8. Further, in 2 Maccabees of the Old Testament of the *Jerusalem Bible* the case is given of Eleazar who chose to be martyred rather than eat pork. It is recorded that

> 'Eleazar, one of the foremost teachers of the Law, a man already advanced in years and of most noble appearance, was being forced to open his mouth wide to swallow pig's flesh. But he, resolving to die with honour rather than to live disgraced, went to the block of his own accord, spitting the stuff out, the plain duty of anyone with the courage to reject what it is not lawful to taste, even from a natural tenderness for his own life.' (2 Maccabees, 6. 18-21)[2]

The attitude of Jews and Moslems to eating of pork is widely believed to stem from the fact that pigs can harbour the cysts of tapeworms and other parasites in their flesh. These can be transmitted to man if the flesh is not properly cooked or frozen. However, Burton suggests that

> 'Perhaps, a better explanation lies in the nature of pig farming. As already mentioned, pigs are kept by settled people and nomadic tribes traditionally feel superior to farmers, an attitude still held by the Bedouin Arabs. They would consequently feel it below their dignity to eat farmers' animals and this attitude has become enshrined in the religion of the Hebrew pastoralists and their spiritual descendants.'[3]

By contrast pork has traditionally been held in high esteem in Europe and China, where wild boar in particular is regarded as a delicacy. The human disease risk factor appears not to have been paramount consideration. Indeed, pigs were sacred to the Norse gods and other heathen gods in Europe and the consumption of pork was associated with festivals such as Yule.[4] Throughout Melanesia (New Guinea and surrounding areas) the pig is held in high esteem and plays an important part in religious ceremonies.[5]

While it is apparent that pigs can carry many diseases and parasites of danger to man and other animals, so too can most mammals. The risks might be higher, however, in the case of an omnivorous animal such as the wild pig, as will become clearer from the discussion below.

Concern about wild (feral) pigs in Australia and New Zealand stems not so much from any danger that they might pose to human health, but their potential to cause economic loss to the livestock industries, by acting as carriers and reservoirs for disease. Particular fear has been expressed that wild pigs could become carriers of diseases exotic to Australia and to New Zealand (such as foot-and-mouth disease) if these diseases happened to be accidentally introduced. It is felt that if wild pigs were to catch one of these diseases that it would make it difficult or impossible to eradicate the disease. Australian policies towards feral pigs have been influenced by this possibility.

The purpose of this chapter is to highlight some general observations (of zoologists and others) about wild pigs as carriers of disease and parasites, outline the diseases and parasites spread by pigs to humans and livestock, and summarize the ways in which infection occurs and means to avoid it. Health and disease risks from feral pigs in Australia will be given special consideration. In particular, government policies and suggested strategies aimed at reducing this risk will be outlined and critically examined.

GENERAL OBSERVATIONS BY ZOOLOGISTS ABOUT WILD PIGS AS CARRIERS OF DISEASES AND PARASITES

Sludskii, after considering the evidence on the wild boar in U.S.S.R. as a vector in the spread of diseases and parasites, came to a number of general conclusions that are worth summarizing.[6] These are:

1. The wild boar is subject to many infectious diseases.
2. These include diseases specific to pigs (swine fever, for example) and some (such as foot-and-mouth disease) that are widespread among other species of animals. Some infections are so contagious for the wild boar (swine fever, foot-and-mouth disease) that temporary (epizootic) infection occurs amongst wild pigs even when they are in excellent physiological condition.
3. Wild pigs constantly exchange infections (a) with domestic pigs, (b) with other domestic livestock, and (c) with wild animals.

4. Because it regularly eats rodents and the carcasses of animals (even when the animals have died from infection) 'the wild boar contracts infections more frequently than other species of wild ungulates, and, since it covers large areas in its wanderings and migrations, it plays a particularly important part in natural outbreaks of a series of diseases (foot-and-mouth disease, swine fever, pasteurellosis, erysipelas, etc.).'[7]
5. It is host to many parasites and plagued by a number of ectoparasites (ticks, pig lice, mosquitoes and gad-flies). The ectoparasites are in turn also sources of infection.

Cross infection between wild pigs and domestic livestock is frequent. This suggests that

(a) domestic pigs and farm animals should be isolated from wild pigs, at least when there is an epizootic outbreak in domestic animals of disease contagious to wild boar;
(b) conversely attempts should be made to isolate wild pigs from domestic livestock when contagious disease is present amongst wild pigs.

It is necessary to prevent livestock from grazing in localities where wild boar are present and carefully remove all carcasses. Pigs are especially fond of carrion and smell this out over long distances. 'When eliminating outbreaks of foot-and-mouth disease among farm animals, together with outbreaks of swine fever, erysipelas, pasteurellosis and other infections, particular importance must be assigned to the part played in them by the wild boar, and, where necessary their numbers must be reduced to a minimum.'[8]

Observations from the United States of America

In countries such as the United States of America, Australia and New Zealand where the wild pig is an introduced animal, there is considerable concern not only about the ability of the wild pig to spread endemic diseases to livestock (and to humans) but at its ability to act as a carrier of (and reservoir for) exotic diseases such as foot-and-mouth disease and African swine fever. Wood and Barrett expressed the position in the United States as follows:

> 'The potential for wild pigs to harbour diseases transmissible to man or domestic animals is considerable. Cholera, swine brucellosis, trichinosis, foot and mouth disease are all diseases that may be transmitted by wild pigs. Cholera is thought, by the USDA, to be eradicated in the U.S. today, although it has been found in wild pig populations in the past. Trichinosis occurs in wild pig populations and poses a health hazard. From 1973 to 1977 there were 24 reported cases of trichinosis in humans that had

> contracted it from wild pigs in the U.S.[9] The fear of foot and mouth disease is considerable in the country and the number of wild ungulate populations that could be reservoirs for the disease should be minimized.'[10]

They also point out that African swine fever occurs in Brazil and the Dominican Republic and that if it should spread to the continental U.S., 'the economic cost of pig eradication to eliminate the disease would be dramatic.'

Furthermore, brucellosis a disease transmissible to man, occurs in feral pigs in South Carolina,[11] Florida[12] and Hawaii[13] and is a source of infection for domestic pigs. The United States Department of Agriculture is concerned that brucellosis is present in wild pig populations because it would like to eradicate swine brucellosis from the country.

In addition pseudorabies has been found to occur in wild pigs in Florida.[14] In California, feral pig populations at least in the past harboured bovine tuberculosis.[15] This disease was discovered on a Californian ranch in 1965 and wild pigs were eradicated on the ranch over a period of two years in an attempt to control the disease.

Another expression of the concern in the United States about the health and disease risks from wild pigs comes from Belden and Pelton. They point out that

> '... concern has been expressed regarding health considerations involving the possible contamination of water courses or sources and the potential of this species to serve as a reservoir for the transmission of parasites and diseases to man, livestock and wildlife.'[16]

This concern led Belden and Peltman to study the wallows of the European wild boar in the Great Smoky National Park and in the Tellico Wildlife Management Area of the Cherokee National Forest, Tennessee. They felt that this wallowing could pollute the drinking water of visitors to these areas. They concluded that

> 'With over 50 per cent of the wallows occurring in or within close proximity to small streams the potential for human consumption of water from the same sources is feasible; this is particularly true considering the almost exponential increase in numbers of hikers and backpackers using the Park in the past five years. Preliminary analysis of water samples from the study area indicates hogs may be contributing to the bacterial loads of streams which they root and wallow.'[17]

There appears to have been no study in Australia of the impact of rooting and wallowing by feral pigs on water quality but the effect is potentially significant (especially since pigs can carry leptospirosis) and they occur in some areas (such as Barrington Tops, N.S.W.) that are catchment zones for municipal water supplies. In addition, there may be risks to hikers and others dependent on local water supplies.

The perceived health threat of feral pigs in Australia was recently summarized in *The Agricultural Gazette of N.S.W.* by Keith Hart, a veterinary inspector on the south coast of New South Wales. He says 'feral pigs have emerged in recent years as an important menace to agriculture ... feral pigs represent a threat to our livestock industries because of their potential for spreading exotic diseases such as foot and mouth disease.'[18] He stresses

> 'It is important to remember that feral pigs can act as a reservoir for exotic diseases. Pigs are susceptible to foot and mouth disease, swine fever, African swine fever, and trichinosis. All these diseases can be transmitted to pigs if they feed on contaminated meat or meat products. Pigs are also susceptible to other exotic diseases such as rinderpest and Aujeszky's disease.'[19]

If there should be an outbreak of any of these exotic diseases amongst domestic livestock with subsequent transmission to feral pigs on the east coast of Australia, eradication of the disease might be impossible since many feral pigs live in inaccessible terrain. However, it is also possible that the disease could first occur in the feral pig population. If, for example, contaminated meat or animal products happened to be imported to Australia, the waste dumped and this then eaten by feral pigs, an outbreak of an exotic disease could occur. 'Diseases such as foot and mouth disease, vesicular exanthema and swine fever are caused by viruses that can survive in meat and other animal products for a very long time.' In some towns, feral pigs have access to (council) garbage tips so it is feasible (as emphasized by Hart) that contaminated scraps could be obtained from this source.

Feral pigs carry a number of established diseases in Australia. These include leptospirosis and sparganosis, diseases dangerous to both humans and livestock, and several other diseases. The nature of these diseases is outlined below.

Concern about pigs and other feral animals in Australia as possibly serious economically damaging carriers of disease was expressed in the recent report of the Board of Inquiry into feral animals in the Northern Territory. The recommendations of the Board included the following:

> 'The evidence clearly shows the nature and extent of existing and potential problems. The Board considers that all Australians should be more aware of this threat. To overcome a situation which has arisen over many years will require new attitudes, money and effort. There are some who regard the presence of brumbies, wild pigs, wild buffalo and other feral animals as a nostalgic symbol of the great outback, or as a good subject for camera or target practice. The romantic merits of these animals must be weighed against

their cost to the nation in terms of disease, environmental damage, and loss of production.'[20]

In particular the Committee recommended that

'The situation which has built up in the Top End [of the Northern Territory] over the past 100 years should not be allowed to continue. Prudence demands that positive programs be developed and carried out as quickly as possible to eliminate animals which are susceptible to foot-and-mouth disease and which cannot be brought under reasonable systems of management within a period of say 5 years.'[21]

In New Zealand, it is also recognized that feral pigs pose a risk factor in the possible introduction of exotic diseases. For example, J. T. Murray points out that '...The wild pig has the greatest potential as a vector of foot and mouth disease because of its wide ranging habits, frequent close contact with domestic stock, and its high susceptibility to the disease. Other species may also be of considérable importance of disease vectors. The question of transmission by the Cervidae (deer family) has been investigated in some detail in Europe and there seems little doubt that deer could become an important reservoir of foot and mouth disease.'[22]

Let us consider some of the particular diseases and parasites that can be carried by wild pigs. It is convenient to cover first those diseases exotic to Australia and then those endemic to or occurring in Australia.

DISEASES AND PARASITES EXOTIC TO AUSTRALIA CARRIED BY WILD PIGS[23]

Wild pigs (*Sus scrofa*) are subject to (most of) the formidable array of diseases and parasites that can infect domestic swine. It is not possible to describe or mention *all* of these diseases here. (A comprehensive coverage is available, for example, in H. W. Dunne, *Diseases of Swine*.[24]) The diseases considered here are those that have evoked the most attention in Australia. The exotic diseases include foot-and-mouth disease, rinderpest swine fever, African swine fever, trichinosis, vesicular exanthema and vesicular stomatitis.

Foot-and-mouth Disease[25]

It has been estimated by the Australian Bureau of Agricultural Economics that the introduction of foot-and-mouth disease to Australia 'would cost $2500 million in one year in direct losses of rural output.'[26] It would result in the closure of important export markets for Australian meat - for example, the United States' market.

Foot-and-mouth disease is a virus disease highly infectious to cloven-footed animals, both domestic and wild. Other groups of animals are usually not affected (although rats and hedgehogs may help spread the disease). It is rare for humans to contract it. Cattle, pigs, sheep and goats (in that order) are most frequently affected, so the disease is of particular concern to the pastoral and dairying industries.[27]

The occurrence of the disease in animals is characterized by vesicular eruptions in the membranes of the mouth and the muzzle and the skin around the feet. Some authorities claim that death in infected animals is uncommon[28] (and is reported to seldom exceed 5 per cent of infected stock[29]) but emaciation generally occurs in diseased animals (due to mouth lesions and lameness) and this alone is sufficient to cause great economic loss. Abortion and other complications may also occur.

There are a number of types and strains of the virus and the susceptability of different species of animals to these can vary. Consequently, if a country has some types and strains present but not others, it may pay it to take measures to prevent the entry of additional strains from abroad.

Foot-and-mouth disease regularly affects animals ('is enzootic') in parts of Europe, Asia, Africa and South and Central America but is not established in a number of regions, including North America, Australia, New Zealand and Japan. It occurs sporadically in Great Britain. Although introduced to Australia several decades ago, it was successfully eradicated.[30]

The virus is easily transmitted either by infected animals, by products from infected animals or by contact with materials, clothing or animals that have been in the area of infection and to which the virus has adhered.

Of special concern to countries such as Australia free of foot-and-mouth disease is the fact that the virus can survive for long periods in animal products such as meat, hides, milk and milk products. Even though importation of such products from abroad is restricted, illegal importation does occur. This is a particular risk with fast international air transport. However, the virus can remain stable for long periods of time. For example in skim milk (despite processing) it can survive up to 1½ years under suitable conditions.[31] It may persist in animals' bodies for up to 1 year and has reoccurred on farms up to a year after its apparent eradication.[32]

Control of the disease is achieved by three main means: (a) Policies such as quarantine of animals from abroad and embargoes on the importation of appropriate foodstuffs and other commodities that could be contaminated by the virus. (b)

Stamping-out policies or slaughter and destruction of livestock (and appropriate wildlife) whenever an outbreak occurs. (c) Vaccination of livestock in areas where the disease is enzootic or in some cases epizootic. Policies (a) and (b) have been adopted in Australia, the United States and Great Britain but, in addition to other policies, some countries in Europe also use policy (c).

The stamping-out or slaughter method involves '(1) quarantine of premises and rigid inspection of the areas involved, (2) slaughter and disposal by burial or burning of all infected and susceptible animals exposed to the disease, (3) thorough cleaning and disinfection of the premises and all possibly contaminated materials, and (4) testing of the premises by careful restocking with susceptible stock' to check whether the disease has in fact been eradicated. Clearly this method is likely to be less costly and more effective the more quickly an outbreak is detected. It would call for the destruction of wild pigs either in contact with infected stock or in contact with wild pigs that have contacted the disease. Movements of wild pigs (as well as other susceptible animals) into and out of affected areas needs to be stopped.

An attempt has been made to analyse the economic benefits and costs of some of these alternative policies in Britain. Power and Harris found in their study of control policies for foot-and-mouth disease that the slaughter policy was more economic for Great Britain than vaccination policies, if one were to judge only on the basis of quantifiable factors.[33] A similar study does not appear to have been done for Australia as yet.

African Swine Fever[34]

African swine fever is a highly contagious viral disease specific to pigs. While the symptoms are similar to those for swine fever (hog cholera), the viruses are different. African swine fever is generally fatal for European pigs (*Sus scrofa*). Infected European pigs 'usually die so quickly that loss of condition is uncommon.' However, in Africa it occurs naturally amongst warthogs (*Phacochoerus*) and bush pigs (*Potamochoerus*) producing in them an inapparent infection.

Originally the disease was confined to Africa. Its effect there is such that it prevents the commercial raising of pigs in some regions, in West Equatorial Africa, for example. However, the disease has in recent years occurred in Portugal, Spain, France and Italy where slaughter policies have been adopted to control it.

The risks of international spread of the virus are considerable since it can survive for long periods of time in meat or contaminated articles. The virus is present in infected pigs 'in the blood, tissue fluids, internal organs, and all excretions and secretions of infected animals.' It has been found that at room temperature the virus can survive in blood for several weeks and cool storage for 6 years, so one has to be especially wary of imports of pigmeat from infected areas. Even on soil, boards and bricks in the open it can survive for 80 days or more.[35]

The disease could be introduced to Australia by contaminated pigmeat products being imported and the uncooked scraps fed to domestic pigs or left on garbage heaps to which feral pigs have access. Alternatively, it could come in by the import of an article in contact with the virus which subsequently comes into contact with pigs in Australia. The consequence would be severe for the domestic pig industry as well as for anyone having a commercial or amateur interest in wild pigs in this country. Because no effective vaccine has yet been developed against the disease, apart from policies aimed at excluding the virus or keeping pigs away from possible sources such as garbage, the only available control policy appears to be that of slaughtering all pigs likely to have come in contact with the virus - that is, a stamping-out policy.

Japanese encephalitis[36]

This disease occurs in Japan and some other Asian countries including Singapore. Humans and horses are especially susceptible to the disease, the virus of which is carried by mosquitoes. Given Australia's proximity to Asia and the presence of pigs in the Top End of Australia, mosquitoes are a potential carrier of the disease to Australia. The disease causes pregnant pigs to lose their young or give birth to abnormal progeny. Its introduction would therefore have economic consequences for the pig-raising industry, as well as for horse-breeding in Australia.

Pseudorabies and Rabies[37]

Pseudorabies (also known as Aujeszky's disease) is a virus that can cause death of pigs of all ages, including the abortion and death of foetuses. Infection occurs most often in pigs, cattle, sheep, dogs and cats. While the disease is of concern in the U.S.A., it is reported that 'it is not a major cause of revenue loss generally' to swine producers.[38] However, the species mentioned above other than pigs (cattle, for example) are more susceptible to the disease than pigs.

Control involves isolating infected animals and suspected carriers of the disease. Contaminated garbage or meat should not be fed to species susceptible to the disease. In particular wild pigs should be kept away from such garbage.

While rabies may occur in pigs, it is reported to be rare.[39]

Rinderpest[40]

The virus disease rinderpest can cause high mortality in cattle and ruminants. Mortality rates in susceptible species may exceed 90 per cent and morbidity in susceptible species is high. It is enzootic in most of equatorial and northern Africa, Asia and Asia Minor.

The virus is present in all the tissues and fluids of infected animals, and appears to be transmitted by contact with 'virus-laden aerosols which may arise from infected nasal secretions, expired air, urine or fresh fecal material.' Therefore, quarantine would appear to be a relatively effective means of control.

The susceptibility of wild pigs to the disease is uncertain. It is reported that European breeds of pigs are usually resistant to the disease but that it occurs in wild pigs in the Middle East and in Thailand. The extent to which Australian feral pigs would be susceptible is unknown but they could become carriers of the disease if it happened to be successfully introduced. The disease is reported to have occurred in Western Australia near Fremantle in 1923 but it did not become established.

Swine Fever (Hog Cholera)[41]

Swine fever (or hog cholera) is also caused by a virus. Originally believed to have been confined to America, the disease now occurs in other parts of the world - for example, in Europe[42] and Asia. Outbreaks have occurred in Australia.[43]

Dunne reports that in the United States 'the virus of hog cholera is still the cause of more swine deaths than any other infectious organism.'[44] The disease, which appears to be specific to pigs, results in high mortality (and morbidity) amongst unvaccinated stock. Where the virus is of low virulence the mortality rate tends to be lower for older animals. The virus is highly infectious amongst pigs. It is transmitted by direct or indirect contact with infected animals. Contaminated feed, water or equipment is a common source. In particular the feeding of uncooked garbage containing pork scraps should be discouraged since it is a common source of infection.

In relation to control of the disease, Dunne says 'The need for effective quarantine of hog cholera infected animals is paramount. Effective measures are necessary to prevent transportation of infected animals to slaughter, to community sales, or to other points of public dispersion. It has been common practice to reduce financial losses by shipping exposed hogs quickly to market. Such animals provide infected meat scraps for wide distribution. Movement of swine in areas where the disease is present must be prohibited to effect eradication of hog cholera.'[45] All trucks and railway wagons used in the transport of infected animals and contaminated pig sties and buildings need to be thoroughly disinfected.

The virus can survive for considerable time in pigmeats. For example it has been shown that bacon prepared from infected pigs can cause swine fever after 27 days.[46] The virus can survive in cured hams and is not destroyed by refrigeration or freezing.[47] Therefore, there is a risk to Australia of a virulent form of the virus entering through (illegal) import of pork products brought in, for example, by plane travellers. In addition to the embargo on such imports, it is important in order to minimize the chances of occurrence of the disease that feral pigs not have access to garbage and that if any garbage is fed to domestic pigs it is well cooked.

Teschen Disease (Encephalomyelitides)[48]

Teschen disease is a viral disease affecting the nervous system of pigs, occurs in Europe but is not established in America and Australia. It results in the paralysis of the infected animals and high mortality rates. No fully effective vaccine appears to be yet available so the main method of control of the disease is by quarantine and the slaughter of infected stock.

Vesicular exanthema[49]

Vesicular exanthema, a viral disease specific to pigs, occurs within the United States. Mortality amongst older pigs is low but can be high in young pigs. It is spread by contact with infected pigs and by the feeding of contaminated raw garbage to pigs. Clearly it can infect feral pigs.

It is reported to cause considerable economic loss in the U.S.A. Stewart Madin says,

> 'Vesicular exanthema is of great economic importance, since it causes serious weight losses in fat hogs, slow gains in feeder stock, deaths in suckling pigs, abortions in pregnant sows, and impaired lactation in nursing sows. In addition it is clinically indistinguishable from foot-and-mouth disease (FMD) and vesicular stomatitis (VS) in swine, thus requiring expensive quarantine procedures.'[50]

The disease is controlled by segragation and quarantine and by ensuring that pigs are not fed uncooked garbage.

Vesicular stomatitis[51]

This viral disease is confined to the Americas. Apart from pigs it can naturally infect horses, cattle and deer. Inapparent infections can occur in humans and in various wild animals.

The disease is 'clinically indistinguishable from foot-and-mouth disease and vesicular exanthema' but it is less contagious than these diseases, less serious to animal health and of limited economic importance. Hanson says, 'While vesicular stomatitis in swine ready for market could be costly to the livestock owner in delay or shrinkage, under most conditions he is not concerned as the animals return to normal appearance and weight within 2 weeks.'[52] Many animals also develop natural immunity to the disease. The main problem associated with the disease is distinguishing it from foot-and-mouth disease and vesicular exanthema, in case of an outbreak. In the case of mistaken diagnosis, costly and unnecessary quarantine and slaughter action may be taken in areas where the virus is present.

The methods of the spread of the disease are uncertain. While it can be spread by contact, mosquitoes and biting flies could also play a role in its spread. If the disease happened to be introduced to Australia it would be difficult to eradicate given the available feral animal reservoirs and the presence of mosquitoes.

Screw Fly (an External Parasite)[53]

The Board of Inquiry into Feral Animals in the Northern Territory reported that the risk of this external parasite being introduced to Australia is high.[54] The problem arises because the adult screw fly lays its eggs on 'any fresh abrasion such as a wire cut, scratch, wound tick or fly bite, and especially to the navel opening in the newborn.'[55] The resulting maggots then feed as a colony on the fresh tissues of the animal. Its presence would be a greater hazard than the blowfly menace in Australia.

The Northern Territory Board of Inquiry said the Bureau of Animal Health tells us that 'Screw Worm infestation occurs throughout South-East Asia and in Papua New Guinea, and the risk of its introduction to northern Australia is increasing. Screw worm would probably make cattle raising in the Top End of the Northern Territory uneconomic. Infestation in all feral animal species may occur, although *feral cattle* and *buffalo* would be the major reservoir species.'[56]

The Committee of Inquiry ventured that the most likely means of the entry of screw fly to Australia 'would be by island hopping, possibly assisted by small boat traffic across the Torres Strait to the Cape York Peninsula. Thence it could proceed through the Gulf Country to Arnhem Land. The presence of feral goats, deer, cattle, buffalo and horses in those parts of the Territory would hamper detection and aid spread.'[57] The Board felt that this consideration added weight to its recommendation that feral animals be exterminated in northern Australia. This recommendation will be taken up later.

Trichinosis (an Internal Parasite)[58]

Trichinosis is caused by a parasitic round-worm (nematode) *Trichinella spiralis*. Although the disease occurs widely (for example, throughout America and Europe) it is not known to occur in Australia. The main hosts of *Trichinella spiralis* are pigs, rats and humans.

Infection occurs in rats, pigs and humans by their eating uncooked meat containing the larvae (cyst form) of the trichina worm. These cysts develop into worms in the small intestine. There the fertilized female worm produces new larvae which enter the bloodstream of the host and eventually invade its muscles, creating new cysts.

The most common way for man to be infected is by eating uncooked or inadequately cooked pork. Infection in pigs generally occurs as a result of the feeding of raw garbage containing infected pork or as a result of pigs eating infected rats. While pigs are said to be relatively tolerant to the parasite, in man its cysts (in the muscles) can give rise to distressing symptoms not unlike rheumatism and in some muscle locations can cause serious breathing difficulties.

With less feeding of raw garbage to pigs, the incidence of this disease is declining in the U.S.A. While in the 1950s 25-50 million Americans were estimated to have trichina larvae in their muscles,[59] recent estimates indicate that the current figure is closer to 5-8 million.[60] However, this is still a considerable number.

Trichina are destroyed by the adequate cooking of pork or by freezing it. To control the disease it is necessary to process pork in a way that will destroy any trichina present. The destruction of rodents in piggeries also assists control and, of course, raw garbage should not be fed to pigs.

The disease occurs in wild boar and is present in feral pigs in Hawaii and California.[61] It could be expected to spread to feral pigs if introduced to Australia - for example, in inadequately processed pork.

Passing Observations on the Prevention of Entry of Exotic Diseases to Australia

It is clear from the above that animal quarantine import regulations would not in themselves be sufficient to prevent the entry to Australia of many of the exotic diseases listed above. The risk of their import in contaminated foodstuffs remains considerable even though the import of such foodstuffs is regulated. Some diseases such as foot-and-mouth disease and African swine fever are liable to be introduced on contaminated products in addition to food. Entry could be by many routes, the postage of illegal items, for example, or via food brought in by air travellers.

To reduce risks of the establishment of exotic diseases, garbage (at least in its raw state) should not be fed to pigs and pigs should not be permitted to come into contact with garbage. In particular, feral pigs should not be allowed access to garbage tips.

In a few cases insects are suspected of being vectors in the spread of pig diseases. The current Australian practices of destroying insects on aircraft arriving from overseas helps to reduce the risk of entry by such a vector but shipping is subject to less precaution. However, in the case of screw worm, because of the proximity of screw fly to the Australian coast (for example, in New Guinea) the risk of entry of this insect by island hopping is high.

Some exotic diseases have broken out in the past in Australia. Swine fever (hog cholera) has been reported as occurring in Australia in 1903, 1927-28, 1942-48 and 1961 but is not established here.[62] An outbreak of rinderpest was reported as taking place near Fremantle, Western Australia, in 1923. These incidences underlie the need to identify quickly an outbreak of an exotic disease in Australia so that rapid measures can be taken to contain and eliminate it.

DISEASES ENDEMIC TO AUSTRALIA CARRIED BY WILD PIGS

It is, of course, not possible within this book to cover all diseases and parasites endemic to Australia (that is, existing within Australia) that are transmissible by wild pigs. I shall therefore concentrate on those that seem to have been given the greatest attention because of their implications for livestock or for man.

Wild pigs in Australia can and do carry a number of diseases and parasites of economic importance to the livestock industries. Some of these are also infectious to man. However, the extent to which feral pigs carry and transmit diseases in Australia has not been fully explored and no estimates have been made of the economic losses entailed. The latter are dependent upon the availability of adequate biological information about the transmission of diseases by pigs in Australia. However, from the information available it seems that the role of feral pigs in carrying specific diseases varies according to regions in Australia. For example, tuberculosis appears to be carried by feral pigs in the Northern Territory but not in New South Wales.

First the possible role of feral pigs in the occurrence of leptospirosis, anthrax, erysipelas, brucellosis, tuberculosis, and Murray Valley encephalitis will be considered. Then attention will be focused on the internal parasites spargana, hydatids and pork tapeworms, *Taenia solium*. Finally attention will be given to external parasites of wild pigs, such as sarcoptic mange, lice and ticks.

Leptospirosis[63]

This disease is caused by various species of leptospirae. Apart from occurring in a number of wildlife species, the disease infects dogs, cattle, pigs, sheep and goats and is transmissible to man. The bacteria invade the bloodstream of the affected animal and eventually become localized in its kidneys. For about two months - that is, while the infection is in its kidneys - the animal sheds large numbers of the bacteria in its urine and this is a source of infection for other animals.

The disease can cause permanent stunting or death of young animals, a loss of weight in older animals and abortions amongst cattle, sheep and goats. According to Siegmund, leptospiral abortions in sheep have not yet been reported. However, mortality rates in lambs 10 to 12 weeks old can exceed 15 per cent, with a similar percentage of lambs stunted by the disease. Morbidity usually approaches 100 per cent. The disease is a common source of abortions in cattle. In young calves mortality from the disease is 5 to 15 per cent but deaths are rare amongst adults. When an outbreak occurs almost all calves in a herd are likely to contract the disease and about 75 per cent of older stock are likely to be infected. Thus considerable economic loss can occur.

The extent of the economic loss from this disease is probably not appreciated since it may be present without an owner of livestock being fully aware of it. Cattle and pigs are the prime carriers of the disease and (at least outside Australia) pigs are considered to be the main reservoir of the disease.

As far as Australia is concerned, Collins points out that the significance of the disease has only recently been recognized and that it probably affects all states. 'In New South Wales the disease is of considerable economic importance, and in Western Australia in which 60 per cent of calves have been affected with a mortality of almost 100 per cent.'[64]

There is little doubt in light of the research by Keast, Littlejohns, Rowan and Wannan[65] that feral pigs, at least in New South Wales, are significant reservoirs of the disease. These researchers concluded from their sample of feral pigs in New South Wales that 'the high proportion of leptospiral reactions, which were all of the pomona type, emphasise the important role of the feral pig as an active source of infection and this must be considered in the overall control of leptospirosis ...'[66]

Pigs are frequently carriers of *Leptospira pomona*, a variety highly infectious to cattle, sheep, goats and horses. The only indication of infection in pigs may be an increase in abortions in late pregnancy and poor weight gain amongst younger pigs.

The disease is spread by contact with infected urine, including the intake of urine-contaminated food or water. Feral pigs with the disease may, for example, infect the drinking water of cattle or other animals as a result of urinating while wallowing or feeding in wet areas. The risk of infection by this route is high since it has been demonstrated that *Leptospira pomona* can survive in cracks, water-holes and swampy areas for 10 to 15 days.

As a step towards controlling the disease, veterinarians recommend that calves and pigs be not allowed to run together. In fact pigs should (if they are likely to be infected) be kept separate from cattle, sheep and goats. In the case of infected wild pigs such separation can be difficult to achieve. However, in some areas it may be worthwhile from an economic standpoint to consider means to exclude feral pigs from livestock if the former are suspected carriers of leptospirosis.

Yet, it must be stressed that although feral pigs appear to be an important reservoir for *Leptospira pomona*, other animals such as rodents and deer are also carriers. Furthermore, in many instances the carriers are not wild animals but domestic livestock, in some of which the disease may take an inapparent form.

Anthrax[67]

Like leptospirosis, anthrax occurs in most countries of the world. However, according to Collins, it is now limited in Australia to small outbreaks in restricted areas of New South Wales and Victoria.[68] The disease is caused by *Bacillus anthracis* and affects a wide range of animals including cattle, horses, sheep, goats and wild herbivores. Pigs and humans can be infected although both show some resistance to the disease.

The disease can spread in several ways and spores of the bacillus can lie dormant in damp or swampy soils for many years. Feral pigs help spread the disease by feeding on the carcasses of animals that have died of the disease, they themselves dying in a few days or so and often after having travelled some distance. But this is not the only way in which the disease can spread. For example, carnivores may spread the disease in a similar way. Animals that have died of the disease should be cremated on the spot or buried deep, under a layer of quicklime.

It might be noted that the bacillus can withstand boiling, so care has to be exercised in ensuring that no infected animals are used, for example, in the processing of pet food or other food, for ingestion of the bacillus is a common method of infection. Furthermore, care must be taken to ensure that blood and bone meal (which is often used in animal foodstuffs) is not contaminated. Care therefore needs to be taken in utilizing feral pigs and other susceptible animals for such purposes.

Erysipelas[69]

In Australia this disease occurs sporadically amongst pigs and is a cause of their unthriftiness and occasional death, particularly in pigs 3-9 months old. The disease is caused by the bacterium *Erysipelothrix rhusiopathiae*. The bacterium also causes nonsuppurative polyarthritis in lambs and post-dipping lameness in sheep. Infection in man, known as erysipeloid, can occur via wounds.

While one could expect the disease to be carried by feral pigs, there are also numerous other natural hosts for the disease and much uncertainty surrounds its epizootiology. Although keeping feral pigs away from susceptible livestock could reduce the incidence of the disease, the effect is unlikely to be marked, for as R. Shuman and R. Wood point out,

> 'When one considers the widespread distribution of *E. rhusiopathiae*, its association with a wide variety of animals (both furred and feathered), its apparent ability to survive under seemingly adverse conditions, and the absence of reliable means of detection, the eradication of swine erysipelas seems very remote.'[70]

Brucellosis[71]

Various types of bacteria can cause brucellosis, or contagious abortion. In cattle the main agent is *Brucella abortus* but *Brucella suis* and *Brucella melitensis* are occasionally responsible for infection. In pigs *Brucella suis* is the usual causal agent but it may also arise infrequently from the other two types of *Brucella*. Man may catch the disease, sometimes called undulant fever, by contact with infected carcasses or organs of infected animals. Abattoir workers, veterinarians, butchers and farmers are most at risk. In animals infection is possible by several methods including direct contact and ingestion of materials such as food and water contaminated by the brucellae. Not only does the disease lead to abortions in animals but it may result in their infertility.

No evidence has been found of brucellosis in feral pigs in New South Wales, according to a recent report of a veterinary officer.[72] In line with this statement Keast and colleagues found no evidence of the disease in their sample of feral pigs from New South Wales.[73] The report of the Board of Inquiry into Feral Animals in the Northern Territory stated that wild pigs are not known to be a factor in the spread of brucellosis at present.[74] Nevertheless there is always a risk that feral pigs could become carriers of the disease in Australia. Furthermore *Brucella suis* has been isolated from feral pigs in Queensland.[75]

Tuberculosis[76]

No evidence has been found that tuberculosis occurs in feral pigs in New South Wales,[77] but the disease is common in wild pigs in the Northern Territory and one suspects therefore also in northern Queensland. Although pigs can be infected by the bovine, avian or human types of tuberculosis, infection by the bovine type *Mycrobacterium bovis* is most common.[78]

While it had been thought that possibly up to 80 per cent of wild pigs in the Northern Territory had bovine tuberculosis, it is now believed as a result of laboratory tests on 726 feral pigs by the Primary Industry Division of the Northern Territory that the percentage is closer to 10 per cent.[79] 'It is also apparent that the disease is not uniformly spread through the wild pig population, but tends to show up in pockets. There is undoubtedly a link between the pig and the buffalo in this respect. As a scavenger, the pig readily takes advantage of buffalo carcasses left after death through bogging, shooting or other causes. Many of these derive from old or infected animals which are more likely to be suffering the effects of generalised tuberculosis.'[80]

Furthermore, it was thought that wild pigs were a source for reinfestation of buffalo and cattle (the primary hosts of *Mycrobacterium bovis*) it is now believed that pigs are an end host. 'That is [according to *Feral Pigs in the Northern Territory*], while pigs may contract the disease from eating infected buffalo meat, it is unlikely that they will contaminate the environment by excreting the organism, leading to infection in grazing buffalo.'[81]

Humans can be infected with tuberculosis by eating the inadequately cooked flesh of pigs suffering from the disease.

Murray Valley Encephalitis

Following examination of the sera from 617 feral pigs from New South Wales, Gard, Giles, Dwyer-Gray and Woodroofe concluded that pigs may play an important part in the life cycle of Murray Valley encephalitis, as they do 'in Japan with the related

Japanese encephalitis virus.'[82] Antibodies to Murray Valley encephalitis were found in the sera of 58 per cent of the feral pigs in their sample. Occasionally Murray Valley encephalitis occurs in man. Because the sample of feral pigs was collected between epidemics of the disease, it was felt possible that feral pigs act as a reservoir for it.

The Overall Role of Feral Pigs in the Spread of Endemic Diseases in Australia

From the above, it appears that feral pigs in Australia significantly increase the incidence of leptospirosis but possibly have a small impact on the frequency of anthrax and erysepilas. There is no evidence to suggest that feral pigs in Australia are important vectors of brucellosis even though it has been isolated from some feral pigs captured near Ayr in Queensland. As for bovine tuberculosis, feral pigs in the Northern Territory are infected with this disease but they are considered to be end hosts and not a source of reinfection of cattle and buffalo. However, feral pork in the Northern Territory should be well cooked to reduce the chance of tuberculosis infection in humans. In south-eastern Australia, there is no evidence of tuberculosis in feral pigs. However, it is possible that feral pigs in this region act as a reservoir (but not necessarily the only one) for Murray Valley encephalitis, a disease that occasionally breaks out in epidemic proportions.

PARASITES OF WILD PIGS IN AUSTRALIA

In Australia, as in all countries where wild pigs occur, they are hosts for a number of internal and external parasites. Most interest in Australia has been focused on the ability of pigs to carry sparganosis and hydatis, both of which can pose a public health problem. Relatively little concern seems to have expressed about the ectoparasites of wild pigs such as mange, pig lice and ticks. Let us consider some of these parasites.

Sparganosis[83]

Sparganosis is caused in pigs and in humans by the invasion of body tissue by the spargana of the cestode *Spirometra erinacei*. Pigs may be infected by eating an infected intermediate host of the cestode, such as frogs, toads, small reptiles and small mammals and in turn man may acquire sparganosis by eating inadequately cooked or inadequately pre-frozen pork containing the spargana.

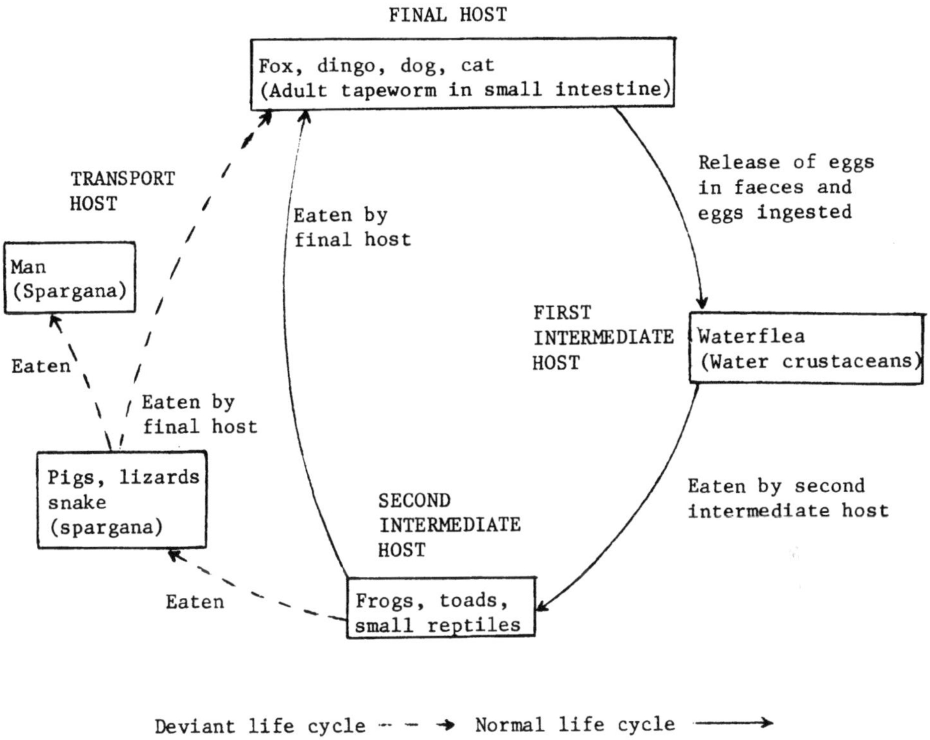

Fig. 8.1 The life cycle of Spirometra erinacei. *This tapeworm is the source of sparganosis in pigs and in man.*

The life cycle of *Spirometra erinacei* is indicated in Figure 8.1. The primary hosts of the cestode are the fox, dingo, dog and the cat. Neither man nor pigs are essential for it to complete its life cycle and are described as transport hosts. The tapeworm has two intermediate hosts - first the water flea (an aquatic crustacean such as *Cyclops*) and then a frog (such as *Hyla aurea*) or a small reptile or mammal. If a primary host such as a dog should eat an infected frog, the tapeworm develops in its small intestine. But if the frog is eaten by a pig, the parasites on reaching the pig's intestine migrate to its body tissue, usually the connective tissue of its muscles especially of the abdomen. There they form spargana. These are about 5 to 8 cm in length and about 25 mm wide and white in colour. They are easily mistaken for nervous or fatty tissue and thus it is difficult for a meat inspector to detect their presence.

Spargana have little or no effect on the health or growth and condition of pigs. Most concern is about the public health hazard posed by spargana.

Appleton and Norton explain in the *Queensland Agricultural Journal*, 'Sparganosis, a parasitic infection of many feral pigs, is a potential health hazard to those who eat infected pork. Many hunters shoot, slaughter and dress their own feral or wild pigs while other feral pigs are processed in local slaughter houses in Queensland.'[84] They explain that when man eats spargana in undercooked pork,

> 'the spargana migrate from the gastrointestinal tract of man to become resident in the loose fibrous tissues under the skin or between the muscles of the stomach and the nearby surrounding regions. Patients usually become aware of a small lump in this region, which can vary in size up to 3 cm and often moves about as the parasite migrates.'[85]

Sparganosis in feral pigs is widespread in Australia and appears to occur throughout Queensland, New South Wales and Victoria. However, it is not a human health hazard if infected pork either is well cooked or frozen for 24 hours. The spargana are destroyed by either process. Gordon, Forsyth and Robinson claim that

> 'The parasite is not very readily detected by health inspectors unless the infestation is very heavy or the inspection is extremely thorough. The spargana do not survive storage of carcasses in the chilling room at -8°C for 24 hours.'[86]

The precaution ought to be taken of freezing feral pig carcasses (as is required in abattoirs in some Australian states) or ensuring that the meat is well cooked. In the case of domestic pigs (where neither adequate cooking or freezing is likely to occur), they should be kept away from areas such as swamps where infection is likely to result.

Hydatis[87]

Pigs (including feral pigs) may act as an intermediate host for hydatid tapeworm *Echinococcus granulosis*. But so too do sheep, cattle and horses, hence the pig is *not an essential link* in the life cycle of this parasite. However, the dog and dingo are important as primary hosts. The eggs of the parasite may be accidentally ingested by man. It has been said that the resulting disease, hydatis, is the most serious parasitic disease affecting man in Australia.

The life cycle of *Echinococcus granulosis* is illustrated in Figure 8.2. In the primary host, usually a dog or dingo in Australia, this small tapeworm exists in the small intestine from whence its eggs are shed in the faeces of the host. These eggs are accidentally ingested (in food or water) by an intermediate host such as the sheep, pig, ox or man, as well as other animals. These develop in the intestine of the host and penetrate the body through the intestinal walls and migrate to a suitable part of the body, frequently the liver or the lungs, to form

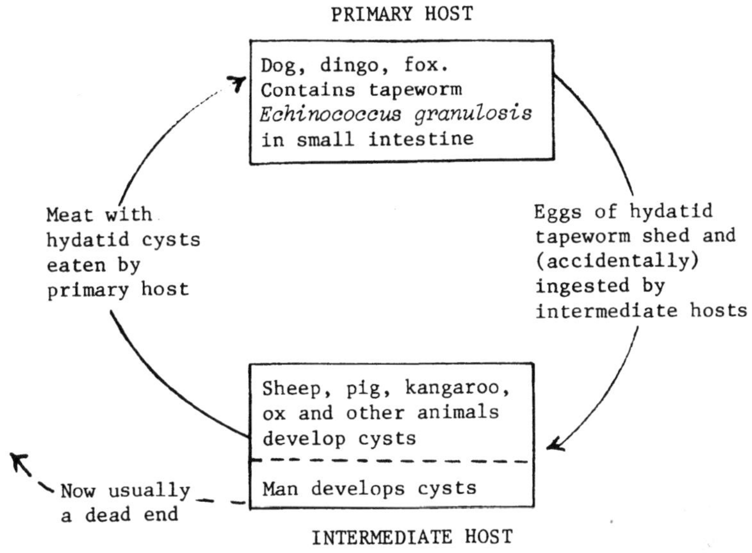

Fig. 8.2 The life cycle of the hydatid tapeworm Echinococcus granulosis. *This tapeworm can cause hydatis in man, pigs and other animals.*

cysts. When inadequately cooked or uncooked meat containing these cysts is eaten by a primary host, the tapeworm develops in the intestine and the cycle repeats itself.

The economic loss caused by hydatis in livestock is uncertain but it does not appear to be a serious source of economic loss, even though infection rates may be high: 30 per cent in sheep in New South Wales, 20 per cent in sheep and cattle in Victoria. However, in man the condition is often fatal.

It is clear that the feral pig is not an essential intermediate host. Eliminating it would be likely to have little if any effect on the incidence of the disease. However, pigmeat or offal from wild pigs to be fed to dogs should clearly be well cooked to reduce the chances of infection in dogs.

Pork Tapeworm[88]

This tapeworm (*Taenia solium*) is uncommon in Australia.[89] Man is the primary host of the tapeworm and the pig is its intermediate host. In the primary host *Taenia solium* attaches itself to the intestine and sheds eggs that are passed out and may be ingested by pigs. If this happens cysts develop in the pig. When inadequately cooked meat containing these cysts is eaten by humans the tapeworm develops in the intestines and the cycle repeats itself. Clearly, the incidence of infection can be reduced by improved human sanitation and ensuring that all pork is adequately cooked. One suspects that feral pigs are more likely to be carriers of the parasite than domestic pigs.

Discussion of Internal Parasites

Other internal parasites may be carried by feral pigs. For example, wild pigs are subject (as are domestic pigs) to kidney worms *Stephanurus dentatus*, and the occurrence of the worm is said to be common in feral pigs.[90] However, as far as man is concerned the risk of infection from internal parasites in pigmeat seem as though they can be eliminated by cooking the meat well or by freezing it for long enough.

Ectoparasites of Feral Pigs

Feral pigs can become infected by some ticks - for example, the bush tick *Haemaphysalis bispinosa*[91] - and sometimes carry sarcoptic mange *Sarcoptes scabiei* var. *suis*.[92] Pig lice *Haemotopinus suis* are reported to be commonly present on feral pigs in the Northern Territory,[93] are found on feral pigs elsewhere and may infest domestic pigs.

The economic significance of these parasites and role of feral pigs in the maintenance of their populations is unclear. Populations of most of these parasites might be expected to continue in the absence of feral pigs. Ticks, for example, have other hosts, and sarcoptic mange and pig lice would possibly continue to exist in domestic pigs.

POLICIES TO REDUCE AUSTRALIAN DISEASE RISKS FROM FERAL PIGS

The purpose of this section is twofold. It is to discuss generally (a) measures that may reduce the occurrence of endemic diseases carried by feral pigs in Australia, and (b) methods of lowering the risks of those exotic diseases being introduced to Australia that can be communicated by wild pigs. As will be seen, a number of measures that decrease the likelihood of infection from endemic diseases also make the successful introduction of exotic diseases less likely. Particular consideration will be given to the proposition that feral pigs should be eliminated in Australia or in particular regions of Australia because of the health and disease risk posed by them.

Measures to Reduce the Risks of Endemic Diseases from Wild Pigs

From the above discussion of endemic diseases transmissible by wild pigs in Australia, it is clear that a number of measures can be taken to reduce the risk of infection to livestock and to humans. These measures can also be expected to be of assistance in other countries in reducing the spread of diseases carried by wild pigs. The following measures will assist:

(1) Garbage, especially if it contains uncooked meat scraps, should not be fed to domestic pigs or be available to feral pigs. Consequently, feral pigs should be fenced out of municipal garbage dumps or excluded in other ways from food at such areas.

(2) If garbage is fed to pigs, it should be well cooked.

(3) Domestic livestock, especially domestic pigs, should be separated from feral pigs if possible. This is especially important in areas where (or at times when) there is a high risk of cross infection. There have been suggestions that the running of domestic pigs on range should be prohibited.

(4) It may be desirable to control movements of feral pigs by road or other transport to ensure that if new feral colonies are to be formed or new stock introduced the introduced stock is disease free. Hunters occasionally introduce wild pigs to new areas in Australia to build up feral herds without any check on the health of the animals involved.

(5) Farmers should exercise care in introducing feral pigs for fattening to areas where domestic pigs are housed.

(6) Feral pig meat for human consumption should be well cooked. As an extra precaution, the meat may also be frozen for 24 hours or more.

(7) Any feral pig meat that is to be fed to dogs (or other farm animals) should be well cooked.

(8) Cooking (or freezing) may not destroy anthrax bacteria. Care should be taken to ensure that animals with anthrax are not processed or consumed. However, the disease is infrequent in Australia and appears to be confined to particular localities.

Implementation of the above measures would involve some cost but at least from a public interest point of view could be worthwhile since most would also lower the risks of successful introduction of exotic diseases to Australia.

A more drastic and costly measure to eliminate the risks from feral pigs as carriers of disease would be to attempt to eradicate them. The question of whether an eradication policy is likely to be justified is given particular attention later in this chapter.

Measures to Reduce the Likelihood of Successful Introduction of Exotic Diseases

Although Australia has a system to reduce the likelihood of the successful introduction of exotic diseases, the system is not foolproof. For this reason, a number of the measures suggested above for the control of endemic diseases would also help to reduce the likelihood of the successful introduction of exotic diseases. These include keeping feral pigs (and other animals) away from garbage tips, not feeding garbage or feeding only well-cooked garbage to pigs, and reducing the contact of domestic livestock with feral pigs.

The risk of contamination through garbage is a real one. Although garbage from overseas aircraft is incinerated at Australian airports, products may be brought in illegally by passengers and, as mentioned earlier, postal items pose a risk. However, garbage from ships is an even greater problem. Garbage may be dumped overboard, sent to tips either for burning or burial, or burnt in an incinerator.

In a recent review of quarantine arrangements, the Department of the Prime Minister pointed out that

> 'During recent times, considerable concern has been expressed at the potential dangers from exotic animal diseases that could be caused by ships that dump their garbage overboard whilst close to shore or during the night when actually berthed.'[94]

While there are regulations prohibiting such dumping, it is not always possible to enforce these. For one thing, the actual dumping has to be observed by enforcement authorities.

In some cases too, garbage is taken from ships to municipal tips. Whenever this occurs (and it is unsatisfactory that it should occur) it is clearly necessary to keep animals away from these tips. The importance of this is underlined by an observation of the *Review of Australian Quarantine Arrangements* that

> 'Tips and burial are not considered to be satisfactory methods of quarantinable garbage disposal. We were advised of an incident near Geelong where garbage was placed in a shallow pit with a two inch layer of earth over it, cattle were observed eating the garbage some time later.'[95]

Amongst the Australian measures to reduce the chance of exotic diseases being introduced are:

(1) Animal quarantine.

(2) Prohibitions or restrictions on the import of various animal products.

(3) Disinsection of aircraft. However, ships are not automatically rid of insects or other pests such as rats. A ship is only fumigated if advice is tendered

by the Senior Quarantine Inspector that this is necessary and the Inspector can make out a case for it. One reason for this is that the cost of fumigation is high.[96]

Early detection and identification of any exotic disease that accidentally enters the country is important. Where an eradication policy is followed, this should make it less costly and more probable that the disease will be stamped out. Slaughter-out programmes have been drawn up in Australia for the control of various exotic diseases if they should enter. These strategies have been formulated by the Exotic Diseases Sub-Committee of the Australian Agricultural Council.[97]

Eradication of Wild pigs: The Ultimate Solution?

It has been seriously suggested that feral pigs in Australia should be eradicated to eliminate the health and disease risk posed by them. Indeed, the report on *Feral Animals in the Northern Territory* recommended that a feral-animal-free zone be established along Australia's northern border (Top End) to reduce the risk of entry of exotic diseases from Asia.[98]

Policy suggestions have been considered for the elimination of feral pigs (a) throughout Australia and (b) for their elimination only in particular areas - for example, in areas of high risk.

In assessing such recommendations one needs to consider: Is eradication feasible? If so, at what cost and in what span of time? Even if eradication is possible will it eliminate the health risk? As pointed out above, other animals are capable of transmitting a number of exotic diseases and domestic pigs can transmit all those carried by feral pigs. If the disease risk is not eliminated is it significantly reduced by eradication of wild pigs and does the benefit from this (and other benefits such as reduction in agricultural damage) exceed the cost of the measures taken? Are there any alternative policies such as partial control of the population of feral pigs that would give higher benefits in relation to costs?

To answer these questions would need a major cost-benefit study involving co-operation between economists and natural scientists. However, there is a case for some quantitative assessment before launching on programmes that can involve heavy economic costs.

Localized or corridor elimination of feral pigs in Australia is likely to be seriously entertained, either as an end in itself or as a step in complete eradication. But any policy along these lines has to take account of the fact that wild pigs are relatively mobile animals and can recolonize areas naturally. Thus for eradication in localized areas to have a permanent effect it would be necessary for there to be geographic barriers to the entry of feral pigs from surrounding regions, or a feral-pig-exclusion fence or barrier would need to be built. In addition, care would need to be taken within the eradication zone to ensure that domestic pigs there do not escape and form new feral pig colonies.

Attempts have been made in the past to eliminate wild mammals or deplete these numbers greatly to reduce the chances of disease or insect invasions. This was, for example, extensively tried in Africa in an attempt to control the spread of the tsetse fly *Glossina morsitans* by depriving the species of natural hosts. In Botswana, for instance, an attempt was made to reduce mammal populations along a corridor abutting the Okavango Swamps. Child, Smith and von Richter,[99] after studying this programme, concluded that 'the hunting programme was not achieving its objectives and that several species were maintaining themselves or increasing in the effective area hunted and that these trends were in general, in accordance with those in surrounding areas.'[100] They also strike a note of caution that is indirectly relevant to the issue of whether to eradicate feral animals in Australia. They point out that

> 'This production [of meat from wildlife shot in control operations], when considered in the light of the ineffectiveness of past control operations, emphasizes the need to be very clear on the cause/effect relationships between tsetse and game, before mounting tsetse control operations ...'[101]

They also go on to point out that the management and harvesting of wildlife populations is sometimes the most economic form of land use. Considerations such as these have led to the suspension of hunting as a means to control tsetse fly in Botswana.

While the problem of controlling feral animals in Australia is not the same as the control of wildlife in Africa, the caution sounded by Child, Smith and von Richter is relevant. Cost-benefit studies need to be done before deciding on feral pig eradication, allowance needs to be made not only for the resources actually used in elimination but also for the value of recreational hunting *forgone* and in some areas, as in Queensland, the loss of the commercial harvest of wild pigs.

CONCLUSIONS

While the feral pig is a carrier of communicable diseases, its role in spreading established diseases in Australia and as a likely vector for the successful introduction and dispersal of exotic diseases can easily be exaggerated. As pointed out above, simple measures can be taken to reduce the health risks from feral pigs. Possibly the public should be better informed about these measures.[102]

As for exotic diseases there are also a number of simple measures that can be taken, as also pointed out above, to reduce this risk. However, the elimination of feral pigs will *not* eliminate these risks since some of the diseases can be carried by other animals and all can be spread by domestic pigs. The extent to which eradication would reduce the risk needs to be studied and weighed against the cost and the alternative strategies (for example, localized eradication or control areas, etc.). In the absence of scientific investigation of these matters there is a danger that the feral pig will become a 'scapegoat', the victim of a costly eradication programme that may do little to reduce the chances of the introduction of an exotic disease such as foot and mouth disease. Compared with wild pig herds in most other parts of the world, Australian feral pigs are relatively disease free. This factor should help to aid the export of this game meat from Australia. While there may be a case for establishing zones in Australia that are free of feral animals, each case needs to be carefully evaluated. There is an urgent need for such evaluation.

CHAPTER 9

Methods and Economics of Control of Wild Pigs

When the wild pig is regarded as a pest - as, for example, it mostly is by Australian agriculturalists - the methods of controlling it and the economics of their use need to be considered. One purpose of this chapter is to outline and compare some of these control methods, concentrating on techniques employed in Australia, most of which could, however, be adopted elsewhere to control pestilent pigs. The chapter concentrates on physical, chemical, environmental and biological means of control. (It does not analyze social mechanisms such as amateur hunting and commercial utilization of wild pigs as control policies; these were discussed in Chapters 3, 4 and 5.) Although a number of control techniques have been considered in the discussions so far, the actual techniques have not been outlined in detail and dealt with as a group.

A further purpose of the chapter is to consider the relevance of simple economic models of pest control to decisions about the control of wild pigs. This is taken up in the second half of the chapter.

PHYSICAL TECHNIQUES OF CONTROL

Just how one should classify the techniques available to control wild pigs is an open question. I have divided these into physical, chemical, environmental and biological but the distinction between these categories is not always clearcut.

Physical techniques of control are either usually intended to kill pigs or to exclude them from particular places. Physical methods of control include shooting, dogging, bowhunting and trapping, and techniques of exclusion such as pig-wire fencing and electric fencing.

Shooting, Dogging, etc.

Shooting is commonly employed in Australia as a pig-control measure, often in conjunction with dogs intended to flush out pigs or to hold them until they can be shot. The shooting of pigs may be:

(a) a *casual* activity undertaken by farmers (for example, some farmers carry a gun on their normal farm activities and shoot any feral pig that is incidentally sighted), or

(b) a *systematic* activity from the landholder's point of view (regular drives may be organized by farmers against wild pigs, frequently in co-operation with neighbours or amateur shooters and using dogs).

The shooting of pigs is a very labour-intensive activity and involves the cost of bullets, the maintenance of firearms, the cost of transport to the hunting area and the expense of keeping dogs where these are used. The economics of the operation is enhanced when labour is provided free, as by amateur shooters (many of whom are experienced), and indeed may meet all costs, as well as pay the landholder a fee for hunting rights, as discussed in Chapter 4. As for the individual farmer, he should reckon the cost of his labour as being equal to the value of other opportunities which he forgoes to hunt pigs. For this reason many agriculturalists limit their participation in pig shooting to periods of slack farm operations.

In general, officers of the New South Wales Department of Agriculture believe that shooting is a relatively ineffective method of controlling feral pigs. For example, Hart says,

> 'This method generally accounts only for small numbers of pigs on the coast, mainly because of the difficult forest and swamp country in which they live. Shooting will often disperse a mob of pigs, making effective control of large numbers difficult. However, some shooters with well-trained packs of dogs have become very efficient at hunting pigs, and have achieved good localized control over feral pigs in some areas.'[1]

In a bulletin for New South Wales' farmers, Hone and O'Grady advise that

> 'Unless shooting is extremely intensive it is only effective on small populations where the pigs are accessible, such as in a region with few watering points. A variety of weapons are used to shoot feral pigs. Large calibre bullets, such as .308, are best at long range but only small calibres or shotguns are needed at short range. Dogs are useful to flush pigs from areas of thick cover. Do not shoot in areas where poisoning or trapping is about to be carried out; the shooting may disperse the pigs and lower bait take.'[2]

Nevertheless, as pointed out earlier (in Chapter 3), amateur shooting does account for a large number of feral pigs in Australia. Furthermore, it is possible for shooting pressures to become sufficiently intense to exert control. For example, in Europe these pressures are so great that many countries have closed seasons on wild pig hunting.

Helicopters are occasionally used to shoot feral pigs in Australia. Their successful use, however, depends upon relatively open country and they are considered to be of greatest value in inaccessible swampy or marshy country. However, even in these areas helicopter-shooting fails to destroy all pigs and costs per pig killed can be high.[3] Because of costs and terrain restrictions, their use is likely to be limited.

While these methods are not used for control purposes, there are of course other ways in which wild pigs can be killed. For example, bowhunting has its enthusiastic amateurs in Australia. In Malaysia beating is sometimes used. As Diong explains: 'This method is traditional in India where bands of men beating drums and gong drive pigs out into the open, to be shot at by men riding on horse-back.'[4] The pigs may also be driven into nets where they are entangled and then killed by the hunters. None of these methods is likely to be used on a wide scale in Australia.

Trapping and Snaring

Trapping is one of the methods of control favoured by the New South Wales Department of Agriculture. The traps recommended for use are in fact (comparatively) easily dismantled pens made of wire mesh with a one-way opening. The wire mesh is secured by wire ties to standard steel posts located on the outside of the pen. The posts are driven in with a post driver or a sledge hammer.

The traps may either be of a square panel form or of a round silo form.[5] A panel trap is illustrated in Figure 9.1. Each of the four *fixed* sides are secured to three steel posts and these panels may also be tied together at the corners. In addition there is a gate or moveable panel that in conjunction with one of the fixed sides forms a funnel. When secured by a spring, tubing or another suitable counterweight this permits the pigs to push their way into the pen, and then closes automatically so that they are trapped. It may be sufficient merely to tie the moveable panel with rope towards the top provided that tension in the wire is sufficient to ensure that the opening closes after a pig pushes its way into the pen. Thus the pen can be constructed by using five pieces of mesh each 2 metres long, nine steel posts, some thick tie-wire and a spring, tubing or a piece of rope. The Department of Agriculture estimated that these materials cost about $60 in 1980.

Pigs have to be attracted to this type of trap by an attractant, and various types of food can be used for this purpose. Wheat, pig pellets, sorghum, corn (maize) and other grains, meat, bread and vegetable wastes (such as apples and lettuce) are just some of the lures that can be used. About 7 kg of grain per night is sufficient. Traps should be set in areas that pigs are known to frequent such as near their pads, feeding points or drinking or wallowing areas. Food can be left out before the trap is constructed to test for the presence of pigs (detected by their hoof marks and the taking of bait) and their attitude to the bait. Once the trap is constructed it is recommended that the gate or entrance be

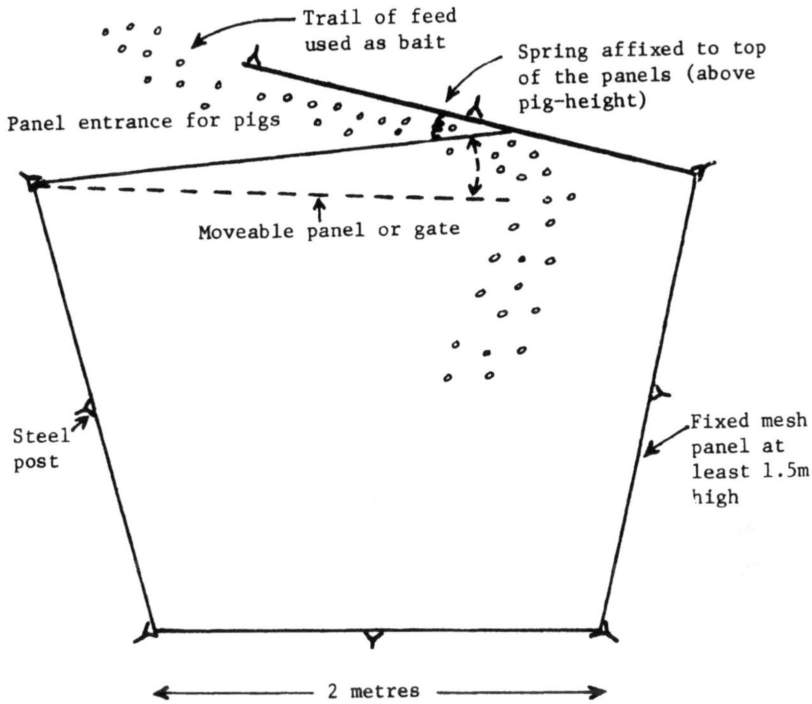

Fig. 9.1 This type of panel trap is commonly used to catch feral pigs in Australia. (Aerial view)

left open for a couple of nights and the pigs be free fed. The bait should trail through the entrance, as indicated in Fig. 9.1, and extend out from the entrance by several metres, so as to increase the chances of the pigs' finding it. Once it is clear that the pigs seem to be taking the bait, set the trap by attaching the spring to the opening panel.

The trap should then be checked daily and any captured pigs destroyed (by shooting, for example). One set of authorities[6] suggest that the dead pigs may be left in the trap where they will act as bait for other pigs the next night (because pigs are cannibals). However, another authority says that 'it has been found that other pigs will not be attracted if pig carcasses are left in a trap.'[7] In any case, it is clear that if the pig carcasses are left outside and near the trap, other pigs may eat them rather than the bait.

Once it is found that no more pigs are being caught in the trap, the entrance should be left open for several nights and possibly a different bait tried. If pigs start taking this bait, the trap can then be set again. The procedure is repeated until it is clear that further pigs are unlikely to be caught or until it is uneconomic to leave the trap in view of the diminishing catch and the potential catch elsewhere. The trap is then moved to another suitable locality.

The main advantage of this type of trapping is that it is comparatively specific to pigs in Australia and poses little environmental hazard, unlike poisoning. Amongst the disadvantages of this control measure are the following:

(a) It is relatively labour-intensive especially since traps have to be visited each day. On large properties and on sites distant from bases, as in the case of some national parks and forests, this can involve considerable travelling time.

(b) Its success depends upon whether pigs are attracted *to* the trap. This can be difficult to achieve. It is practically impossible to attract pigs to a trap when grain is in head or when other, possibly more palatable food is available. It may also be difficult to locate suitable sites for the traps. Increased success using this trapping method is dependent upon the development of improved attractants. There is room for further research here on the types of foods, odours, etc., that may attract pigs. At the Department of Biological Sciences, University of Newcastle, N.S.W., some research has been carried out into sexual attractants (pheronomes) that might entice pigs to the traps but studies in this area have yet to be fully developed.

(c) In soft soils, it is also possible for pigs to dig under this type of trap.

The economics of this trapping procedure might be improved if captured pigs can be used commercially. Furthermore, travelling overheads are likely to be spread and other economies achieved if a number of traps can be operated simultaneously.

Other types of traps are possible. For example, C. H. Diong describes and illustrates a trap that has been successfully used in Malaya.[8] This is a special wire-noose trap.

Portable or semi-portable traps are used for catching wild pigs in the United States. Semi-portable types have been described by Matsche[9] and sectional portable types by Williamson and Pelton.[10] Pigs are attracted to these traps by food such as corn.

Sludskii points out that in the past pit traps were used in the U.S.S.R. to catch pigs.[11] A pit was dug on the pad of wild boars and covered. Sludskii also describes various snares and booby traps that have been used in the past.[12] But none of these can be regarded as safe or effective control methods - for example, guns or bows and arrows set up with trip-wires.

Exclusion by Fencing, Including Electric Fencing

Netting (pig-wire or ringlock) has been used to exclude feral pigs on some

Queensland sugar cane fields, and is sometimes used to enclose domestic pigs in large runs. The wire must be well secured (especially at the bottom which should be on the ground) otherwise pigs root their way under it, as of course they will do in any case if the ground is very soft. Posts need to be comparatively close together and usually a strand of barb wire is added to the bottom of the fence to reduce the likelihood of pigs rooting under the fence. Alternatively the bottom of the netting may be staked into the ground. The fence is relatively expensive to construct but in some circumstances is economic in reducing damage to high-value crops. In light or soft soils, however, it may not be effective because pigs may root under it.

Electric fences are being increasingly used or considered in Australia as means to exclude feral pigs from lambing paddocks and from crops.[13] Fences may be especially built with this in mind or existing fences may be modified for this purpose. It is important that the construction of the fence be such that the pig is restrained by the fence while in contact with a live wire and an earth wire (or earth), otherwise if it is halfway under the fence and gets a shock it may merely continue through.

For new fences one possible construction, considered as suitable by the New South Wales Department of Agriculture, is shown in Figure 9.2. The fence restrains both cattle and sheep and given the addition of the earth wire at 2-5 cm above ground level is also expected to be effective in excluding pigs. If hardwood posts are used, insulators are not necessary in the construction of such fences, the power for which may be supplied by a power energizer operated from electricity mains. But solar units to power energizers are also available[14] as are batteries. The cost of materials for such a fence was estimated in 1980 by the New South Wales Department of Agriculture to be $230-250 per kilometre including the costs of the energizer.[15] The extra cost of adding an additional wire at 2-5 cm height to help exclude pigs is slight.

According to Plant, it is difficult to satisfactorily electrify existing fences.[16] Suggestions have, however, been made for dealing with the problem of existing fences. Pharaoh, for example, suggests that a live wire could be set outside 15 cm from the existing fence on small posts at a height of 5-20 cm. In this case a low earth wire may need to be added to the existing fence or to the small outside posts to ensure that the progress of the pig is impeded and effective earthing of pigs occurs on contact with the live wire.[17] Alternatively the live wire may be offset from the existing fence. However, arrangements of this kind may work best if the existing fence is of ringlock or of pigwire because these fences cause the pig to stop.

While electric fencing can be effective and economically worthwhile in some

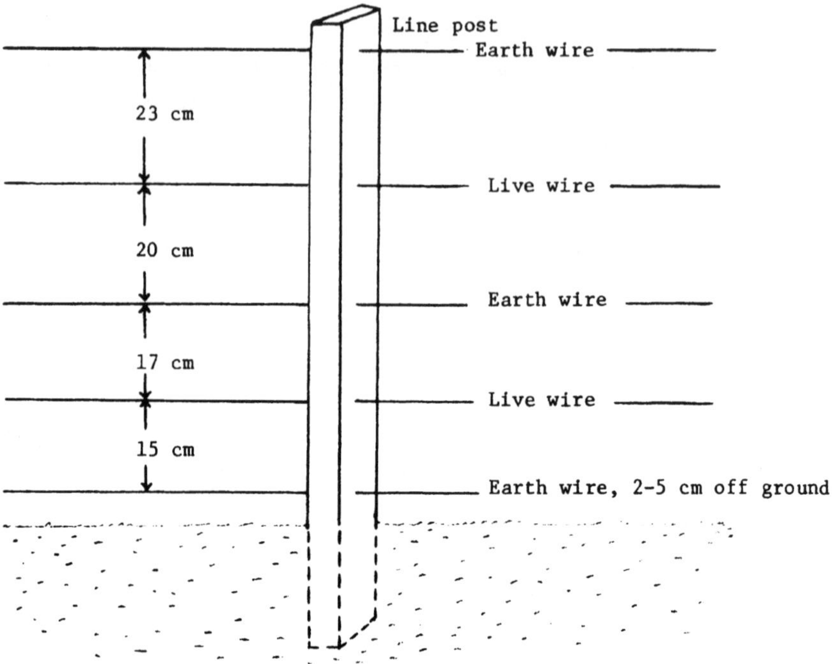

Fig. 9.2 A cattle and sheep fence modified by the addition of an earth wire 2-5 cm off the ground to help exclude feral pigs by ensuring contact of the pig with a live wire and an earth wire.

circumstances, it is not without its limitations. These include:

(a) The fence may be broken by animals such as kangaroos and emus and some pigs are liable to get through in any case.

(b) The effective earthing of the fence can be difficult to achieve in some dry areas of Australia.

(c) Grass and undergrowth must be kept away from the fence to stop it from shorting out.

(d) Care must be taken not to fence feral pigs in, as may occur if a large area including scrub or swamp is enclosed.

Nevertheless where a valuable crop is being grown or for lambing paddocks, electric fencing can prove to be economically worthwhile, even if it does not offer complete protection from feral pigs.

CHEMICAL TECHNIQUES OF CONTROL

The main form of chemical control used at present is the poisoning of food left out for feral pigs. Indeed poisoning using sodium monofluoroacetate (1080) followed up by other control measures is a means favoured by some officers of the New South Wales Department of Agriculture. Although other available poisons may be used in food, usually they are not recommended for use because of their serious side-effects on other species. In the United States experiments have been taking place to develop chemical sterilants for wild pigs. Let us consider these control techniques in turn.

Sodium Monofluoroacetate (1080) Poisoning

The compound sodium monofluoroacetate (1080) is in its pure form a white, odourless and practically tasteless powder. But the commercial product may be coloured, may not be quite pure and hence may have a slight 'vinegarish' smell.[18] The salt occurs naturally in the South African weed 'Gifblaar' (*Dichapetalum cymosum*) which poisons stock if eaten.[19]

Originally it was developed during World War II in the United States as a rodenticide.[20] In the 1950s it was also found to be an effective poison in Australia and New Zealand for rabbits.[21] Since then it has also been used to poison other mammalian pests such as opposums in New Zealand, dingoes and, in recent years, feral pigs in Australia. It is an extremely toxic substance, the lethal dose varying with the type and size of the animal. Its toxicity to other animals will be discussed later.

The poison (given to animals with food-bait) interferes with the use of citric acid in the body of the victim. As a result of poisoning:

> 'Carnivores generally show central nervous symptoms and die in convulsions. In herbivorous animals, however, the symptomatology is different and the clinical picture is that of cardiac failure with ventricular fibrillation as the terminal event. As would be expected, of course, mixtures of the two clinical patterns are seen. In the human subject symptoms may occur after a latent period which is stated to vary from a few minutes to several hours. Usually they commence with vomiting, excitability, convulsions, and irregularity of the heart. Exhaustion and cardiac and respiratory failure are the terminal events.'[22]

It has been said by the New South Wales Department of Agriculture that

> '1080 is a relatively useful poison in dealing with large populations of noxious animals as there is no taste or smell, it is slow acting, painless and economical compared to other substances.
>
> As the properties of the baits are readily eaten and symptoms occur from 30 minutes to several hours later, animals do not associate the illness with

the bait or disturb other animals still feeding on the bait.

For these reasons wild pigs do not become bait shy and if used properly has a kill rate of more than 80 per cent of the population in an initial effort.

No other forms of control give this high success in one operation.'[23]

The poisoning procedure for wild pigs is usually as follows: Pigs are free fed on bait (unpoisoned for 3-7 days). Grain or pellets (pig pellets or bran-pollard pellets) are usually used as bait. The use of meat in New South Wales for bait is illegal. Pig numbers are estimated from the free feed eaten and pig tracks, and an appropriate amount of poisoned bait is put out in the evening. This may be laid in furrow to help attract pigs (see Figure 9.3). All domestic stock (especially dogs) must be kept away from the poisoned bait. The next morning the site is checked and any dead pigs and remaining bait buried or burnt.

In New South Wales landholders using 1080 poison have to comply with a number of requirements. They must inform all landholders within an 8-kilometre radius

Fig. 9.3 Furrow made to receive bait to poison feral pigs with 1080.

seven days before using it, and must place warning signs on all access points to their property and the paddock in which it is laid. Furthermore, the poisoned bait is only available from a Pastures Protection Board, where a trained officer mixes the poison with the bait.

From an environmental view the use of 1080 has some advantages. Unlike DDT it does not accumulate in the food chain and it breaks down in the soil to harmless substances as a result of fungal and bacterial action. However, the substance is extremely poisonous to dogs (including dingoes), and they may die if they eat the carcass of a pig that has been recently poisoned by 1080. The compound decomposes in the soil. But under dry conditions and in winter temperatures if baits remain intact, the poison can remain potent for a long period, beyond 4 weeks, for example. Rain, however, leaches the poison from the bait and higher air temperatures assist the speed of decomposition of the poison. It is reported that under normal conditions in New Zealand about half of the poison is leached out of the bait in each successive fortnightly period.[24] Rapid leaching and decomposition of the poison might be expected on the relatively wet eastern coast of Australia (decomposition being more rapid in the warmer north) with slower degradation of the poison occurring in the drier interior of the country.

As mentioned earlier, the toxicity of the poison to various animals varies. Furthermore there is always the problem that non-target species may be killed by poisoning.

Table 9.1 indicates the approximate toxicity of sodium monofluoroacetate to selected types of animals. By these standards, dogs are very susceptible; rabbits, pigs and humans and some marsupials reasonably susceptible; and birds, lizards and frogs relatively resistant to the poison. Fish are also reported to be relatively resistant to it. However, since the food consumption of birds may be high in relation to their body weight, incidental poisoning of birds by 1080 is not uncommon.[25] Consequently, research has concentrated on ways to make bait less attractive to non-target species such as birds. For example, it has been found that if the bait is dyed dark green that it is less likely to attract birds. It is also advisable to cover any bait left out by day (with tree branches, for example) so that birds are less likely to consume the bait. The type of food used as bait also influences the risk to non-target species. For example, if meat is used as bait, dingoes are susceptible and scavengers including scavenging birds and possibly reptiles could be at risk.

At various times conservationists, hunters and some farmers in Australia have expressed concern about the use of 1080 because of its spillover or side-effects on non-target species. In the light of this, an assessment by an officer of the National Parks and Wildlife Service of New South Wales may be pertinent. He has

TABLE 9.1 Approximate Minimum Lethal Doses of Sodium Monofluoroacetate (1080) (in milligrams per kilogram of body weight)

Animal	mg/kg	Animal	mg/kg
Dogs	0.1	Marsupials	1-10
Rabbits	0.6	Birds	4-20
Humans	0.7	Lizards	100
Pigs	0.8	Frogs	500

Source: Figures reported by J. R. Giles at Regional Wild Pig Advisory Council Seminar on Wild Pigs, Tamworth, March 1979, and based on CSIRO research.

said,

> 'Principal advantages of the poison are its high toxicity to some vertebrate pest species, and its rapid rate of breakdown in the field. Dogs are highly susceptible to "1080" (minimum lethal dose 0.1 mg/kg [body weight]), rabbits and pigs are quite susceptible (minimum lethal dose 0.6-0.9 mg/kg), whereas birds, most marsupials, reptiles, amphibians and fish are more resistant. This variation in susceptibility of species to the poison, combined with selection of suitable baits and knowledge of the ecology of target and non-target species makes "1080" one of the most useful and environmentally safe aids available for control of some vertebrate pests, when used properly and in carefully constituted control programmes.'[26]

Despite the apparent relative superiority of 1080 poison to known alternative poisons, a cautious attitude still remains as far as the effect of 1080 on non-target species is concerned. Batcheler suggests that 'many studies have yet to be undertaken to pin down the importance of the deaths to survival of the [non-target] species, particularly to the rarer ones which may not be able to sustain an extra nudge beyond normal winter mortality, before they are headed for local extinction.'[27] Batcheler suggests that birds, apart from being killed as a result of directly eating bait, may be poisoned by eating insects that have fed on the bait.

Yet one cannot judge the impact on non-target species of a poisoning programme for pigs merely by the numbers of non-target animals killed. The whole ecological balance needs to be considered. To the extent that pigs are competitive with non-target species, the benefits to the species of a reduced pig population may more than outweigh deaths caused by incidental poisoning of some non-target animals. More research into the side-effects of 1080 poisoning programmes on non-target species is needed.

Other Poisons

Other poisons have been used to kill pigs. These include yellow phosphorus, organophosphate pesticide poisons, arsenic trioxide, potassium cyanide and strychnine alkaloid.

In Australia, yellow phosphorus was sold as a pest destroyer under the registered trade name 'S.A.P. Rabbit, Pig and Crow Poison.'[28] Phosphorus was also used to control feral pigs in New Zealand in the 1950s, a phosphorus-carbon bisulphide mixture (supplied by the Department of Agriculture) was distributed in carcass baits.[29] These phosphorus poisons have undesirable side-effects on non-target species. Birds are particularly susceptible.

Organophosphate poisons have also been tried in the past but their use in New South Wales is illegal. However, Benson after his survey of farmers in New South Wales, commented that 'The poison 1080 was seen by those who were poisoning but not using 1080 as being disadvantaged by its inconvenience against other poisons, especially S.A.P. Since S.A.P. went off the market there has probably been a big increase in the use of organophosphate poisons such as Lucijet and Phosrin.'[30] Lucijet is the registered trade-name for fenthion-ethyl, a concentrate added to sheep dips. Poisons of this nature are available from the shelf, but 1080 is available only from Pastures Protection Boards and users are required to take special precautions.

Organophosphate poisons are a threat to wildlife. Benson emphasizes this in the following way: 'These poisons do a great deal of damage to wildlife, Lucijet in particular used in the undiluted form on carcasses was noted by many landholders to have killed birds with their claws still in the carcass and some have discontinued its use.'[31]

As for the other poisons mentioned above: Arsenic trioxide was used in New Zealand as a pig poison and in the mid-1950s it was preferred to phosphorus because of the inflammability of the latter substance.[32] Strychnine is used occasionally for poisoning pigs in New South Wales but its use is legally restricted. It is illegal in New South Wales to use potassium cyanide for poisoning noxious animals[33] because of the risk posed to wildlife.

A problem with all poisons is that they are usually not confined in their effects to the target species. Research continues in trying to make poisoning programmes more specific to wild pigs - for example, by variations in the bait used, and by choosing poisons such as 1080 restricted in the range of animals to which toxicity is high.

Chemical Sterilants

The use of chemicals to break the reproductive cycle of mammalian pests is appealing but not without its difficulties. The chemical has to be administered to the animal and this normally means the employment of an oral contraceptive. In turn this means that bait containing the contraceptive substance must be taken and presumably taken on a relatively regular basis.

Smith and Diong are evaluating this as one possible method of pig control in a national park in Hawaii. They are, however, cautious about the prospects of the control method. They point out that

> 'The effectiveness of chemical sterilants in field situations will be reviewed since this approach is ostensibly a very humane way of dealing with the problem. However, the impact of these chemicals on all life-forms would need to be assessed because the substances could be released into the streams and ground-water systems in the [Kipahulu] Valley. Since the State of Hawaii still retains certain water rights within the valley, the impact of such a control program on their rights would also have to be evaluated.'[34]

ENVIRONMENTAL AND BIOLOGICAL MEANS OF CONTROL

Environmental and biological methods of controlling wild pigs include destruction of their habitat, denying them access to water, the encouragement of natural predators of pigs (or at least evaluating the impact of the destruction of such predators), the possible introduction of an exotic disease specific to pigs and alterations in the composition of farm production so that commodities of less attractiveness to pigs are produced. Let us consider each of these possibilities in turn.

Habitat Change and Denial of Water

The survival of pigs in some hot areas of Australia depends upon their access to shade and shelter - provided by such shrubs as lignum *Muehlenbeckia cunninghamii*. The clearing of shrub cover reduces the chances of survival of wild pigs - they cannot escape the heat and they become relatively easy targets for hunters. Pigs also need regular access to water. In some of the drier parts of Australia the number of watering points are limited. If pigs can be denied access to these for one or two weeks in summer their numbers can be considerably reduced.[35] However, both strategies can have an undesirable impact on wildlife. Furthermore clearing of shrub or cover can be costly and it may add to erosion risks. Hone, O'Grady and Pedersen express the view that although habitat destruction is a relatively minor method of control of wild pigs at present, 'it offers considerable potential as careful changes in the pig's habitat could bring about a permanent reduction in feral pig numbers. The habitat components that can be changed are vegetation and water supply.'[36]

Natural Predators

Where these predators continue to exist, the most important predators of wild pigs are wolves, tigers and leopards. According to Sludskii, the chief natural predator of wild pigs in the U.S.S.R. today is the wolf and this is capable of exerting considerable control on pig populations.[37] In the past when tigers were more widely present in the U.S.S.R., they also intensively hunted wild boar and continue to do so today in the Soviet Far East. In Asian countries where it is present, the tiger continues to be an important predator on wild boar. The leopard *Felix pardus* is also mentioned by Sludskii as hunting wild boar in the U.S.S.R. but because its numbers are so depleted there it has little influence today on wild boar populations.[38] In South Africa the leopard is regarded as an important predator of the bushpig *Potamochoerus porcus* and increases in the population of bushpigs have been attributed to a decline in the population of leopards.[39]

In Australia, there are no large natural predators of pigs. However, dingoes and wild dogs may play a useful role in limiting the numbers of wild pigs. Opinions about the importance of dingoes as predators of feral pigs differ but some foresters and wildlife officers, as pointed out in Chapters 6 and 7, believe that dingoes are important predators of pigs and that policies to reduce the population of dingoes has resulted in increased feral pig populations in Australia. However, no scientific investigation appears to have been done of the relationship between the two species.

Judging from reports by Sludskii for the U.S.S.R., it is likely that the fox *Vulpes vulpes* is a minor predator of pigs in Australia. Predation is likely to be confined to piglets or weak young animals. Sludskii believes their impact on pig populations is negligible.[40]

It is also not unlikely that the Australian wedge-tailed eagle *Aquila audax* and feral cats *Felis catus* take a few piglets. The dingo *Canis familiaris dingo* and feral dogs are likely to be the most important predators of wild pigs in Australia but their impact on pig populations has not been scientifically studied.

Predator-prey relationships raise many interesting questions, especially when *both the predator and the prey are considered to be pests*. Reducing the population of a predator-pest (such as the dingo) may result in an increase in the population of a prey-pest (such as the wild pig) and the economic costs of an increase in this population *may* be greater than the economic advantages reaped from a reduction in the population of the predator-pest. The optimal economic balance between these two populations can be determined in theory, but insufficient empirical scientific evidence is available at present to apply the theory to the relationship between numbers of dingoes and the feral pig population. However, sight should not be lost of ecological interdependence.

Disease Introduction

The introduction of an exotic disease to control feral pigs in Australia is unlikely to be seriously entertained because it could also devastate the domestic pig industry. However, swine fever, and particularly African swine fever (see the previous chapter), could be expected to reduce feral pig numbers sharply if introduced to Australia. Nevertheless these diseases would not be certain to eliminate feral pig populations. While Australian authorities are unlikely to use swine fever to control feral pigs, Pakistan government authorities have spread swine fever amongst wild boar to reduce populations there.[41]

Adjustment of Economic Activity

As mentioned in Chapter 2, one environmental means of coping with wild pigs is to change land use so that the produce of the land is less palatable to pigs. For example, a farmer may change from sheep-raising to beef-production and grow crops or varieties of crops that are known to be less attractive to wild pigs. From the farmer's point of view, as explained in Chapter 2, this can be more economical than not adjusting and attempting to control feral pig numbers. However, it may not be optimal for all to adjust and even after adjustment of economic activity, control of feral pig numbers may be economical.

GENERAL DISCUSSION OF AVAILABLE CONTROL TECHNIQUES AND STRATEGIES

Usually a combination of methods is used to control pigs. A programme favoured by some officers of the New South Wales Department of Agriculture is poisoning of pigs, 'followed by repoisoning or trapping in 1 to 2 months if feral pigs are still present in reasonable numbers. It is essential for the initial poisoning to be followed by further control efforts. If the initial work is satisfactory, control efforts should be concentrated around watering points or areas of thick vegetation. Trapping or shooting is the usual follow-up.'[42] When trapping rather than poisoning is used as the basic control method, shooting (and dogging) is the usual follow-up method. In the absence of follow-up control, pig populations are likely to quickly return to carrying capacity levels. In Australia feral pig populations can exhibit an increase of 60 per cent per year or more.[43] Depending upon the population of the pig population removed, populations can easily return to pre-existing levels within a year or two if there is no follow-up control. (This is ignoring the possibility of immigration of wild pigs.) Hone, O'Grady and Pedersen suggest that 60 per cent of pigs in a population must be removed throughout a year (on a continuous basis) to keep the population stationary.[44] Giles indicates from his research that if 70 per cent of wild pigs are removed in one operation and if no further controls are undertaken, the population (in the absence of changed conditions and immigration) usually returns to its pre-existing level within two years. He points out that in Australia female wild pigs can breed throughout the year and a population can more than double in size in a year.[45]

It has been argued that if a wild pig population is to be controlled, it is desirable to reduce its population *substantially* in the first control operation (possibly to the ultimately desired control level) and then hold it there by serially repeating follow-up measures (that is, by continual mopping-up of the population growth at the new population level).[46] If a substantial reduction in the population is *not* achieved and mopping up of subsequent population growth is to be the rule, the rate of increase of the feral pig population may be a maximum sustainable one (or nearly so). Consequently to hold the population level, a greater number of pigs need to be removed on a continuous basis than would be the case if population were lower. This is illustrated by Figure 9.4. There the curve OABC shows the relationship (hypothetical) between the rate of increase of a wild pig population and the number of wild pigs. The maximum sustainable level of the feral pig population (the ecological carrying capacity of an area) is P_E. The feral pig population increases at its fastest rate (gives maximum sustainable yield) for a pig population of P_M, which in the case shown is half the ecological carrying capacity. If the population P_E of feral pigs is reduced to approximately P_M (by about half) the amount of mopping-up of pigs required to sustain this population will be much greater than if the population is reduced to say P_D, about a quarter of P_E.

However, it cannot be assumed that the most economical policy is to reduce pest populations to levels that give a low yield. It depends, for one thing, on the cost of reducing the population level and this may rise steeply the greater is the attempted reduction in the population of a pest. Furthermore the cost of

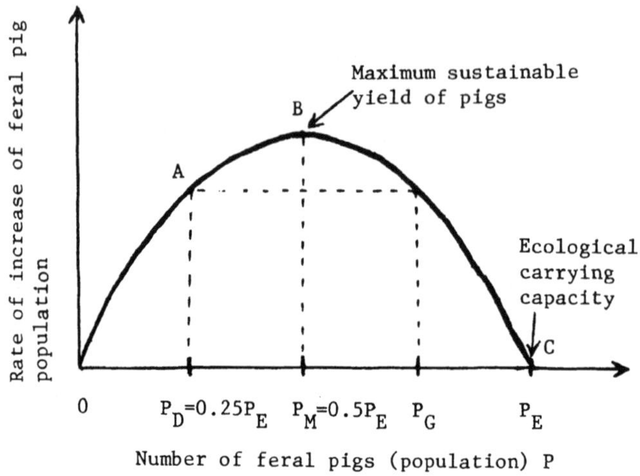

Fig. 9.4 The rate of increase of wild pig populations is likely to depend on the density of wild pigs. Assuming the usual (quadratic) relationship shown above, the number of pigs that need to be continuously 'cropped' to keep a feral pig population at a control level depends on the population level after control.

holding a population at a particular level does not depend merely on the off-take needed to keep the population stationary. Thus the cost of maintaining a pest population at $P_D = 0.25P_E$ is likely to be higher than keeping it at $P_G = 0.75\ P_E$ even though the off-take needed to keep each population level stationary is the same. In the latter case the density of the population is greater and thus less control activity (fewer traps, less poisoning, etc.) is likely to be needed for the same catch. This is not only because of the likely proximity of more pigs to any trap or control device but also because there is more competition amongst them for resources, for instance, for food. I shall return to consider the most economical policy later.

However, it should be noted that there is a whole array of possible control strategies. Apart from the strategy of reducing a pest population to some level and holding it there by continual 'mopping-up', an alternative is to engage in non-continuous control. Control action may be on a large scale but be confined to periods well spaced in time. Other alternative patterns can also be evaluated.

Furthermore, there are other complications to consider: the efficacy of controls depends on the age of animals removed and is influenced by changing ecological conditions. If older animals are removed, these might have died anyway, so control measures are therefore of little effect. Similarly, if a drought period is commencing it is possible that animals destroyed during its onset would have died anyway.

In Australia, feral pig populations can fluctuate greatly because of variations in seasonal conditions. Drought, in particular, can reduce their numbers dramatically. This raises the question of whether control programmes can be made more effective by taking advantage of natural fluctuations. During drought, for instance, feral pig populations are likely to be especially vulnerable and it is sometimes argued that this is the ideal time to step up control operations because recovery of pig populations after the drought will then be slowed. It could well be that this is an ideal time in which to concentrate control efforts, but it must be remembered that the economics of the control depends upon more than the reduction achieved in population levels. It depends upon the amount of damage produced and the *value* of production lost as well as control costs. In order to make further progress with this problem, we need to consider some simple economic models of pest control.

ECONOMICS OF PEST CONTROL IN RELATION TO WILD PIGS

While the benefits and costs from any set of measures to control feral pigs can as a rule be determined and compared to decide whether or not that set of measures is likely to be profitable (and such calculations have in fact been done by the New South Wales Department of Agriculture for a number of situations[47]), analysis and discussion of control policies are facilitated by considering models of pest control. The purpose of this section is to introduce some of these models. The simplified Headley model[48] is outlined first and then some implications and modifications of it are discussed. Modifications are made to it - for example, to allow for the influence on control of the mobility and fecundity of a pest. The private profitability of control of a pest population is likely to be less the more mobile is the pest (other things equal) and the greater is its fecundity. The social profitability of control is also likely to vary inversely with the fecundity of the species, as is explained below. Since the explanation of these models involve technicalities, some readers may wish either to skim this section for main points or turn directly to the concluding section of this chapter.

Headley's Pest Control Model

Headley's model of the economics of pest control is a very simplified one.[49] He assumes that the amount of damage caused by a pest population increases at an increasing rate with the population of the pest, once the pest population exceeds some particular threshold. Indeed, he represents the amount of damage caused by a pest population by the positive values of the accelerating branch of the quadratic function

$$D_t = bP_t^2 - A \tag{9.1}$$

where D_t is the amount of damage in t, P_t represents the level of the pest population (for example, a wild pig population) in t, and A is a constant which defines the damage tolerance level. This is Headley's damage function. The constant A may indicate the maximum amount of damage which may occur before it is of any significance for production.

Examples of this damage function are illustrated in Figure 9.5. For instance, the curve OHJ is a case in which A = 0 and b = 1. The curve KLN is a damage function in which A = OK and b = 1. In the first case damage occurs when a pest population exists but in the second case damage occurs only if the pest population exceeds $\hat{P}_t$ in size. Because of this threshold factor, the damage function in reality is $O\hat{P}_tLN$, if negative damage values are assumed to be impossible. One

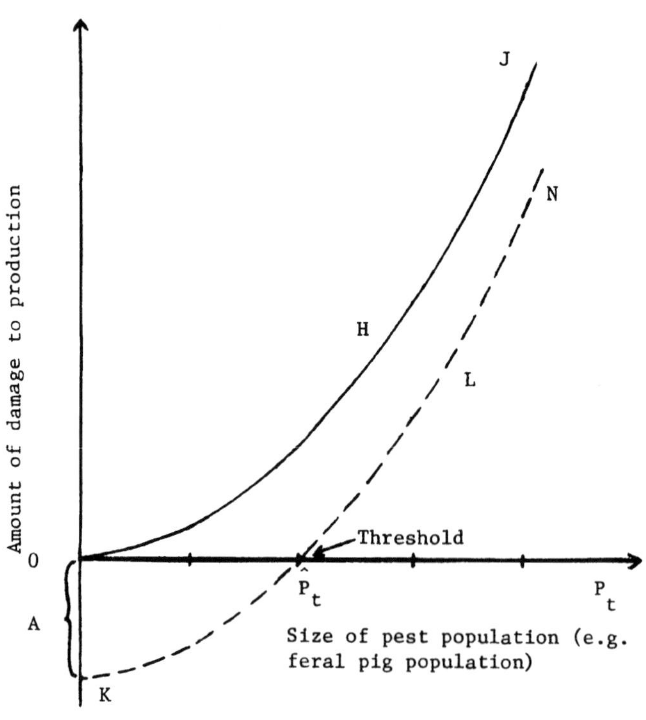

Fig. 9.5 Two examples of Headley's damage function. In the case where the function is represented by KLN, no economic damage occurs unless the est population exceeds a threshold, $\hat{P}_t$.

problem with Headley's model is that he does not indicate how damages are to be measured.

The level of production in this model is represented by

$$\left.\begin{aligned} x_t &= K - cD_t && \text{for } D_t > 0 \\ &= K && \text{for } D_t \leq 0 \end{aligned}\right\} \qquad (9.2)$$

where x_t is the output in period t of the product under consideration, K is a constant indicating the level of output when no effective damage occurs, and c is a parameter.

Letting $\hat{P}_t$ represent the threshold level of pest population at which damage becomes effective, function (9.1) can be used to express relation (9.2) in terms of the level of pest population at time t. The production function becomes

$$\left.\begin{aligned} x_t &= K - c(bP_t^2 - A) && \text{for } P_t > \hat{P}_t \\ &= K && \text{for } P_t \leq \hat{P}_t \end{aligned}\right\} \qquad (9.3)$$

This yields a production function of the form shown in Figure 9.6 by curve RST. The hatched area represents the reduction in output caused by the pest. Assuming that the productive unit under consideration (for example, farm or region[50]) does not influence the price of the final product, the total revenue function for the productive unit expressed as a function of P_t is of the same form as the production function.

'The final element in this model is a function that relates control costs to population - that is, which tells the cost of control to maintain a population of a certain size. Assume that this cost relation is continuous and decreases as the population is allowed to be larger and that the cost decreases at a decreasing rate, which is in keeping with the findings of entomologists that the cost of reducing a population increases at an increasing rate.'[51] Although this may not *always* be the relevant relationship and an alternative is considered later, let us explore some implications of Headley's model. He assumes specifically that the cost of population is a hyperbolic function of the following type:[52]

$$C = \frac{\theta}{P_t} \qquad (9.4)$$

where C represents the cost of varying and holding population to level P_t and θ is a constant.

Given Headley's assumptions, the marginal revenue foregone as a result of an increased pest population can be represented by a curve like $O\hat{P}_tC$ in Figure 9.7, and the marginal cost of reducing the pest population below its equilibrium or natural level, P_E, is like that shown by curve HJ. Given the revenue and cost conditions expressed by these curves, the profit-maximizing level of the pest population is $\bar{P}_t$, the level at which the appropriate marginal conditions are satisfied. At this level, the marginal cost (extra cost) of reducing the pest population

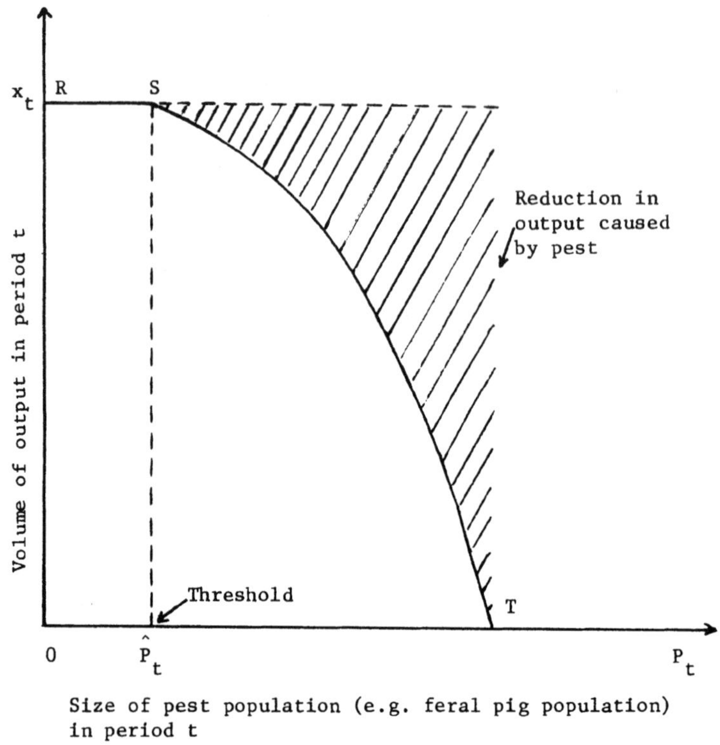

Fig. 9.6 Headley assumes a simple relationship between the loss in production caused by a pest and its population level. The resulting production function is like RST in the above figure.

is just equal to the marginal gain (extra revenue) obtained by reducing the population.

The implications of the Headley model include the following:

(a) If elimination of a pest is not profitable but some control is profitable, profit is maximized by varying the pest population to a level *above* the threshold $\hat{P}_t$ that causes damage. This is so if additional costs must be incurred to reduce the pest population.[53]

The profit-maximizing policy is likely to be one of allowing a pest population that causes some economic damage. This is analogous to the proposition that it is frequently not economical to eliminate all pollution that causes social damage.[54]

(b) The greater the price received for the product subject to damage by the pest the more profitable it is to reduce the pest population. Conversely the lower

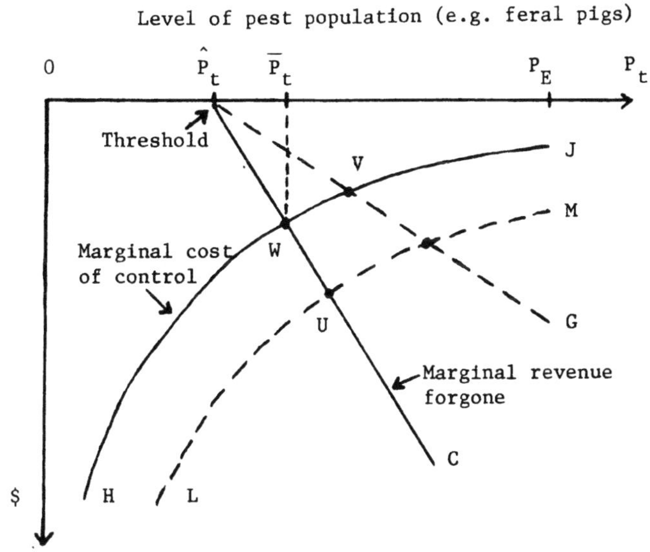

Fig. 9.7 In the Headley model the optimal population size (profit-maximizing one) of a pest after pest control is that for which the marginal revenue forgone as a result of an increase in the pest population is equal to the marginal cost of reducing that population. In the above case this implies a pest population of $\bar{P}_t$ so the population is reduced to the level from its pre-existing level P_E.

the price received for the product, the less profitable is control of the pest and the more likely is it optimal to allow a greater pest population.[55] Thus in Figure 9.7, a lowering of the price of product x causes the marginal revenue curve from control to shift from $O\hat{P}_tC$ to $O\hat{P}_tG$. Whereas in the first instance it was profitable to reduce the pest population to $\bar{P}_t$, the reduction in the product's price means that it is only profitable to reduce the population to the level corresponding to the point V. Thus, other things equal, one might expect greater control to be exerted on wild pig populations during times of high prices for production (boom times) and in regions where the value of the product is high. Yet, as discussed in Chapter 4, this is not always certain to be so in practice.[56]

(c) Other things equal, the greater the (marginal) loss in production caused by each member of the pest's population (the more destructive it is) the greater can be expected to be the profit-incentive to reduce its population, and conversely. When the destructiveness of the pest is greater (and the threshold unaltered) the effect is (as can be seen from the equation given in note 55) to cause the marginal revenue curve to decline more steeply. Thus in Figure 9.7 an increase in the destructiveness of a pest might shift the relevant portion of the marginal revenue curve from $\hat{P}_t G$ to $\hat{P}_t C$, thus lowering the level of the pest population to be tolerated from that corresponding to V to that corresponding to W. If at the same time, the threshold $O\hat{P}_t$ is reduced (as is likely), this reinforces the tendency. It follows that in regions or on farms where the marginal loss in production due to pigs is greatest more control of the population is optimal than where less destruction occurs, other things equal.

This observation and that in (b) can be combined in a familiar economic proposition. Together they imply that greater control of a pest population (feral pig population) is profitable the greater is the *value of the marginal product*[57] from such control, other factors assumed constant.

(d) The greater is the marginal cost of reducing a pest population, the smaller is likely to be the profit-maximizing reduction in the pest population. The converse, of course, also holds. In Figure 9.7, for example, an increase in the marginal cost of control from that indicated by curve HWJ to that shown by curve LUM, implies that if the marginal revenue curve is $\hat{P}_t C$, a pest population corresponding to U rather than W maximizes profit.

Limitations of Headley's Model

While Headley's model provides a useful introduction to this difficult subject, it does have a number of limitations that need pointing out. Headley assumes that the level of production aimed for by the farm (farms) x_t is constant and not subject to influence by the presence of the pest. The farmer's attempted supply of the product is completely inelastic and any reduction in that supply is merely due to the depredations of the pest. In turn, this suggests that the resources allocated to the production of product are constant (a profit-maximizing purely competitive firm may do this if its marginal cost curve of production is of a reversed L-shape). It is of course an empirical matter to determine the extent to which this assumption is satisfied. However, theoretically, it might be more satisfying to develop a model that includes the pest population as a partially controllable input in the farmer's production function with a *negative* impact on output. For the sake of simplicity, however, it is worthwhile developing Headley's model further.

Headley does not allow for the mobility and fecundity of pests in his model. Mobility of a pest between farms (or between regions) may increase the marginal cost to a farmer (or region) and/or reduce the revenue from control when compared to the absence of mobility. This reduces the profitability of control from the point of view of an individual farmer or region. Similarly, other things equal, the greater the fecundity of the pest the smaller the most profitable reduction in its population because the additional costs of reducing the pest population and holding it at the reduced level are likely to be higher. Let us consider these matters using a more general form of Headley's model.

A More General Pest Control Model

In order to put the Headley model in a more general form, assume that the population of a pest is of a size P_E. It is thus in its natural or sustainable equilibrium. Feral pigs in parts of Australia have possibly achieved this status. Assume as Giles[58] does for feral pigs that the control strategy consists of reducing the basic population from its natural level to a lower one and adopting measures to hold it there - that is, to hold the population in a steady state. Assuming a planning period of any specified length, the discounted costs of these control measures and the discounted gains or benefits for the planning period can be computed in principle and the corresponding marginal curves can be derived.[59]

Typically the marginal discounted cost curve of population reduction (and measures to maintain the new population) might be as indicated by curve OAB in Figure 9.8. The discounted marginal benefit curve for the planning period might be like curve CGD. These curves are of the same nature as assumed by Headley except that now they are related directly to the reduction, Z, achieved in the pest population. Assuming that these curves refer to private benefits and costs, private net benefit from reducing the pest population is maximized in Figure 9.8 when the pest population is lowered by Z_2 and the remaining population is $P_E - Z_2$.

The amount of reduction in the pest population required to maximize net benefits is sensitive to shifts in the discounted marginal benefit curve and to shifts in the discounted marginal cost of control curve. If the marginal benefit curve shifts downwards, the optimal reduction (that needed to maximize net benefit) of the pest population declines, other things equal. Similarly, this occurs if the discounted marginal cost curve of control shifts upwards. Using these relationships, consider the effect of the mobility and fecundity of a pest on its optimal control.

Mobility of a Pest (Such as the Wild Pig) and its Optimal Control

Wild pigs, like many pests, are extremely mobile.[60] They have a considerable daily range. Under stationary environmental conditions they usually remain in a large home range. However, seasonal migration occurs in some regions (for example, from high areas to lower regions in the winter period, with return to higher areas in the summer) and permanent migration over long distances can occur, especially when

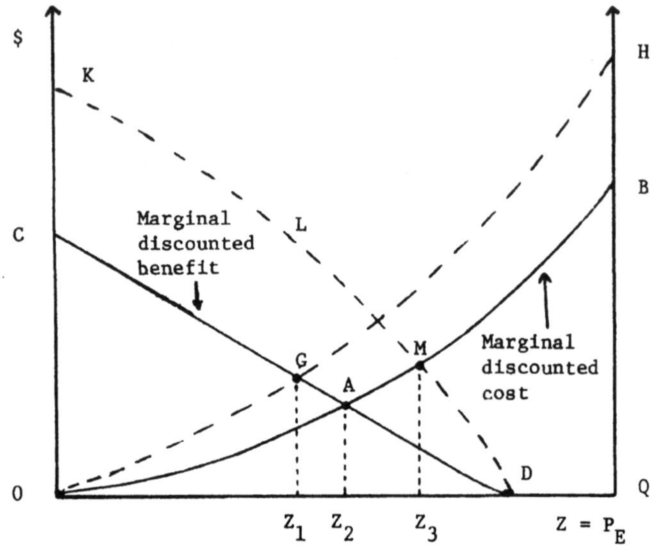

Fig. 9.8 Given the marginal discounted benefit and marginal discounted cost curves shown by the solid lines above, discounted net benefit is maximized when the pest population is reduced by Z_2, permanently.

wild pigs are under environmental stress.

In Australia, the Working Party on Feral Pigs of the Australian Standing Committee on Agriculture reported that under normal conditions, adult male feral pigs have a home range of 10–50 km^2, dry sows a range of 5–20 km^2 and lactating sows ranges of 5 km^2. The Committee reports that 'When a population has an adequate supply of food and water in a survival area, it is unlikely that the animals will travel more than 5 km outside this area except when there are mature crops in the vicinity. In the grain belt of New South Wales there are often decimal forays of up to 10 km from bedding areas to crops which can be an important source of food.' However, migration and dispersal of pigs under stress conditions can occur. Evidence on the spread of pigs in New South Wales indicates that a pig population can extend its range by 50 km per year. Feral pig populations have been known to migrate up to 20 km in 48 hours when suffering from food shortages.[61] Hone, O'Grady and Pedersen point out, 'Movement patterns after control work are also important because feral pigs may move into areas where control work has been undertaken. The rate of immigration will depend on many factors ... The first feral pigs seen moving into an area are usually adult males which is to be expected because of their larger home range.'[62]

Experience with movements of the wild boar in the Soviet Union is similar to

Australian experience as far as diurnal movement of the feral pig is concerned, but there is more evidence from the U.S.S.R. about migration of wild pigs. Sludskii points out that 'during feeding wild boar cover diurnally distances of up to 9 km, sometimes swimming across large rivers every day'[63] but that to escape unfavourable environmental changes (such as severe floods, fires, snow and lack of food generally) 'wild boar undertake exceptionally long migrations, covering in the course of one season up to 500 km.'[64] Migrations of 100-200 km 'as the crow flies' are comparatively common. Also various regular seasonal migrations have been observed from the mountains (of boar living there) to lower country in the winter, with the reverse trek occurring in the summer. As a result of this seasonal migration the wild boar may move a linear distance of 100-150 km one way. Other seasonal migrations also occur from north to south with changing seasons, and vertically from river beds and swamps, for example, to surrounding low hills. The wild boar is extremely mobile and, according to Sludskii,[65] A. N. Nasimovich considers it to be the most migratory ungulate in the Western Caucasus. Because of this great mobility, especially in the span of a season if stress occurs, there is considerable intermixing of wild pig populations and wild pigs are a very effective force in the spread of diseases.

Similarly, studies in the United States of America[66] and in New Zealand[67] confirm the mobility of the wild pig, and results seem consistent with other findings. Barrett concludes after his study in the United States:

> 'These findings are similar to those of other studies which have suggested that wild pigs normally have home ranges of 5 to 30 km^2. Wild pigs have been known to travel longer distances up to 210 km, and they exhibit distinct seasonal migrations. Most authors have concluded that wild pigs are irregularly sedentary animals. They remain in an area with suitable cover only so long as sufficient food and water are available, but will shift their home range as a result of drought, flooding, or severe disturbance.'[68]

What are the economic implications for control of a pest population of such mobility? The main implication is that, where the wild pig is regarded as a pest, private self-interest cannot be relied upon to ensure its socially optimal control. Less than an optimal amount of control can be expected to occur where individual landholders take relatively independent control action. This is so for two reasons: (a) No landholder appropriates the full gain from eradicating pigs present on his property. Other landholders who might have expected to have the same pigs on their properties at some future time also gain. These spillover benefits are ignored by the individual landholder in determining his optimal control strategy. (b) Because wild pigs are likely to be present on adjoining properties, some may in time be expected to move into a property where control is contemplated. This reduces the profitability of control on the property where control is being considered, especially if the speed of inflow of new pigs is related to the density of pigs on the property - for example, if a reduction in feral pig densities speeds the inflow of pigs from outside the property.

An *outflow* of pigs from a property means that the marginal discounted benefit curve for an individual landholder of his control of a pest is below the social or collective gains from it. If in Figure 9.8, CADQ represents the private marginal discounted benefit of feral pig control on a farm (that appropriated by the farmer) and OAMB is the marginal discounted cost curve of control, the farmer maximizes his benefits by reducing the population of pigs by Z_2. But the social marginal discounted benefits of his control (allowing for its beneficial consequence on other landholders) might be as represented by curve KLMDQ. From a social point of view, it is consequently optimal for the farmer to reduce feral pigs on his property by a greater amount, namely by Z_3, the amount for which the social marginal discounted benefits of control equal the marginal discounted costs of control.

Consider the implications of an *inflow* of feral pigs to a property. An inflow of pigs to the property implies that the cost of reducing and maintaining the feral pig population on the property at a constant level is greater. The marginal discounted cost of control on the property can be expected to be higher than in the absence of such an inflow. Thus rather than curve OAMB applying in Figure 9.7, curve OGH might apply. In that case the landholder finds it profitable to reduce the pest population only by Z_1, a smaller amount than Z_2 as is the case where no inflow occurs.

Both inflows and outflows of pests can occur on a property simultaneously, but provided that either movement occurs it reduces the private profitability of pest control. Some hypothetical patterns of possible pest (feral pig) movements in relation to a property (shown as a rectangle) are illustrated in Figure 9.9. In case (a), the property is a breeding ground for the pest and only outward movements of the pest occur. In case (b), only inflows of the pest are experienced and rate of inflow could depend on the density of the pest on the property. In case (c), the pest population circulates - members of it being on the property sometimes but also moving to other properties and returning. In case (d), the pest population keeps migrating through the property. In cases (a), (c) and (d), private control is likely to be less than socially optimal. And in case (b), less private control is likely to occur the stronger in the rate of inflow of feral pigs in response, say, to attempts to reduce the level of population on the property.

The socially optimal control of many pests, including wild pigs, requires that account be taken of their mobility. Collective effort is needed to control them. However, on those properties where the private marginal benefits of control are less

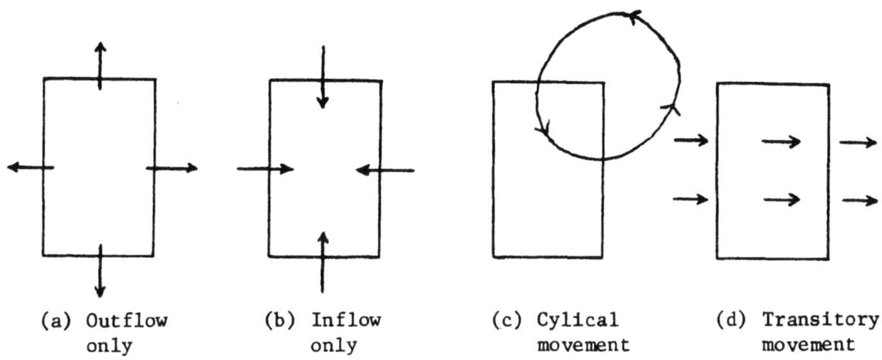

Fig. 9.9 Some hypothetical patterns of pest (e.g. wild pig) movements shown by arrows in relation to a property indicated by a rectangle.

than the social benefits from control, the question arises of who should pay for control beyond the level that is privately most profitable.[69] It is possible that when a pest recirculates as in diagram (c) of Figure 9.9 the *mere co-ordination* of control effort by affected landholders is sufficient to ensure optimal control of the pest. This will tend to be so when all of the landholders on their properties, have a similar density of the pest population, the members of which regularly move to the other properties to be replaced by members from these properties. A simple case is like that in diagram (a) of Figure 9.10. There are two concentrations, (1) and (2), of the pest population which circulate at a constant rate and in a regular fashion between two properties A and B. If landholders A and B face an identical economic situation, and *simultaneously* launch like measures to control the pest both will gain equally. But if these measures are not synchronized the effectiveness of control may be reduced. For example, if concentration (1) of the pest is

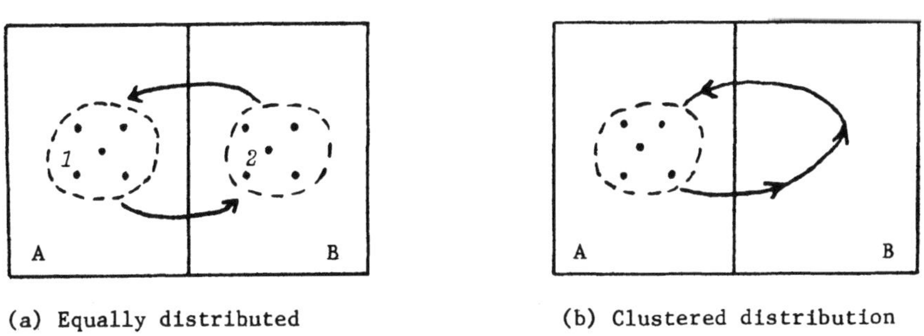

Fig. 9.10 Not only the pattern of movement of the pest population but its degree of clustering has implications for the optimality of its control.

attacked by landholder A first and then control measures are taken later by B when (1) has moved to his property, this may mean a smaller reduction in the whole pest population for the outlay made because concentration (2) is not controlled.

In the case shown in diagram (b) of Figure 9.10, the pest population is spatially bunched. Here serial control by the property owners rather than simultaneous control is likely to be optimal. However, the socially optimal amount of control effort by the first landholder to begin measures can be expected to be in excess of that needed to maximize his private gain. In this case, landholder B might contribute to the initial effort by A, and A might contribute something to the follow-up effort by B, and so on. Collective (co-operative) accommodation needs to be made between the landholders. Public policy may assist such co-operation especially when many landholders have to be organized.[70]

It is clear that with mobile pest species (and most are), considerable attention needs to be given to their movement patterns in determining optimal control measures. Many patterns of movement are conceivable. For example, another case akin to that illustrated in (a) of Figure 9.10 is one in which after removal of any *portion* of the pest population, the density of the whole population tends to equalize over the whole of its range even though it continues to circulate. It is clear from the available evidence that wild pigs are subject to a variety of patterns of movement. Under stable environmental conditions recirculation may be common but transitory movement can occur if stress causes migration. There is a need to develop the theory of spatial movements of pests and its implication for economic control of pests to a higher level.[71] However, the general implications of such movements is that they give rise to spillover effects from private pest control and that collective co-ordination and planning of control efforts can in principle result in more effective or more economical control (from a social viewpoint) of a pest population.

Fecundity of a Pest (Such as the Feral Pig) and its Optimal Control

As pointed out earlier, the rate of increase of a wild pig population is usually high. In Australia sows can have two litters per year and breeding is almost continuous. Consequently in Australia, an (instantaneous) rate of increase of 0.6 (60 per cent) for a feral pig population is considered to be realistic by Hone, O'Grady and Pedersen.[72]

In Europe and the U.S.S.R. the rate of increase of wild boar populations may be slower because due to the severity of the winters sows may have only one litter per year. However, according to Olaff wild boar populations in the Solling mountain range (in deciduous forests) of West Germany can show a substantial multiplication in one year, if the increase in the number of pigs one year old or older are counted in the spring. In years of poor food supply this population is estimated to increase by 50 to 60 per cent, in years of average food supply 80 to 100 per cent, and in years of plentiful food supply by 150 per cent or more.[73] The Forestry Commission of Danndorf, West Germany, believe that even higher rates of increase apply than estimated by Olaff.[74] Even in the Soviet Union, where winters can be very severe, large rates of increase in wild boar populations are recorded. Heptner, Nasimovic and Bannikov point out that within 1-2 years the number of pigs in a wide area can fall to 10 per cent of the pre-existing population and then increase two or threefold within the next year, even disregarding any migration of wild pigs into the area.[75]

Wood and Barrett claim that the wild pig is the most prolific large mammal in the United States and that sows may produce two litters per year.[76] Pigs become sexually mature at 6-8 months. However, the mortality rate of piglets tends to be high - for example, 50 per cent in South Carolina and 70-90 per cent in California. Nevertheless 'theoretically, even at 90% mortality rate, if adult sows produce an average of 5 young per litter and breed twice per year, the breeding population will still increase 33 per cent annually.'[77] Barrett summarizes the position as follows:

> 'This relatively high reproductive potential added to their unsynchronized breeding habit enables feral pigs to withstand mortality rates well above those tolerated by other ungulates. Feral hogs can quickly respond to seasonal food supply and rapidly repopulate an area after severe population reduction. This situation poses both the opportunity for a large annual yield, and the necessity for a substantial harvest frequently enough to control the population and prevent environmental damage.'[78]

All researchers are agreed that the wild pig has a high net reproduction rate and can increase its population rapidly. This makes it more difficult to hold its population below the ecological equilibrium level because the higher the rate of increase of the population of a pest the more members that have to be continually removed to hold the population level stationary. The total costs are raised of reducing the equilibrium population *and* holding it at the same reduced level. Consequently in terms of Figure 9.8 the marginal discounted cost curve of permanent reduction may move upwards. Hence marginal costs and marginal benefits are equal for a smaller reduction in the population. Thus the apparent paradox arises that it is usually less economic (from the points of view of both an individual farmer and a whole region) to reduce a pest population the higher is its fecundity, assuming that the pest population is already established. The effect is similar to that when high rates of immigration of a pest prevail. But *elimination* of the pest can become a *relatively* more attractive strategy when the net reproduction rate of a pest is high. This is because control of the population (without elimination) becomes less profitable with a greater rate of increase of a species, other things equal. Hence, when elimination is not the optimal strategy and the pest species is established, we can expect less reduction in its population to be optimal (given the available techniques of control) the greater is the fecundity of the pest. However, where a pest is established only in a limited area and there is a possibility of its spreading to other areas, the socially economic gains to be made by eliminating or containing it in the area where it is established are likely to be higher, the greater is the rate of increase of its population.

Multiple Control Optima

In each of the above models, there is only *one* (internal) optimum for reduction in a pest population or for the level of the pest population after control, namely the level at which the appropriate marginal cost curve equals the appropriate marginal benefit curve of control. However, the marginal cost curve of pest control could have a different shape and this could give rise to multiple local optima. For example, a situation like that shown in Figure 9.11 cannot be ruled out *a priori*.[79] There the marginal discounted benefit curve is indicated by FGQ and the marginal discounted cost of control curve is shown by the curve passing through OABCD. Marginal cost equals marginal benefits for three levels of reduction, namely Z_1, Z_2 and Z_3. Z_1 and Z_3 provide relative maxima of gain and Z_2 gives a relative minimum of total net gain. It is necessary to compare total net gain for reductions Z_1 and Z_3 to decide which of the two is absolutely optimal. Shortsightedness can result in the choice of reduction in the pest population of only Z_1, when Z_3 provides greater gains.

Other Complications Involved in Determining the Optimal Control of a Pest Population

Many other complications can arise in determining the optimal reduction (from both a social viewpoint and a private viewpoint) of a pest population, such as the population of wild pigs. Economic and environmental conditions are rarely stationary and they may alter with some uncertainty. Ideally this should be taken into account in deciding on the optimal level of control of a pest population.[80]

Furthermore it is assumed above that it is optimal to reduce the pest population to its desired level and hold it there by frequent follow-up action. But that

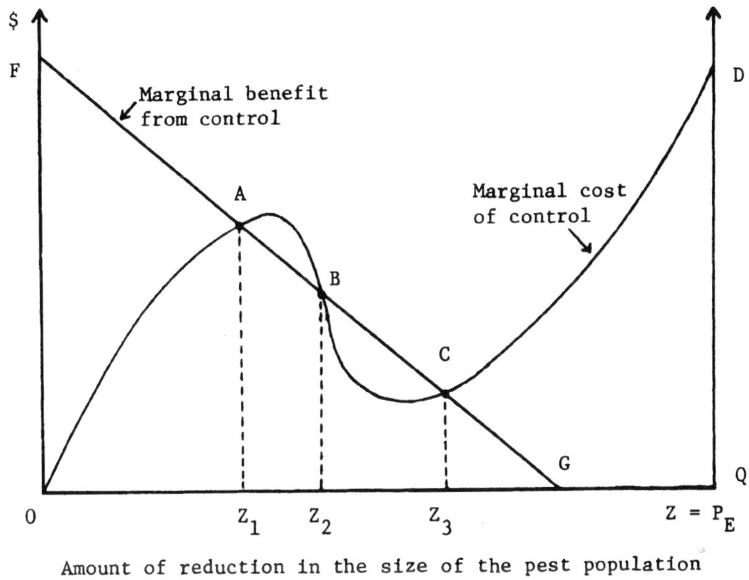

Fig. 9.11 Finding the level of pest reduction that maximizes total net gains may not be straightforward. In the above case two relative maxima of net gain exist. One corresponds to point A and the other to point C. Given the particular curves above, Z_3 in fact gives a greater net gain than Z_1.

is just one possible pattern of control. In some circumstances, intermittent or 'bang-bang' control might be less costly, especially if economies of scale are achieved with increased size of each control campaign or of each 'pulse' of control effort.[81]

As mentioned earlier, from a social point of view account needs to be taken of the externalities or spillover impacts from different control techniques on non-target species (for example, the impact of the poisoning of pigs on non-target species). The social value of non-target species may exceed their value as perceived by individual landholders, and public action may be needed to control the use of pest control techniques with unfavourable spillovers of this kind.[82]

Other types of spillovers may also call for public control of pest eradication. The target species may develop resistance to a control technique[83] - for example, by natural selection or *by learning* from experience. These effects may depend upon the *widespread* use of the technique and may be virtually independent of its use by an individual landholder. Hence any individual landholder is likely, acting in self-interest, to ignore the collective effect of his use of the technique. The technique is liable to be used therefore at a greater rate than is socially optimal. This can be explained in terms of a matrix as shown in the accompanying note.[84] In the context of feral pig control, it is sometimes said to be one of the advantages of 1080 poisoning that pigs do not learn to avoid the poison and there is little sign of growing biological resistance to it.

The above 'complications' cannot always be ignored. Considerably more research on and consideration of their economic consequences are called for. But to do so here would involve going beyond an introduction to the subject.

CONCLUSIONS

There are several techniques for controlling wild pigs. Commonly used methods in Australia include poisoning, especially with sodium monofluoroacetate (1080), trapping and shooting often combined with the use of dogs. Some use is also made of exclusion-fencing. Little use has been made so far of biological and environmental means of control.

Naturally application of all of the available techniques involve costs, and none ensures perfect control of a wild pig population. The considerable *mobility* of pigs and their *high rates of population increase* hamper the control of this species. These factors, as explained earlier, may result in less control of wild pig populations than otherwise. The presence of various spillovers from wild pig control (some of which arise from the mobility of pigs and others from the side-effects of the different techniques such as the poisoning strategy) suggest that the (social) efficiency of the control of wild pigs *could* be improved by government or public intervention in control activities.

However, there is still much research needed to decide even in *theory* upon optimal patterns and strategies for controlling pests including wild pigs. Although central (government) guidance of control efforts can be economically advantageous, government strategists have to act with some degree of ignorance. So caution is required in recommending and implementing collective policies.

The wild pig is of course not regarded universally as a pest. Some consider it to be an asset, at least in reasonable numbers for hunting and other purposes, and only consider it to be a pest when its numbers are high. Heptner, Nasimovic and Bannikov say, for example, that the danger is not the pig but great numbers of pigs.[85] In many countries, there is a need to balance the claims of those who regard the wild boar as an asset, such as hunters, against those who consider it to be a pest, such as some farmers. Wild boar populations are thus managed to sustain populations at a level that takes account of these conflicting interests.[86]

Australian attitudes towards the feral pig are still developing but on the whole it is regarded as a pest, particularly because the wild pig is not an indigeneous species. Many (but not all) Australians would like to see it eradicated, an issue to which I shall return in the next chapter. However, there seems to be little prospect of eliminating feral pigs in Australia in the near future using present techniques, especially taking account of the fact that feral pigs are well established in some remote and inaccessible places. However, historical experience indicates that it is possible to exterminate the wild pig.

The wild boar was exterminated in Britain and in Ireland in historic times. While it is difficult to pinpoint the exact reason for this, it seems from Harting[87] that the most important factors were the increased hunting of it and the clearing and economic development of the land (an environmental change). Harting concludes his study by saying that

> 'Lake and moor have become fields of yellow grain; forest has been changed into morass, morass into moor, and moor again into forest, until finding nowhere to rest in peace, the bear, the beaver, the reindeer, the wild boar, and the wolf, have become in Britain amongst the things that were.'[88]

In Australia, one suspects that it will still be some time before the economic development of the land and the density of the human population reaches levels where the continued existence of the feral pig is threatened by environmental changes and hunting.

CHAPTER 10

Concluding Observations

Even when the wild boar or pig remains within its original geographic range, there is usually a need or demand to manage it collectively. This is partly because it is a mobile species with favourable spillovers or collective benefits for some and with unfavourable spillovers or collective disbenefits for others in a society. The demand for its regulation is also influenced by the volatility of its population levels - for instance, its ability to increase its numbers rapidly by reproduction if environmental conditions are favourable and if hunting pressures are inadequate. In Europe and parts of Asia, hunting by man is the major control on the level of wild pig populations. In the absence of such hunting, wild pig populations would be much higher and considerable economic damage could be anticipated.

Possibly in some regions, economic development has increased the *equilibrium* level of the population of wild pigs that would be sustained in the absence of the hunting of them by man. The provision of more numerous and more reliable sources of water and of a greater quantity of palatable crops in agriculture would contribute to this. In addition, a very significant contribution to this result flows from the reduction in the population of natural predators of pigs, such as wolves and tigers. Offsetting influences, however, would be the destruction of habitat suitable for wild pigs as a result of economic development of the land. Shrubs, forests and reeds may be destroyed and marshy or wet places drained. Such destruction accompanied by heavy hunting pressure from man can lead to the extermination of wild pigs as, for example, apparently occurred in Britain.[1]

The intensity of hunting in Europe and other densely populated countries where the wild pig is sought for recreational hunting and/or human consumption is more than sufficient to control its numbers and policies are adopted to ensure that these remain within 'acceptable' positive ranges of population. In many countries, closed hunting seasons and bag limits apply and overall policies are adopted for managing the wild boar stock.

In Australia, however, the position is quite different. While recreational hunting of wild pigs is very popular, the human population is much smaller in relation to the population of wild pigs, which often occur in areas remote from the main settlements, in rugged or inaccessible terrain and in vegetation that provides them with considerable protection. Recreational hunting pressures in most regions are inadequate to threaten the survival of feral pig populations. No attempt is made to conserve the species officially. On the contrary, wild pigs are officially regarded as a pest[2] and their eradication is considered to be a desirable if as yet unattainable goal. This view is reinforced by the fact that wild pigs are of feral origin in Australia and are an exotic species.

ECONOMIC EVALUATION OF WILD PIGS AND OF POLICIES TO MANAGE THEM

Given our modern desire for precise answers, it would be useful if the social costs and benefits of wild pigs and policies to control them could be expressed in dollars and a conclusion reached about the optimal social policy for their control. However, given our present state of ecological and other knowledge, such a cost-benefit analysis would give a misleading impression of accuracy. Apart from that, it would also be based upon a set of social principles *not* acceptable to a large group of individuals.[3] For this reason no such final evaluation of this type is made here, but the work has involved a search and one that needs to continue for the elements that would need to be taken into account in such an evaluation. The study represents an environmental impact statement rather than a definitive cost-benefit analysis.

Even if we had the information to complete a definitive cost-benefit analysis, a number of philosophical questions would arise in using it for policy decisions. As a rule, this type of analysis is based upon the proposition that if the gainers from an economic or social change can compensate losers from the change and remain better off than before it, then it is desirable.[4] Actual compensation need not be paid according to this test of a social improvement which is known as the Kaldor-Hicks test. This creates problems of evaluation when some gain and some lose as a result of policy, and in relation to the management of wild pig populations some citizens are likely to gain and others to lose. For example, as discussed in this book, hunters may lose if measures such as poisoning programmes are adopted to eradicate pigs or reduce their numbers to low levels, although many landholders may gain.

Furthermore an assessor may arrive at a different evaluation of such a policy depending upon which set of parties is to be compensated or bribed. The Kaldor-Hicks test says that the policy is desirable if *gainers* can bribe or compensate losers. Thus if landowners are the only gainers from wild pig eradication and hunters the only losers, eradication is judged desirable by this test if landowners could financially compensate hunters for it. An alternative test would be to suggest that the social change or policy would be desirable if *the losers* could not bribe the gainers to desist from the policy. If hunters could not financially compensate or bribe landowners in this case to adopt other than an eradication policy, eradication would be judged a socially desirable policy on the basis of this test. Depending, for example, upon which group has the lower income, these alternative ways of approaching the problem can give different results.[5] Some authors suggest that the result depends on who is assigned property-rights. Do farmers have a right to pig-free properties or do hunters have a right to a stock of wild pigs? The matter of who should compensate whom revolves around such questions[6] and economists have no ready answer to them.

Cost-benefit analyses are also heavily influenced by the dominant values of the time and the individualistic evaluations of those alive at the time.[7] This is of course an unavoidable constraint and creates a problem for social rationality. It suggests not that social rationality should be avoided as a goal but that aids to it, such as cost-benefit analysis, should be treated with scepticism. One needs somehow to take account of the fact that values may and do change and so does our information. Thus in social decision-making a premium should be placed on the keeping open of options.[8]

Over a long period, society's attitudes towards different species and natural environments change. In Australia, for example, in this century compared to last, attitudes towards the dingo (native dog) seem to have become more favourable whereas those towards most introduced species have become unfavourable. Indeed, in the last century naturalization societies were active in Australia and New Zealand introducing such species as deer, foxes and rabbits from Europe.[9] Introduced species are held in less regard than previously because many have caused considerable environmental disruption in Australia and Australians place higher value on native species than previously. We cannot be sure that values will not alter again and this should make us hesitant about putting too much faith in current cost-benefit evaluations.

Methods for determining the actual amounts that individuals would be willing to pay for the preservation or eradication of a species such as wild pigs face a number of difficulties. How can individuals be induced to reveal their true preferences for the conservation or eradication of a species? They may try to overstate their preferences for strategic reasons.[10] There is also a danger in relying on individual preferences when individuals are badly informed or actually misinformed, say, about the effects of a species, as in fact many individuals are in relation to wild pigs. An actual evaluation of wild pigs in Australia requires many difficult evaluations. For example, what value should be placed on saving a rare species that might be threatened by pigs?[11] Are in fact individuals prepared to pay anything to save relatively unknown rare but unspectacular species, and if not does this imply that such species are not worth saving? A recent Australian study found that individuals seemed to place little value on the saving of a threatened species, namely the ground parrot *Pezoporus wallicus*.[12] If cost-benefit analysis is to be used sensibly to evaluate the control of feral pigs in Australia, we have yet to resolve a number of issues about the evaluation of natural ecosystems[13] that are affected by pigs.

As mentioned earlier, the Kaldor-Hicks test of a desirable policy does not require that compensation be actually paid. However, if compensation is paid then adopted policies involve no one being made worse off and some individuals being made better off. A so-called Paretian improvement occurs. Possibly the French scheme of hunters contributing funds to indemnify landholders against damages caused by game come close to being a policy of this type. Game are accorded some protection (for example, in closed seasons) 'in return' for this indemnity.

CONTROL POLICIES AND METHODS

The optimal social policies suggested by the tests mentioned above are likely to be ones of compromise. For example, where landholders and farmers are in conflict about the optimal stock of wild pigs to maintain, the social optimum as determined by the above tests is likely to be between the level most desired by landholders and that wanted by hunters.[14] However, such compromises would be compatible with the elimination and exclusion of pigs from some areas, such as from some national parks or even from certain regions.

Nevertheless, if wild pig populations are to be regulated to certain levels, both social and technical or physical means are required to achieve this end. As mentioned earlier, in parts of Europe recreational hunting pressures are more than adequate to reduce wild pig populations to desired levels and the problem is to ration or limit hunting by the use of closed seasons and other regulations. But in Australia recreational hunting, although popular, has been insufficient because of the low population density of Australia to reduce wild pig populations to levels widely considered to be desirable. Despite the lack of official restrictions on the hunting of the Australian wild pig, it has remained a net pest.[15] Its numbers exceed levels commonly considered to be desirable.

In order to reduce the population of wild pigs, two types of policy measures have been adopted: (a) the declaration of the wild pig as a noxious animal, and (b) the payment of bonuses or royalties on pig snouts.

The noxious animal declaration obliges landholders to take action to eliminate wild pigs from their properties or face penalties for not doing so. However, judging by experience in New South Wales these regulations have been relatively ineffective.[16] No prosecutions have been launched in New South Wales for failure to take action to control wild pigs. Successful legal action could be difficult to obtain because a defendant could argue that the wild pigs in question have only recently strayed to his property from elsewhere. Furthermore, the question arises of what is an adequate attempt to eliminate wild pigs. The difficulty of detecting breaches and enforcing the noxious animal regulations tends to make it ineffective for the control of wild pigs.

Bonuses have also been officially paid in Australia and New Zealand in the past for the destruction of wild pigs.[17] Opinions differ about the effectiveness of such bonuses or subsidies in controlling pest populations. It has been argued, for example, by S. Whitehouse that bounty schemes designed to control pest populations are relatively ineffective.[18] However, where there is common-access to the species or it is virtually common property as is (or has been) the case with feral pigs in Australia, the effectiveness of the subsidy or bounty will depend upon the elasticity of supply of hunting effort in relation to costs. Where this is comparatively inelastic, as it might be for recreational hunting, the bounty will have little impact on the kill of feral pigs. The opposite is the case if hunting effort is very elastic in relation to costs. The applicable relationship needs to be determined empirically.

If hunting is responsive to the subsidy, it is still possible that problems can arise with the bonus scheme. Cheating is always a possibility. Several examples are given by Whitehouse.[19] Apart from this it would probably be optimal to vary the bonuses by regions making them higher in regions where wild pigs cause the greatest marginal damage. But if bonuses vary by region, an incentive exists to destroy pigs in areas where they are most easily taken and take their snouts to areas paying the highest bonuses.[20]

Again when bonuses are paid on the production of physical evidence, such as snouts, this tends to favour some control techniques over others. Shooting, dogging or trapping of wild pigs is favoured compared to poisoning. When poisoning of pigs is used as a control method the carcasses are rarely found.

When a market exists for the meat and other products of the wild pig, commercial harvesting of wild pigs is encouraged and this may, as discussed below and considered earlier in this book, regulate the number of the species.

It has also been pointed out that regulations to control human transport of feral pigs and the free-ranging of domestic pigs might assist in ensuring against the re-emergence of feral pigs in areas where they have been eradicated provided that these areas are separated by some physical means from those areas where feral pigs occur. In order to guard against possible transmission of diseases, it is desirable that wild pigs be excluded from garbage dumps and that the meat from wild pigs be either frozen for sufficient time or well-cooked before it is eaten. In addition, domestic pigs should not be allowed to roam in the same area as feral pigs.

A number of different technical methods to destroy wild pigs such as shooting, dogging, trapping and poisoning of various kinds have been examined. No technique was found to be best in all circumstances and a combination of techniques is sometimes called for. In assessing techniques, account needs to be taken of spillover effects from them. Trapping appears to be the most expensive technique compared to shooting and/or dogging and to poisoning with sodium monofluoroacetate (1080). This is so if one judges on the basis of the cost per pig taken. But trapping does not have the environmental spillovers of the other two techniques, such as the impact of 1080 on non-target species. It is one of the few available techniques in densely timbered areas of high rainfall.

Exclusion of wild pigs by electric fencing or pig-wire fencing is another technique with few spillovers. Some other possible techniques were mentioned but many are still in the development stage such as the use of sexual sterilants and of pheronomes to trap wild pigs. The spread of diseases such as African swine fever and of swine fever would exert considerable control but to introduce these to Australia and use them for this purpose might prove to be the death-knell of the domestic pig industry. Furthermore, over a long period, biological selection and adaptation might be expected to occur amongst wild pigs.

For any programme of control of feral pigs to have a noticeable or somewhat lasting effect in an area, a *substantial* reduction in the wild pig population is required over a considerable distance. Otherwise the pig population is quickly replenished by natural population growth and by immigration of wild pigs from other areas. As pointed out, the fecundity of the pig is considerable and it is also a mobile animal.

COMMERCIALIZATION

While some Australians see the commercial use of wild pigs as a means to control wild pig populations at little or no expense to the landholder, some authorities such as the N.S.W. Department of Agriculture have been opposed to commercialization on the grounds that it might result in wild pigs being preserved for a profit rather than being eradicated. The effect of commercialization depends on the circumstances. Under some conditions, it lowers the population of wild pigs but raises this under other circumstances.

If pigs were perfectly mobile (and if, say, their density tended to equalize almost instanteously everywhere) they would virtually be common property. There would be no incentive to any landholder to conserve them and in fact the higher the current price received for wild pigs (or for permission to hunt them on properties) the greater would be the permitted hunting pressure. In these circumstances, commercialization would be an effective way to lower wild pig populations and they could be driven to extinction if their commercial value became high enough.[21]

However, while pigs are mobile, they are not perfectly so and do not necessarily equalize densities quickly to 'natural' levels. As observed in the survey in the Hay district (reported in Chapter 4) hunting pressure on a particular property when little hunting occurs nearby can lower pig populations on the property concerned because pigs learn to avoid the property or leave it for the safer haven nearby. Thus landholders selling hunting rights may moderate hunting pressures on their property to conserve their pig numbers and maximize their gains from the sale of (recreational) hunting rights. Nevertheless, as pointed out in Chapter 4, control of the wild pig population may be greater than if the sale of hunting rights is not permitted.

The utilization of wild pigs in the game-meat industry in Australia is a recent development. The effect of this development on the population of wild pigs remains to be seen. However, the industry is so organized by government regulations that there usually is only a single freezer unit or game processor in any region. Thus professional hunters usually have a single buyer for their products. Because of this (the existence of a monopsony), game (including pigs) are likely to be conserved in each area because each area is large. The single buyer is able to appropriate any gains from conservation in the area. Yet, as pointed out, greater control of the wild pig population could be obtained than in the absence of their utilization in the game-meat industry. Nevertheless the chances of feral pigs being eradicated are reduced and a new lobby may grow up based on the game-meat industry. The latter may come into conflict with recreational hunting interests. Recreational hunters may have less game, be excluded from more properties and or be forced to pay higher fees for hunting rights if feral pigs are utilized in the game-meat industry. Already there are signs in Australia of emerging conflict between recreational hunters and the game-meat industry.

DAMAGES ATTRIBUTED TO WILD PIGS

It may be helpful to recount some of the damages that have been attributed to wild pigs in Australia. These include considerable damage to agriculture and it was estimated (in Chapter 2) that the value of agricultural production lost in 1979-80 was about $73.3 million. Cereal crops, oilseeds, sugar cane, potatoes and other crops may be trampled, rooted up or eaten by pigs. Pastoral losses can also be high and are often greater than suspected. Some wild pigs kill lambs and it has been observed that lambing rates fall markedly in pig-infested areas. By rooting and eating grass, wild pigs compete with cattle and sheep for food. In addition, pigs damage fences, dams and watering points, and roads and promote erosion by their wallowing and digging habits. Their cloven feet add to erosion in a country that had no hoofed animals before the comming of British settlers.

The extent to which wild pigs spread diseases injurious to the pastoral industries is uncertain. In Australia, however, there is evidence to suggest that they are vectors in the spread of leptospirosis. They may also play a role (even if it is not a major one) in the occurrence of anthrax, erysipelas, brucellosis, tuberculosis and Murray Valley encephalitis. They are not, however, the only wild animals capable of carrying these diseases. Yet they are considered to be high-risk animals because of their mobility and their strong preference for carrion (including that of their own kind). In many cases carrion is from animals that have died of disease and pigs can smell out dead animals over long distances. Furthermore, pigs frequently congregate with domestic stock such as sheep and cattle so that any disease that they might have is brought to the proximity of such animals and so increases their chance of infecting domestic stock. Their mobility is another factor that makes wild pigs significant carriers of disease. These animals increase the likelihood of the introduction of several exotic diseases such as foot-and-mouth disease to Australia and would make these diseases more difficult to contain and eradicate if they should enter Australia.

As for risks to human health, the principal risk in Australia is a parasite problem, that of contracting sparganosis by eating pork either not well-cooked or not frozen for sufficient time. If wild pigs containing spargana (of *Spirometra erinacei*) are eaten in these cases, cysts form in the flesh of humans resulting in a condition that is uncomfortable or painful but normally not fatal. This risk is easily guarded against by either freezing meat from the wild pig for a sufficient period or cooking it well.

It was observed that foresters are unsure about the impact of feral pigs on Australian forests. In New South Wales, the value of timber lost as a result of pig activity was considered to be negligible. However, in the rainforests of northern Queensland considerable loss of timber was reported as a result of pigs helping to weaken the root system of trees which subsequently topple over, and it is believed that feral pigs help to increase the occurrence of the fungal dieback disease *Phytophthora cinnamomi*, which attacks the root and kills trees. Pigs were also said to increase erosion in forests and to cause damage by rooting up fire-trails, the edges of roads and dams supplying water for fire-fighting, thus adding to the costs of maintaining forests. Those leasing grazing rights in the forests were reported to be affected adversely by the presence of pigs in forests. Furthermore, neighbouring property-holders sometimes accuse forest authorities (and national parks authorities) of harbouring wild pigs which visit these neighbouring properties and do damage. The movement of pigs between forested areas and agricultural land seems to be common.

Practically no scientific work has been done in Australia to determine the impact of feral pigs on native flora and fauna. In this respect Australia compares unfavourably with the U.S.A. and New Zealand. However, it is known that damage does occur, for example, to terrestial orchids, bulbous plants and ferns. The pig is also believed to play an important role in spread and propagation by rooting of several introduced plant pests, for example, Spanish broom, Noogoora burr, Bathurst burr and blackberries. Again the effects of the feral pig on native animal species has not been scientifically studied but it has an impact through habitat destruction, predation and food competition. Some water birds such as the brolga may be especially affected, along with small mammals with habitats on swamp or creek edges or in the floor of rainforests, such as the rare rat-kangaroo. The brush turkey and the brush hen may suffer from the destruction of their nests (which are mounds on the ground) by pigs. Pigs may also eat the eggs of these birds. As pointed out in the earlier chapters, reptiles may also be affected and frogs in some areas such as in sphagnum bogs could be under particular threat. Even the spawning of native fish species could be affected.

There is a need for scientific work on the impact of feral pigs on Australian wildlife if a rational programme of pig management is to be devised. Australian research dealing with feral pigs has concentrated on their impact on agriculture and methods of controlling them in agriculture with some study of the spillovers from these methods, for example, on native wildlife. Thus the emphasis of research has been on immediate economic or commercial gains and reflects the funding of much research through primary industry levies. It does not reflect the full value that Australians place on their resources including their native species.[22]

BENEFITS FROM WILD PIGS

The benefits from feral pigs in Australia have to be offset against their damages. It has been observed that they are probably the most important wildlife asset in Australia for recreational hunting. Possibly about 100 000 Australians go pighunting each year and no species is given greater coverage by Australia's main shooting and hunting magazines. Annual expenditure on pighunting might amount to $5-15 million per year and the hunters' surplus received from it could exceed $20 million. Unfortunately, however, our current data are inadequate to pinpoint these values accurately.

Feral pigs have also become a useful resource for the developing game-meat industry in Australia. In Queensland the mainstays of the industry are considered to be feral pigs, feral goats and kangaroos. It has been estimated that at current commercial harvesting rates of 60 000 head per year and current export prices, feral pigmeat valued at $50 million will be exported annually. Higher sustainable commercial harvesting rates seem possible, so even greater export income might be earned in the future.

In a few areas of Australia, such as the Hay region, landholders have been able to supplement their income by selling rights to the recreational hunting of feral pigs. Especially when the prices of agricultural produce have been depressed in the past, this extra income has been important for the survival of their farming enterprises. As the commercial utilization of feral pigs spreads in Australia, it is possible that landholders will be able to charge higher fees for permitting hunting on their properties.

Some foresters pointed out in the surveys that were reviewed in Chapter 7 that feral pigs could have minor benefits for Australian silviculture. Marginal benefits might come from eating the larvae of leaf-eating beetles, and their rooting activity is seen by some foresters as assisting the regeneration of cypress pine. The rooting may provide seedbeds for beneficial tree species especially around swampy areas. While feral pigs were seen as a rule not to benefit natural ecosystems, some respondents pointed out that feral pigs help to open up swampy and reedy areas and may thereby increase plant diversity. They were also claimed to help control some exotic weed pests, such as Paterson's curse *Echium lycopsis*.

GENERAL QUESTIONS

Despite the desirability of coming to a definitive cost-benefit conclusion about feral pigs in Australia, our present knowledge is too inadequate for us to put much faith in such a calculation. Furthermore, it is doubtful whether such analysis would be helpful in settling social conflicts that arise from the presence of wild pigs and policies towards them.

Possibly we need also to question the status accorded to feral animals. Currently feral animals are accorded a much lower status by conservationists than animals of natural origin, and feral animals are regarded as particularly inferior when descended from species introduced from abroad. The World Conservation Strategy proposed by the International Union for Conservation of Nature and Natural Resources (IUCN) with the co-operation of the United Nations Environment Programme (UNEP) and the World Wildlife Fund (WWF) reflect this. Robert Allen, in his book based on this strategy, says 'where an introduced exotic species is threatening the survival of native species, the introduced species should be eliminated if possible.'[23] Undoubtedly the introduction of the pig to new environments has resulted in a reduction in the diversity of species in the world. But how valuable is diversity and what is man's brief in preserving it? If the pig is better adapted to an environment than other species, why should man suppress this 'natural' dominance? When is the principle of the survival of the fittest in nature acceptable and when is it not? I am not arguing that feral animals should be accorded equal or superior status to native species but am pointing out the need for us to elaborate on why we accord species the status that we do. Much more clarification is needed about man's responsibility for nature and the way in which he ought to value it.[24] Wild pigs and feral pigs must ultimately be assessed in this context.

One can indeed give several plausible arguments for attempting to retain diversity in nature. This diversity provides a greater variety of material to satisfy man's curiosity and preserves a larger genetic base thereby keeping more options open for mankind. Diversity of species may also make for greater stability of natural ecosystems.[25] To maintain such diversity, it may be necessary to control the number of feral or wild pigs. However, on the same grounds there would be no case for eliminating wild pigs. Indeed, they and even feral pigs could provide a useful genetic reserve for the domestic pig industry. They might prove a basis for breeding domestic pigs with superior qualities, such as greater resistance to particular diseases. Already research is occurring with this aim in mind.

CONCLUSIONS

There is little doubt that wild pigs or feral pigs are a pest when they occur in large numbers. They then have adverse effects on agriculture, forestry, wildlife and natural ecosystems. In the absence of control by man, populations of wild pigs can increase to noticeably damaging levels because man has reduced environmental constraints on their population expansion or introduced wild pigs to areas or countries where fewer constraints are present.

While the wild or feral pig is an environmental pest when it occurs in large numbers, it can also be a useful economic resource. It is an important asset for amateur or recreational hunters in many countries, including Australia. If its population is sufficient, it can help support a game-meat industry, as in New Zealand and as is now occurring in Australia.

Its advantages have to be offset against its disadvantages. A clearcut case has not been made for eradicating it in Australia. This is so even apart from doubt about whether (with existing control techniques and Australia's low density of human population) it is feasible or economic to eradicate the feral pig. If the proposition is accepted that the feral pig is in Australia to stay, at least for the forseeable future, there is a need to devise policies to manage it rationally. This book, however, reveals many gaps in our knowledge about the feral pig in Australia (for example, about its value to hunters and its impact on native wildlife and on Australian forests). These gaps need to be filled to some extent so that rational policies for managing wild pigs can be formulated. To embark on a national eradication campaign[26] before adequately studying such matters and examining alternatives could be a costly folly.

Notes and References

CHAPTER 1. Status of Wild Pigs: General Features

1. For an interesting discussion of the domestication of animals and of feral animals and problems of classification, see I. Lehr Brisbin, Jr, 'The ecology of animal domestication', *The ASB Bulletin* (Official Quarterly Publication of the Association of Southeastern Biologists) 21 (1974): 3-8.
2. See, for example, *Collier's Encyclopedia,* Crowell Collier Publishing Co., New York, 1962, p. 699. Asian domestic breeds were crossed with European domestic breeds to produce most of today's varieties of domestic breeds.
3. For some general background information on the pig family, see, for example, B. Grizmek (ed.), *Grizmek's Animal Life Encyclopedia,* Van Nostrand Reinhold Company, New York, 1979, Ch. 4; *McGraw-Hill Encyclopedia of Science and Technology,* McGraw-Hill, New York, 1977; *Larousse Encyclopedia of Animal Life,* Paul Hamlyn, Sydney, 1968, pp. 588-91.
4. See, for example, Susan P. Bratton, 'Wild hogs in the United States: Origin and nomenclature', in G. W. Wood (ed.), *Research and Management of Wild Hog Populations,* Belle W. Baruch Forest Science Institute of Clemson University, Georgetown, South Carolina, 1977, pp. 1-4.
5. Grizmek, *op. cit.*
6. See, for example, P. Le S. Milstein, 'The bushpig *Potamochoerus porcus* as a problem animal in South Africa', Proceedings Entomological Symposium, Pretoria, 27 September 1971 (roneoed), pp. 1-29, esp. p. 3.
7. For some interesting general information about the warthog, see D. Mason, 'The fair warthog', *African Wildlife* 32 (1980): 5, 27-30.
8. Grizmek, *op. cit.,* p. 106.
9. V. G. Heptner, A. A. Nasimovic and A. G. Bannikov, *Die Säugetiere der Sowjetunion* (V. G. Heptner and N. P. Naumov, eds), Vol. 1, Gustav Fischer, Jena, 1966, p. 31.
10. Heptner *et al.,* p. 43, point out that while the wild pig can cope with temperatures of -20 to -25 C and even as low as -30 C because of its short legs it cannot live normally in areas with high snow falls.
11. Heptner *et al., op. cit.* p. 31. Translation by Dr M. E. Tisdell.
12. See, for example, G. Barraclough (ed.), *The Times Atlas of World History,*

William Collins, London and Sydney, 1978, p. 37.

13. Professor J. Golson, in a personal communication to me dated July 1981, says 'There is no evidence for pigs in Australia prior to their introduction by Europeans. The faunal records of the final prehistoric period in northern Australia are not admittedly as extensive as one might wish, and they are better for Arnhem land than for Cape York. I have just finished the editing of a manuscript for publication which describes the excavation of a rockshelter near Laura. This has the best and best-studied faunal suite for tropical Queensland (even though the fauna is pretty sparse). There is no pig here and none in the Arnhem Land sites.'

    Pullar also came to the conclusion on the basis of the absence of an Aboriginal word for pig and other factors that pigs did not occur in Australia prior to European settlement. See E. M. Pullar, 'The wild (feral) pigs of Australia and their role in the spread of infectious diseases,' *Australian Veterinary Journal* (1950): May, 99-110, esp. p. 100. See also Pullar, 'The wild (feral) pigs of Australia: Their origin, distribution and economic importance', *Memoirs of the National Museum,* Melbourne, 18 (1953): 7-23.

14. Archeological evidence from three sites in the Central Highlands of New Guinea indicate that pigs were present there 6000-5000 years ago. See J. Golson and P. J. Hughes, 'The appearance of plant and animal domestication in New Guinea' (a revised version of a paper presented in Colloque XXII of 9th Congress of the International Union of Pre and Protohistoric Sciences, Nice, 1976), Department of Prehistory, R.S.P.S., Australian National University, undated, p. 11; to be published in French in *Journal de la Sociēte des Océanistes.* Golson and Hughes found further evidence in support of this from the Kuk site near Mt Hagen (*op. cit.*, p. 9).

    See also O. A. Christensen, 'Hunters and horticulturalists: A preliminary report of the 1972-4 excavations in the Manim Valley, Papua New Guinea', *Mankind* 10 (1975): 1, 24-36.

15. Barraclough, *op. cit.*, p. 48.

16. J. Golson, in a personal communication of July 1981. The only claim in writing appears to be the find of a pig incisor in the Highland rockshelter site of Kiowa at a level dated about 10 000 years ago, as set out in S. Bulmer, 'Settlement and economy in prehistoric Papua New Guinea: A review of the archeological evidence', *Journal de la Societé des Océanistes* 31 (1975): 46, 7-75. Both Golson and Hughes, *op. cit.*, and J.M.B. Smith raise the query as to why pigs did not come to Australia at that time because Australia was connected by a land bridge to New Guinea. Smith says '... why pigs did not establish themselves in Australia until carried here by Europeans is puzzling.

Pigs are known to have been brought to New Guinea as long ago as 10 000 years. Dry land, which could have allowed pig migration, extended from New Guinea to Australia until about 7000 years ago when the sea level rose as continental ice-sheets melted...', J.M.B. Smith, 'Biogeographic history of Australian flora and fauna', in H.F. Recher, D. Lunney and I. Dunn (eds), *A Natural Legacy: Ecology in Australia,* Pergamon Press, Sydney, 1979, p. 26.

17. See, for example, Barraclough, *op. cit.,* pp. 48-9.
18. See G. M. Thomson, *The Naturalisation of Animals and Plants in New Zealand,* Cambridge University Press, Cambridge, 1922, p. 33; and K. A. Wodzicki, *Introduced Mammals of New Zealand: An Ecological and Economic Survey,* Department of Scientific and Industrial Research, Wellington, 1950, p. 4.
19. Generalized distribution of bushpig after Grizmek, *op. cit.* Generalized distribution of wild boar after Heptner, *op. cit,* but modified for later information. Generalized distribution of feral pig based information obtained from various sources.
20. G. W. Wood and R. G. Barrett, 'Status of wild pigs in the United States', *The Wildlife Society Bulletin* 7 (1979): 4, 237-46, p. 237.
21. Heptner, *et al., op. cit.*
22. Wood and Barrett, *op. cit.,* p. 237.
23. *Ibid.,* pp. 237-8.
24. Personal communication of May 1980 from Professor Roberto P. Schlatter, Institute of Zoology, University Austral de Chile, Valdivia, Chile. Information about wild boar and feral pigs in Brazil was supplied by personal communication of August 1980 from Mr Paulo Nogueira Neto, Secretario de Meio Ambiente, Ministério de Interior, Brasilia, Brazil.
25. After T. McKnight, *Feral Livestock in Anglo-America,* University of California Press, Berkeley and Los Angeles, 1964, Fig. 7 updated and extended using additional information. See also G. W. Wood and T. E. Lynn, Jr, 'Wild hogs in Southern forests', *Southern Journal of Applied Forestry* 1 (1977): 2, esp. Fig. 1, 'Distribution of feral hog populations in the southeastern United States by countries'. See also M. D. Springer, 'Ecologic and economic aspects of wild hogs in Texas,' pp. 37-52, esp. Fig. 1, in G. W. Wood (ed.), *Research and Management of Wild Hog Populations, op. cit.*
26. Wood and Barrett, *op. cit.,* p. 237.
27. N. Myers, *The Sinking Ark,* Pergamon Press, Oxford, 1979, p. 35; and F. R. de la Fuente, *World of Wildlife,* Vol. 8, Orbis, London, 1971-74, p. 243.
28. See Milstein, *op. cit.*
29. Another factor limiting the spread of feral pigs in Africa could be the greater

susceptibility of *Sus scrofa* to African swine fever.

30. Thomson, *op. cit.*, p. 33.
31. *Ibid.*, p. 33.
32. *Ibid.*, p. 34.
33. *Ibid.*, p. 37.
34. Wodzicki, *op. cit.*
35. C. N. Challies, 'Feral pigs in New Zealand', in A. H. Whitaker and M. R. Rudge (eds), *The Value of Feral Farm Mammals in New Zealand*, New Zealand Department of Lands and Survey, Wellington, 1976, p. 23.
36. *Ibid.*, Figs 4 and 5. Their distribution is also mapped in A. L. Poole and J. H. Johns, *Wild Animals in New Zealand*, 2nd edn, A. H. & A. W. Reed, Wellington, 1973, p. 99.
37. Challies, *op. cit.* p. 23.
38. The view that Captain Cook introduced pigs to Australia is apparently common in northern Queensland. See the reference in note 59. The view was recently stated in the newspaper article referred to in note 58.
39. See the reference in note 59.
40. J. E. Harting, *British Animals Extinct Within Historic Times*, Trübner & Co., Ludgate Hill, London, 1880.
41. Personal communication from Dr Juliet Clutton-Brock, Department of Zoology, British Museum (Natural History), London, March 1981.
42. Harting, *op. cit.*, p. 206.
43. *Ibid.*, p. 207.
44. Personal communication from Dr Clutton-Brock.
45. Harting, *op. cit.*
46. Quoted in Harting, *op. cit.*, pp. 94-5.
47. Quoted in Harting, *op. cit.*, p. 95.
48. Quoted in Harting, *op. cit.*, p. 95.
49. Jordbruksdepartmentet, *Vildsvin i Sverige*, Statens offentliga utredningar, Stockholm, 1980, 11, p. 7.
50. *Ibid.* Also a personal communication to me from Dr S-O. Lindblom, District Veterinary Surgeon, Jönköping, Sweden, September 1980.
51. Ref. 49.
52. Ref. 49.
53. Ref. 49, pp. 8 and 9, and personal communication from Dr S-O. Lindblom, previously cited.
54. Personal communication from Dr S-O. Lindblom, previously cited.
55. Ref. 49.
56. More investigation of this disappearance is required, for the possibility

cannot be ruled out that African swine fever or other disease played a role.

57. The Environment Agency of Japan kindly supplied me with a grid map showing the distribution of wild boar in Japan in 1978. This indicates that wild pigs are practically absent from Hokkaido. The Agency also pointed out that feral pigs occur on the Ogasawara Islands of Japan.
58. See, for example, 'Plunder in the bush', *National Farmer,* 26 June 1980, p. 38.
59. E. M. Pullar, 'The wild (feral) pigs of Australia: Their origin, distribution and economic importance', *Memoirs of the National Museum,* Melbourne, 18 (1953): 7-23.
60. *Ibid.,* p. 9.
61. Quoted by Pullar, *op. cit.,* p. 9.
62. Pullar, *op. cit.,* pp. 9, 10. See also G. A. Letts, 'Early livestock introductions to the "Top End" of the Northern Territory', *Australian Veterinary Journal* 38 (1962): 282-7.
63. Pullar, *op. cit.,* p. 17.
64. This map has been compiled using information from a number of sources. Much of this information has been assembled on a state-by-state basis by Dr D. M. Flynn. See D. M. Flynn, *Report on Vertebrate Pests and Exotic Animal Diseases,* Australian Bureau of Animal Health, Department of Primary Industry, Canberra, June 1980, appendices D1, D2, D3, D4 and D6. One of the earliest maps of the geographical distribution of feral pigs in Australia was completed by Pullar in the early 1950s. (See Pullar, *op. cit.,* p. 16.) The present distribution of pigs is wider than that indicated by Pullar at that time.
65. T. McKnight, *Friendly Vermin: A Survey of Feral Livestock in Australia,* University of California Press, Berkeley, Los Angeles and London, p. 52. McKnight also maps the geographical distribution of feral pigs in his Fig. 1.2.
66. 'Domestic animals gone bush', *Ecos* (1977): 13 (August), 10-18, p. 11.
67. Flynn, *op. cit.,* p. 32.
68. The fact that pigs are often active only in the early morning or late evening and not during the main part of the day makes estimating their populations difficult.
69. The questions were:
    1. Do wild pigs occur in your country? If so, what species are present and where are they distributed?
    2. Are feral pigs (wild pigs descended from domestic stock) present? If so to what extent are they present? When and how did such pigs come into existence? Are they regarded as a pest? Why?
    3. What damages are being caused by wild and/or feral pigs, for instance to crops, livestock, native wildlife, natural environment? What dangers do

they pose - for example, to health?

4. What measures are being taken in your country to manage or control wild and/or feral pigs?
5. Is any commercial or other use being made in your country (a) of feral pigs, or (b) of wild pigs generally? If so what use?
6. Are wild pigs or feral pigs important for amateur hunting purposes in your country? Please explain.
7. In general, how do *feral* pigs rate in your country as (a) an environmental pest, and (b) an economic resource?
8. In general, how do wild pigs (*other than feral* pigs) rate in your country as (a) an environmental pest, and (b) an economic resource?

70. I intend to consolidate the replies in a research report to be published later.
71. Personal communication from Evangelos Papaevangelan, Head of Game Department, Ministry of Agriculture, Athens, 30 September 1980.
72. Based on information from an original source (unknown) in Yugoslavia and contained in a personal communication from Mr J. Iliev, First Secretary, Embassy of the Socialist Federal Republic of Yugoslavia, Canberra, September 1980.
73. Communication from Eng. Vladimir S. Naumov, Provincial Hunting Inspector, Autonomous Province of Kosovo, Yugoslavia, June 1980.
74. Information provided through the Ministry of Agriculture and Fisheries, Lisbon, Portugal, in personal communication dated August 1980.
75. Information was kindly supplied by Mr Akiyoshi Nawa, Deputy Director, Wildlife Protection Division, Environment Agency, Japan, 6 August 1980.
76. Personal communication, June 1980.
77. Personal communication, July 1980.
78. Personal communication from Dr Setijati Sastrapradja, Director, National Biological Institute, Bogor, Indonesia, June 1980.
79. Correspondence from Turkey received through the Turkish Embassy, Canberra, August 1980.
80. Dated October 1980 and received through the Embassy of Pakistan, Canberra.
81. Communication mentioned in note 80.
82. Personal communication from Mr Paulo Nogueira Neto, Secretario de Meio Ambiente, Ministério de Interior, Brasilia, Brazil, August 1980.
83. Personal communication from Mr C. K. Manu, Chief Game & Wildlife Officer, Department of Game and Wildlife, Accra, Ghana, July 1981.
84. For additional information on the biology and habits of the wild pig, see the references listed in notes 3, 9, and 88; and, for example, A. A. Sludskii, *The Wild Boar: Its Ecology and Economic Importance*, translated by P. Aukland

from the original Russian edition (IZADATEL, Alma-Ata, 1956), CSIRO Translation No. 11383, Melbourne, 1974; and L. Heck and G. Raschke, *Die Wildsauen*, Paul Parey, Hamburg, 1980, Ch. 1 and further references given there. For some interesting material on the management of pigs as a resource in Germany, see, for example, Heck and Raschke, *op. cit.*, and Erhard Ueckermann, *Der Schwarzwildabschuss*, Paul Parey, Hamburg, 1977.

85. J. Hone and J. O'Grady, *Feral Pigs and their Control*, N.S.W. Department of Agriculture, Sydney, 1980, p. 9.
86. See, for example, Ken Cuthbertson, *Pighunting in New Zealand*, A. H. & A. W. Reed, Wellington, 1974.
87. Hone and O'Grady, *op. cit.*, p. 9.
88. R. H. Barrett, 'The feral hog on the Dye Creek Ranch, California', *HILGARDIA* 46 (1978) 9: 283-355, p. 344.
89. Wood and Barrett, *op. cit.*, pp. 242, 243.
90. Compare J. Hone, J. O'Grady and H. Pedersen, *Decisions on the Control of Feral Pig Damage*, Ag. Bulletin 5, N.S.W. Department of Agriculture, Sydney, 1980, p. 9.

CHAPTER 2. Pigs as an Agricultural and Pastoral Pest

1. Personal communication from M. Jean de Chancel, for the Director of Nature Conservation (Ministry of the Environment and Planning), le Neuilly, France, July 1980.
2. E. Kesteloot in *Vu Par Les Belges* (1977): 56, p. 17.
3. Personal communication from Ministry of Agriculture and Fisheries, Portugal, August 1980.
4. *Ibid.*
5. *Ibid.*
6. Personal communication from Prof. Dr Jerzy Krupka, President, Polish Hunting Association, June 1980.
7. Regina Mackin, 'Dynamics of damage caused by wild boar to different agricultural crops', *Acta Theorologica* 15 (1970): 27, 447-458.
8. A. A. Sludskii, *The Wild Boar: Its Ecology and Economic Importance*, translated by P. Aukland from the original Russian edition (IZADATEL, Alma-Ata, 1956), CSIRO, Melbourne, 1974, pp. 27, 32.
9. M. D. Springer, 'Ecologic and economic aspects of wild hogs in Texas', in G. W. Wood (ed.), *Research and Management of Wild Hog Populations*, Belle W. Baruch Forest Science Institute of Clemson University, Georgetown, South Carolina, 1977, p. 41.
10. *Ibid.*

11. G. W. Wood and R. H. Barrett, 'The status of wild pigs in the United States', *The Wildlife Society Bulletin* 7 (1979): 4, 237-46.

12. *Ibid.*

13. Personal communication from Professor R. Schlatter, Institute of Zoology, Universidad Austral de Chile, May 1980.

14. H. Fradrich, 'Swine and peccaries', in B. Grizmek (ed.), *Grizmek's Animal Life Encyclopedia*, Van Nostrand Reinhold, New York, p. 81.

15. C. H. Diong, 'Studies of the Malayan wild pig in Perak and Johore', *Malayan Nature Journal* 26 (1973): November, 120-51.

16. Personal communication from Department of Wildlife and National Parks, West Malaysia, July 1980.

17. Diong, *op. cit.*, p. 128.

18. Personal communication from the Director, Forest Research Institute, ARDA, Department of Agriculture, Bogor, Indonesia, July 1980.

19. Personal communication from Dr S. Sastrapradja, Director, National Biological Institute, Indonesia, June 1980.

20. Personal communication from the Director, Wildlife Conservation Division, Royal Forest Department, Thailand, June 1980.

21. Boonsong Lekagul and J. A. McNeely, *Mammals of Thailand*, Kurusapha Ladprao Press, Bangkok, Thailand, p. 664.

22. Personal communication from the Department of Wild Life Conservation, Sri Lanka, July 1980.

23. Information supplied in personal communication by the Nature Conservation Bureau, Environment Agency, Japan, August 1980.

24. Personal communication from the South Korean Embassy in Australia, May 1980.

25. K. A. Wodzicki, *Introduced Mammals of New Zealand: An Ecological and Economic Survey*, Department of Scientific Research, Wellington, 1950, p. 235f.

26. *Ibid.*, p. 236f.

27. Personal communication from the New Zealand Forest Service, April 1980.

28. See M. D. Murray, 'The role of wild animals in the spread of exotic diseases in Australia', *Australian Veterinary Journal* 52 (1976): December, 547-54, esp. p. 549.

29. See, for example, T. McKnight, *Friendly Vermin: A Survey of Feral Livestock in Australia*, University of California Press, Berkeley, 1976.

30. This follows from the assumption that the utility functions of farmers are strictly concave.

31. See, for instance, J. H. Johnston, 'Public policy on cattle tick control in N.S.W.', *Review of Marketing and Agricultural Economics* 43 (1975): 1, 3-39.

32. Unpublished report by Mr W. Holmes, agricultural economist, Department of

Primary Industry, Charleville, Qld, 1975.

33. Report attributed to Queensland Working Party on Feral Pigs, Australian Standing Committee on Agriculture.

34. J. Hone, 'Effect of feral pig rooting on introduced and native pasture in north-eastern New South Wales', *Journal of the Australian Institute of Agricultural Science* 46 (1980): 2, 130-132. Hone refers to R. H. Barrett, Ecology of the feral hog in Tehama County, California (Ph.D. thesis), University of California, Berkeley, 1971: and A. Lazenby and F. C. Swain, *Intensive Pasture Production,* Angus & Robertson, Sydney, 1973.

35. Queensland Working Party on Feral Pigs, *op. cit.*

36. C. W. Smith and C. H. Diong, *Proposal to Study Feral Pigs in Kipahula Valley Haleakala Park,* Cooperative National Park Resources Studies Unit, University of Hawaii, 1977, p. 9.

37. K. Cook, *Pig,* Schwartz, Melbourne, 1980.

38. See note 2.

39. R. Benson, *Evaluating Feral Pig Damage and Costs of Control in Western New South Wales,* Research Workpaper No. 120, Division of Marketing and Economics, N.S.W. Department of Agriculture, Sydney, 1980.

40. These details are reported in Benson, *op. cit.*

41. Queensland Working Party on Feral Pigs, *op. cit.*

42. Personal communication from Mr Sturgess, Director, Bureau of Sugar Experiment Stations, Brisbane, March 1979.

43. The information about sugar cane losses was kindly supplied by Bureau of Sugar Experiment Stations.

44. See note 42.

45. Queensland Working Party on Feral Pigs, *op. cit.*

46. This incident was related by the grazier, Mr Rup Hanna. The offending boar is shown in Figure 3.11.

47. Reported in Benson, *op. cit.*

48. J. Hone, J. O'Grady, H. Pedersen, *Decisions on the Control of Feral Pig Damage,* Ag. Bulletin 5, Department of Agriculture, Sydney, 1980, p. 3.

49. Benson, *op. cit.*

50. *Ibid.,* p. 9.

51. Report attributed to Working Party on Feral Pigs, Vertebrate Pests Committee, Standing Committee on Agriculture, Australia.

52. *Feral Animals in the Northern Territory,* Report of the Board of Inquiry of Northern Territory Legislative Council, Darwin, 1979, pp. 105 and 106.

53. *Ibid.,* p. 105.

54. J. Hone and others, *op. cit.,* p. 2.

55. A copy of the questionnaire is included as Appendix B in C. Tisdell, *Observations on the Wild Pig Problem in N.S.W.*, Research Report or Occasional Paper No. 41, Department of Economics, University of Newcastle, 1978. ISBN 0 7259 0 304 X.
56. J. Hone and J. Waitham, 'Feral pigs are spreading', *Agricultural Gazette of N.S.W.* (1979): April, pp. 12, 13.
57. Map given personally to author by Giles. It is similar to the one appearing in J. R. Giles, 'Control of feral pigs', *Wool Technology and Sheep Breeding* (1977): June/July, 29-31.
58. C. Tisdell, *Simple Economic Models of Pest Control*, Research Report or Occasional Paper No. 33, Department of Economics, University of Newcastle, 1977. ISBN 0 7259 0265 5. See also Tisdell, *Observations on the Wild Pig Problem ... op. cit.*, pp. 2-5 and Ch. 9.
59. McKnight, *op. cit.* Assuming that Queensland has as many and probably more wild pigs than New South Wales, the N.S.W.'s estimates suggest that there are about 3-6 million feral pigs in Australia. If the N.S.W.'s point estimate of 1.55 million is doubled the Australian figure is 3.1 million feral pigs but a somewhat higher Australian figure is plausible.
60. B. V. Fennessy, 'The impact of wildlife species on sheep production in Australia', *Proceedings of the Australian Society of Animal Production*, 6 (1966): pp. 148-56.
61. M. D. Murray and W. A. Snowdon, 'The role of wild animals in the spread of exotic diseases in Australia', *Australian Veterinary Journal* 52 (1976): 547-54. See also report on *Feral Animals in the Northern Territory*, *op. cit.*
62. CSIRO, *Division of Wildlife Research Report* 1972-74, Canberra.
63. *Ibid.*
64. Tisdell, *Simple Economic Models of Pest Control*, *op. cit.*
65. Tisdell, *Observations on the Wild Pig Problem ...*, *op. cit.*
66. For a discussion of the bounty system, see S. Whitehouse, 'Bounty systems and vermin control', *Journal of Agriculture of Western Australia* 17 (1976): 3, 58-59; and C. Tisdell, 'Wildlife: A national asset or pest to be managed', in Department of Science and Environment, *Environmental Economics*, Canberra, 1979, pp. 79-89.
67. Benson, *op. cit.*, p. 1.
68. Giles, *op. cit.*, pp. 29, 30; Hone and others, *op. cit.*, p. 10; Springer, *op. cit.*
69. Hone and others, *op. cit.*, p. 8.
70. *Ibid.*

CHAPTER 3. Recreational Hunting of Wild Pigs

1. M. Clawson and J. L. Knetsch, *Economics of Outdoor Recreation*, John Hopkins Press (for RFF), Baltimore, 1966, pp. 24–25.
2. For detailed statistics, see Australian Bureau of Statistics, *Preliminary Estimates of Firearm Ownership and Possession, May 1975*, Canberra, July 1975; and *Firearm Ownership, May 1975*, Canberra, October 1975.
3. As advised by N.S.W. Firearms Registry, June 1980.
4. Possibly one cannot place too much reliability on the sample, but 80 hunters in the sample indicated that in their districts 54 per cent of the hunters go pig-hunting. See C. Tisdell, *Amateur Hunting of Wild Pigs in Australia: A Report (Stressing Economic Factors) on a Survey of Hunters*, Department of Economics, University of Newcastle, N.S.W. 2308, May 1979. ISBN 0 7259 0343 0, p. 17.
5. An additional indicator of the importance of recreational hunting in Australia is that the *Australian Shooters Journal* has a monthly circulation of 20 000.
6. For information on the hunting of wild boar in Russia, see A. A. Sludskii, *The Wild Boar: Its Ecology and Economic Importance*, translated by P. Aukland from the original Russian edition (IZADATEL, Alma-Ata, 1956), CSIRO, Melbourne, 1974, p. 260 *et seq.*
7. Information kindly supplied by the Assistant Vice-Consul of the Federal Republic of Germany in Australia, July 1980.
8. *Hunting and Fishing: Spain*, Secretariat of Tourism, Madrid, 1974, p. 5.
9. G. W. Wood and R. H. Barrett, 'The status of wild pigs in the United States', *Wildlife Society Bulletin* 7 (1979): 4, 237–46.
10. Stegeman reports that 15 wild boars were introduced into North Carolina from Northern Germany in 1912 by a hunting club and were kept enclosed for eight years and allowed to increase in numbers. When they were hunted in 1920, they escaped from their enclosure and formed the nucleus of the present U.S. population which has interbred with feral pigs. See L. R. Stegeman, 'The European Wild Boar in the Cherokee National Forest, Tennessee', *Journal of Mammalogy* 19 (1938): 279–91.

    See also Wood and Barrett, *op. cit.;* and, for a less technical account, Perry Jones, *The European Wild Boar in North Carolina*, Division of Game, North Carolina Wildlife Resources Commission, Raleigh, N.C., 1972.
11. For more information on feral pigs and their use for hunting in U.S., see R. H. Barrett, 'The feral hog on the Dye Creek Ranch, California', *HILGARDIA* 46 (1978): 9, 288–355.
12. Personal communication from Professor R. Barrett, College of Natural Resources, University of California, Berkeley, June 1980.
13. See V. G. Heptner, A. A. Nasimovic and A. G. Bannikov, *Die Säugetiere der*

Sowjetunion, Vol. 1, Gustav Fisher, Jena, 1966, p. 51.

14. Personal communication from Dr Roberto P. Schlatter, Instituto de Zoologia, Universidad Austral de Chile, May 1980.
15. H. J. Frith, *Wildlife Conservation*, Angus & Robertson, Sydney, 1973, p. 208.
16. Clawson and Knetsch, *op. cit.*, pp. 161-2.
17. J. L. Knetsch and R. K. Davis, 'Comparisons of methods for recreation evaluation', in A. V. Kneese and S. C. Smith (eds), *Water Research*, John Hopkins Press (for RFF), Baltimore, 1965, pp. 126-7.
18. For further information on the available methods, see Knetsch and Davis, *op. cit.*; and J. V. Krutilla, 'Methods of estimating the value of wildlife resources', in J. A. Bailey *et al.* (eds), *Readings in Wildlife Conservation*, The Wildlife Society, Washington, 1974, pp. 125-36.
19. For more details, see, for example, Knetsch and Davis, *op. cit.*
20. P. P. Pearse, 'A new approach to the evaluation of non-priced recreational resources', *Land Economics* 44 (196): 1, 87-99. The model is applied to evaluate big game hunting resources in East Kootenay, British Colombia.
21. Knetsch and Davis, *op. cit.*, p. 129.
22. C. Tisdell, *op. cit.* In the survey the average number of trips annually was nine and average annual expenditure per hunter was close to $450, implying $50 cost per trip. But in the whole population of pighunters, fewer trips can be expected. The number might be close to the mode in my sample: two trips annually.
23. R. L. Gum and W. E. Martin, 'Problems and solutions in estimating the demand for and value of rural outdoor recreation', *American Journal of Agricultural Economics* 57 (1975): 4, 558-66, Table 2, p. 562.
24. D. W. Seckler, 'On the uses and abuses of economic science in evaluating public outdoor recreation', *Land Economics* 42 (1966): 4, 485-94, p. 485.
25. I would like to thank members of the Newcastle club of the Australian Sporting Shooters Association for their assistance in obtaining the hunting magazines.
26. Frith, *op. cit.*
27. Another influence could be the availability of interested writers on particular species.
28. For details of these trends, see C. Tisdell, *The Relative Value Placed on Species for Amateur Hunting in Australia: Observations from Hunting Magazines*, Department of Economics, University of Newcastle, N.S.W. 2308, December 1978. ISBN 0 7259 0326 0.
29. A copy of the survey form is available as Appendix A of C. Tisdell, *Amateur Hunting of Wild Pigs in Australia: A Report (Stressing Economic Factors) on a Survey of Hunters*, Department of Economics, University of Newcastle, N.S.W. 2308,

May 1979. ISBN 0 7259 0343 0.

30. C. Tisdell, *Observations on the Wild Pig Problem in N.S.W.*, Research Report or Occasional Paper No. 41, Department of Economics, University of Newcastle, 1978. ISBN 0 7259 0 304 X. In view of inflation the cost would now be close to \$10 per pig on average.

31. Where x represents the number of pigs taken and y the cost per pig taken the best fit curves to the observations are:

    For N.S.W., $y = 4.8 + 187.1/x$ $(R^2 = 0.70)$

    For Qld, $y = 1/(0.0304 + 0.002x)$ $(R^2 = 0.77)$

    The decreasing cost relationship is similar to that found by R. Turvey for the shooting of feral pigs by landholders. See R. Turvey, *Approaches to the Selection of Optimal Control Strategies for Feral Pigs: Cost-Effectiveness Analysis of Present Control Measures*, B.Agric.Econ. dissertation, University of New England, October 1978.

    A similar relationship for Victoria is graphed in C. Tisdell, *Amateur Hunting of Wild Pigs..., op. cit.*

32. These relationships are graphed in Tisdell, *ibid.* The possibility should not be dismissed that to some extent the above cost curves are of a mongrel type because of the identification problem.

33. Tisdell, *Observations on the Wild Pig Problem in N.S.W., op. cit.* See also chapter 2.

34. See map by Central Mapping, N.S.W. Government, *Feral Pig Distribution in New South Wales*, October 1977. Map compiled by J. O'Grady, N.S.W. Department of Agriculture.

35. Data supplied to me by Mr O. Sturgess, Director of the Bureau of Sugar Experiment Stations, Brisbane, indicate that feral pigs damaged cane valued in 1977 at more than a quarter of a million dollars.

36. Carried out by Richard Benson, Senior Economist, N.S.W. Department of Agriculture, Dubbo. See R. Benson, *'Evaluating Feral Pig Damage and Costs of Control in Western New South Wales'*, Research Workpaper No. 120 (February 1980), Division of Marketing and Economics, N.S.W. Department of Agriculture.

37. Benson, *op. cit.*

38. Tisdell, *Observations on the Wild Pig Problem in N.S.W., op. cit.* My population estimates also seem to be supported by the Benson study previously cited. See also note 59 of Chapter 2.

39. Estimates are based upon the assumption of a stationary pig population in balance with holding capacity and suppose that the average pattern for replacement of an adult is three months in the uterus plus six months to maturity with zero damage in the uterine stage rising linearly from birth to a maximum

constant level at six months of age. If the population is *below* holding capacity, the removal of adults on average will reduce the effective population (below what it would have been) by an even greater amount.

40. Tisdell, *Observations on the Wild Pig Problem in N.S.W., op. cit.* My estimate in 1977 of damage was $6 per pig but this needs to be updated to allow for general inflation and the rise of agricultural prices since then.
41. For further discussion on subsidies, see C. Tisdell, 'Wildlife: A national asset or pest to be managed?', in G. Gorrie (ed.), Department of Science and Environment, *Environmental Economics,* Australian Government Publishing Service, Canberra, 1979, pp. 79-87, especially pp. 85-7.

CHAPTER 4. Landholders' Commercial Gain from Amateur Pighunting

1. Extract from a letter to me from a senior officer of the N.S.W. Department of Agriculture, July 1979.
2. See Chapter 1.
3. Letter from senior official of New Zealand Forest Service to me, April 1980.
4. See Tisdell, 'Kangaroos: the economic management of a common-property resource involving interdependence of production', *Economic Analysis and Policy* 4 (1973): 2, 59-75.
5. See C. W. Clark, *Mathematical Bioeconomics: The Optimal Management of Renewable Resources,* Wiley, New York, 1976.
6. J. Marchington, *Game Shooting: Management and Economics,* Faber, London, 1976; and M. D. Springer, 'Ecologic and economic aspects of wild hogs in Texas', in *Research and Management of Wild Hog Populations* (ed. G. W. Wood), 1977, pp. 37-46, Belle W. Baruch Forest Science Institute of Clemson University, Georgetown, South Carolina.
7. See for instance, B. W. Bootes, 'Feral Pigs', *Agricultural Gazette of New South Wales,* June 1976, 51; and J. Hone and J. Waithman, 'Feral pigs are spreading', *Agricultural Gazette of New South Wales,* April 1979, 12-13.
8. Incomes for hunting fees were estimated in most cases by taking weekend hunting fees and multiplying by an estimate of the annual number of paying hunters coming on to the property.
9. M. D. Springer, *op. cit.,* p. 46.
10. W. E. Howard, *Control of Introduced Mammals in New Zealand,* New Zealand Department of Scientific and Industrial Research, Lower Hutt, 1965.
11. Brochures and information kindly supplied by Bexley Carmen, Outback Horizons, P.O. Box 2, Nailsworth, S.A. 5083.
12. Personal communication from WIMRAY Pty Ltd, P.O. Box 1634, Darwin, N.T. 5794.

CHAPTER 5. Commercial Use of Wild Pigs for Meat and Other Products

1. C. H. Diong, 'Studies of the Malayan wild pig in Perak and Johore', *Malayan Nature Journal* 26 (1973): 120-51.
2. *Ibid.*, p. 148.
3. *Ibid.*, p. 149.
4. See *China's Foreign Trade* 1 (1979): pp. 24-5, Guozi Shudian, Peking.
5. Personal communication from Deputy Director, Wildlife Protection Division, Environment Agency, Japan, August 1980.
6. A. A. Sludskii, *The Wild Boar: Its Ecology and Economic Importance*, translated by P. Aukland from the original Russian edition (IZADATEL, Alma-Ata, 1956) CSIRO, Melbourne, 1974, p. 275.
7. *Ibid.*, p. 272.
8. *Ibid.*
9. Personal communication from J. Breytenbach, Officer in Charge, Saasveld Forestry Research Station, George, R.S.A., November 1980.
10. Personal communication from D. Mason, Kruger National Park, Skukuza, R.S.A., June 1980.
11. See, for example, K. A. Wodzicki, *Introduced Mammals of New Zealand: An Ecological and Economic Survey*, Department of Scientific and Industrial Research, Wellington, 1950.
12. See, for example, C. Miller, 'Europe goes the whole hog', *The Bulletin*, 14 October 1980, pp. 44-5.
13. *Ibid.*
14. Personal communication from Dr C. N. Challies, Forest Research Institute, Christchurch, New Zealand, March 1980.
15. Personal communication from Mr S. Fokerd, for the Director-General, N.Z. Forest Service, Wellington, New Zealand, April 1980.
16. For some further details about this survey see C. Tisdell, *On the Commercial Use of Feral Pigs in the Meat and Related Trades: Including a Report on a Survey and a Theoretical Note*, Department of Economics, University of Newcastle, N.S.W. 2308, July 1979. ISBN 0 7259 0 351 1. The questionnaire form is included as Appendix B.
17. P. L. Appleton and J. H. Norton, 'Sparganosis... a parasitic problem in feral pigs', *Queensland Agricultural Journal* 102 (1976): 6, 339-43. See p. 343 for feral pig condemnation rates for sparganosis at slaughter at Queensland meatworks.
18. Field days were held by the Regional Wild Pig Advisory Council at Nundle, N.S.W., on 27 and 28 March 1979, and a seminar was held on wild pigs in Tamworth, N.S.W., on 29 March 1979.

19. *Report of Conference on the Feral Pig Problem*, Canegrowers' Building, Brisbane, 1974.

20. For the actual weights of dressed carcasses, see Tisdell, *op. cit.*, p. 8.

21. H. McL. Gordon, B. A. Forsyth and M. Robinson, 'Sparganosis in feral pigs in New South Wales', *Australian Veterinary Journal* (1954): May, 135-8.

22. Appleton and Norton, *op. cit.*, p. 342.

23. For more detailed discussion of sparganosis and other parasites and diseases of feral pigs, see Chapter 8.

24. A rendering works near Scone, N.S.W.

25. For details on commercial value of pig hide, see Tisdell, *op. cit.*, Appendix C.

26. Tisdell, *op. cit.*, pp. 12-13.

27. Australian Bureau of Animal Health, 'Summary of Proposed Amendments to the Exports (Meat) Regulations to Permit the Export of Meat and Meat Products Derived from Game Animals', Canberra, November 1979, p. 1.

28. Australian Bureau of Animal Health, 'Operational Guidelines for the Inspection of Game Meat for Export', Canberra, 5 June 1980, p. 4.

29. *Ibid.*, p. 7.

30. Ref. 27.

31. C. Miller, 'Europe goes the whole hog', *The Bulletin*, 14 October 1980, p. 45.

32. *Ibid.*, p. 44.

33. *Ibid.*, p. 45.

34. *Ibid.*

35. See, for example, Dr D. M. Flynn, *Report on Vertebrate Pests and Exotic Animal Diseases*, Australian Bureau of Animal Health, June 1980, p. 32. See also estimates given in Chapter 2 of this book. These indicate 3-6 million pigs present in Australia.

36. J. Hone, J. O'Grady and H. Pedersen, *Decisions on the Control of Feral Pig Damage*, Ag. Bulletin 5, N.S.W. Department of Agriculture, Sydney, 1980, p. 8.

37. Miller, *op. cit.*, p. 45.

38. See Chapter 2; and Sue Carney, 'Plunder in the Bush', *National Farmer*, 26 June 1980, p. 38.

39. See, for example, C. Tisdell, 'The economic conservation and utilisation of wildlife species', *South African Journal of Economics* 40 (1972): 235-48; and H. Scott Gordon, 'The economic theory of a common property resource: The fishery', *Journal of Political Economy*, 62 (1954): 124-42.

40. As Professor J. Wiseman, of York University, England, pointed out to me landowners may obtain a de facto property-right in wild animals on their property.

41. Personal communication from Mr S. Fokerd, for the Director-General, N.Z. Forest Service, Wellington, New Zealand, April 1980.

42. Some theoretical aspects of this matter are considered in Tisdell, *op. cit.*, Appendix A, 'A theoretical note on the influence of commercial use on the rate of killing of a species which is taken in the absence of commercialisation'.
43. Kangaroo processors and hunters are often given exclusive areas by state government. It has been reported by Sue Carney, *op. cit.*, that Game Management Pty Ltd has negotiated exclusive hunting rights in several south-eastern Queensland shires in return for building a processing works at Goondiwindi to process 100 000 feral pigs per year.
44. See, for example, A. Scott, 'The fishery: The objectives of sole ownership', *Journal of Political Economy* 63 (1955): 116–42.
45. For an outline of the impact of monopsony on the payment for resources, see, for example, C. A. Tisdell, *Economics of Markets*, John Wiley, Sydney, 1974, Ch. 12.
46. Referred to earlier in this chapter.
47. Personal correspondence from a Toowoomba hunter and hunting store proprietor, May 1981.
48. See, for example, *The Deerstalker* 4 (1978): 5 (October), p. 13.
49. Personal communication from M. Jean de Chancel, for the Director of Nature Conservation (Ministry of the Environment and Planning), le Neuilly, France, July 1980.
50. The existence of the kangaroo in reasonable numbers appears to be valued by the general public.
51. Sludskii, *op. cit.*, pp. 284, 285.
52. Sludskii, *op. cit.*, p. 285.
53. Sludskii, *op. cit.*, pp. 279, 280.

CHAPTER 6. Pigs in National Parks and as a Threat to Wildlife

1. F. Rodriguez de la Fuente, *World of Wildlife*, Vol. 8, Orbis, London, 1971, p. 243.
2. H. J. Frith, *Wildlife Conservation*, Angus & Robertson, Sydney, 1973, p. 159.
3. For general statements about this research, see CSIRO, *Division of Wildlife Research Report* 1972–74, pp. 15, 16; 1974–76, pp. 14, 15; 1976–78, p. 14.
4. This research includes the effect of the dyeing of grain bait on non-target species, particularly birds. Personal communication from J. Hone, Veterinary Research Station, Glenfield, N.S.W., June 1980.
5. Harry Butler, *In the Wild*, Vol. I, Australian Broadcasting Commission, Sydney, p. 23.
6. *Feral Pigs in the Northern Territory*, Report of the Board of Inquiry, established

by the Northern Territory Legislative Council, Darwin, 1979, pp. 39, 42.

7. CSIRO, *Division of Wildlife Research Report* 1974-76, pp. 14, 15. Unfortunately no publications on this subject are indicated.
8. J. Giles, J. Hone *et al.*, *The Biology and Control of Feral Pigs*, Australian Meat Committee Final Report, June 1977, N.S.W. Department of Agriculture, Glenfield, 1977, pp. 14, 15.
9. E. M. Pullar, 'The wild (feral) pigs of Australia and their role in the spread of infectious diseases', *Australian Veterinary Journal* (1950): May, 106.
10. Sludskii estimates that in Russia a full-grown wolf eats 50-80 wild boar per year. Wild boar are an important part of the diet of wolves. See A. A. Sludskii, *The Wild Boar: Its Ecology and Economic Importance*, translated by P. Aukland from the original Russian edition (IZDATEL, Alma-Ata, 1956), CSIRO, Melbourne, 1974, p. 147f. In Asia, the tiger was a significant predator on wild boar, before its numbers were much reduced.
11. The following are some general background books on Australian fauna: H. G. Cogger, *Reptiles and Amphibians of Australia*, A. H. & A. W. Reed, Sydney, 1975; H. Frauca, *The Book of Australian Wild Life*, Heinemann, London, 1965; Frith, *op. cit.*; H. J. Frith, *Waterfowl in Australia*, Angus & Robertson, Sydney, 1967; H. J. Frith and J. H. Calaby, *Kangaroos*, Cheshire, Melbourne, 1969; J. M. Leach, *An Australian Bird Book*, 9th edn, Whitcombe & Tombs, Melbourne, 1961; J. D. MacDonald, *Birds of Australia*, A. H. & A. W. Reed, Sydney, 1973; D. Ovington, *Australian Endangered Species*, Cassell Australia, Sydney, 1978; M. J. Tyler (ed.), *The Status of Endangered Australasian Wildlife*, Royal Zoological Society of South Australia, Adelaide, 1979.
12. For information on the adverse effects of feral pigs and wild boar on birds on islands and other continents, see, for example:

    Lord Howe Island: D. Lunney and H. F. Recher, 'National parks: A museum, a garden and an asylum', in H. F. Recher *et al.* (eds), *A Natural Legacy: Ecology in Australia*, Pergamon Press, Sydney, 1979, p. 193. Lunney and Recher point out that the 'Lord Howe Island woodhen *Tricholimnas sylvestris*, has been reduced to 25 individuals by the combined predation of rats, pigs and dogs.'

    New Guinea: Richard Schodde, 'The status of endangered Papuasian birds', in M. J. Tyler (ed.), *The Status of Endangered Australasian Wildlife*, Royal Zoological Society of South Australia, Adelaide, 1979. He mentions adverse impacts on Audubon's shearwater *Puffinus therminieri heinrothi* and on Bruijn's brush turkey *Alectura bruisnii*.

    New Zealand: C. N. Challies, 'Feral pigs (*Sus scrofa*) on Auckland Island: Status, and effects on vegetation and nesting sea birds', *New Zealand Journal of Zoology* 2 (1975): 4, 479-90, and other references listed there. He says that 'Pigs have undoubtedly had an effect on the distribution and number of some bird species but their role is often hard to separate from that of feral cats, or from the past influence of feral dogs (*Canis familiaris*) and man. Auckland Island has become vulnerable for some birds and others have been adversely affected, especially the sea and oceanic species that nest on the island. Pigs in particular have been accused from time to time of digging up the burrows of petrels and taking the eggs and young of albatrosses, mollymawks, penguins and shags.' (p. 486) Challies concludes from his studies that remaining species on the island are now in balance with the feral pig population and most of the flora and fauna is represented elsewhere in the island group. His policy conclusion is very interesting. 'It is difficult to justify control of pigs on Auckland Island solely as a conservation measure. If control must be attempted, the only worthwhile objectives are extermination or a regulated lower density both of which appear to be impracticable at present. There is little point in reducing the density of pigs since, without a sustained effort, their numbers are likely to increase again, and may even temporarily reach high levels. If the population is left undisturbed the present balance between the pigs and their modified environment should continue.' (p. 489)

K. A. Wodzicki, *Introduced Mammals of New Zealand: An Ecological and Economic Survey*, Department of Scientific and Industrial Research, Wellington, 1950, on p. 232 quotes T. Kirk, 'The displacement of species in New Zealand', *Trans. N.Z. Inst.* 28 (1896): 1-27, as claiming that 'pigs together with other introduced animals were responsible for the destruction on the mainland of the tuatara and also most of the kakapos, and in addition has caused a reduction in the numbers of other ground-nesting birds.'

J. C. Yaldwyn (ed.), *Preliminary Results of the Auckland Island Expedition 1972-73*, Department of Lands and Survey, Wellington, 1975.

United States: V. G. Henry, 'Predation on dummy nests of ground-nesting birds in the Southern Appalachians', *Journal of Wildlife Management* 33 (1969): 1, 169-72. Henry suggests that European wild boar are a minor nest predator of turkeys and ruffed grouse.

V. G. Henry and R. H. Conley, 'Fall foods of European wild hogs in the Southern Appalachians', *Journal of Wildlife Management* 36 (1972): 3, 854-60.

M. D. Springer, 'Ecologic and economic aspects of wild hogs in Texas', in G. Wood (ed.), *Research and Management of Wild Hog Populations*, Belle W. Baruch Forest Science Institute of Clemson University, Georgetown, South Carolina, 1977, p. 41. Springer claims that birds were a common dietary item of wild pigs in Texas.

Le Roy C. Stegeman, 'The European wild boar in the Cherokee National Forest,

Tennessee', *Journal of Mammalogy* 19 (1938): 3, 279-90. See pp. 288-9 in particular. He finds that young birds and birds' eggs of the bob-white quail and ruffed grouse may be eaten on occasions and there is food competition with some species. On the whole he believes the impact to be small for the species studied in his area.

G. W. Wood and R. H. Barrett, 'The status of wild pigs in the United States', *The Wildlife Society Bulletin* 7 (1979): 4, 237-46, esp. p. 241 and references given there.

G. Wood and T. E. Lynn, Jr, 'Wild hogs in Southern forests', *Southern Journal of Applied Forestry* 1 (1977): 2. See also Chapter 7 refs. They surveyed 733 foresters, wildlife biologists and land managers in 11 south-eastern states. 'Competition with deer, turkey, and squirrels for food resources was believed to be an important influence by less than half the respondents while about one-third rated turkey nest predation as important.'

Europe and U.S.S.R.: Sludskii, *op. cit.*, and others report predation in Europe and U.S.S.R. on rats and mice and young rabbits in their burrows. See also V. G. Heptner, A. A. Nasimovic, and A. G. Bannikov, *Die Säugetiere der Sowjetunion*, Vol. 1, Gustav Fischer, Jena, 1966, p. 28f.

13. Personal communication from Professor R. G. Barrett, University of California, Berkeley, August 1980.
14. For information on the adverse effects of feral pigs and wild boar on other mammals in the United States, see the references listed above (in note 12). Stegeman, *op. cit.*, found that there could be some food competition between pigs and squirrels, chipmunks, opossums and raccoons and does not rule out the possibility of competition with deer for mast.
15. M. Morcombe, *Australia's National Parks*, Lansdowne, Sydney, 1977, p. 158.
16. Pig predation on reptiles including snakes has been reported overseas. See, for example, Springer, *op. cit.*; Stegeman, *op. cit.*; Wood and Barrett, *op. cit.*; Wood and Lynn, *op. cit.*; and M. D. Springer, Food Habits of Wild Hogs on the Texas Gulf Coast, M.Sc. thesis, Texas A & M University, 1975. He found in his study in the Arkansas National Wildlife Refuge (1973-74) that 'snakes were the only type of vertebrates taken every season'.
17. *The Australian Encyclopedia*, Vol. 5, The Grolier Society of Australia, Sydney, 1977, p. 21.
18. Pullar, *op. cit.*, p. 21.
19. Wood and Barrett, *op. cit.*, p. 241.
20. S. W. Gorham, 'Fiji frogs: Life history data from field work', *Zool. Beitr.* (new series) 14 (1968): 427-46. See also M. J. Tyler, 'The impact of European man upon Australasian amphibians', in Tyler, *op. cit.* Springer in his study,

op. cit., found that the Rio Grande leopard frog was commonly eaten by pigs.

21. In relation to the Great Smoky Mountains National Park, U.S.A., Scott and Pelton report that 'Rooting and wallowing habits of hogs are another concern due to their possible detrimental effects on native brook trout (*Salvilinus fontinalis*) populations through siltation and the possible contamination of water supplies.' See C. D. Scott and M. R. Pelton, 'Seasonal food habits of the European wild hog in the Great Smoky Mountains National Park', *Proceedings of the 29th Annual Conference of the Southeastern Association of Game and Fish Commissioners*, 1975, 585–92.

    Predation on fish and crustacea is reported from several countries. See, for example, Sludskii, *op. cit.*; Springer, *op. cit.*; and C. H. Diong, 'Studies of the Malayan wild pig in Perak and Johore', *Malayan Nature Journal* 26 (1973): 120–51.

    In Australia, native fish lay their eggs in billabongs when these are joined together by floods. After the floods subside, hatchling fish, now separated to some extent from predators, in the discrete billabongs feed off grasses and herbage along the banks of the billabongs. Feral pigs by feeding on and trampling herbage at the margin of billabongs reduce the available food source for the hatchlings. See R. D. McDowall (ed.), *Freshwater Fishes of South-Eastern Australia*, A. H. & A. W. Reed, Sydney, 1980.

22. The problem of controlling feral pigs and/or wild boar in national parks and wildlife refuges is not unique to Australia. In the U.S.A., for example, the problem exists in the south and in Hawaii. See, for example:

    Susan P. Bratton, 'The effect of the European wild boar (*Sus scrofa*) on high-elevation vernal flora in Great Smoky Mountains National Park', *Bulletin of the Torrey Botanical Club* 101 (1974): 4, 198–206.

    Susan P. Bratton, 'The effect of the European wild boar (*Sus scrofa*) on Gray Beach Forest in the Great Smoky Mountains', *Ecology* 56 (1975): 6, 1356–66.

    Susan P. Bratton, 'The effect of the European wild boar on the flora of the Great Smoky Mountains National Park', in G. Wood (ed.), *Research and Management of Wild Hog Populations*, *op. cit.*, pp. 47–51.

    R. H. Conley, 'Management and research of the European wild hog in Tennessee', in G. Wood (ed.), *op. cit.*, pp. 67–70.

    J. R. Fox and M. R. Pelton, 'An evaluation of control techniques for the European wild hog in the Great Smoky Mountains National Park', in G. Wood (ed.), *op. cit.*, pp. 53–63.

    F. J. Singer *et al.*, *Home Range, Movements, and Habitat Use of European Wild Boar*, Report for the Superintendent of Great Smoky National Park, U.S. Department of the Interior, National Park Service, Gatlinburg, Tennessee, May 1979.

C. W. Smith and C. H. Diong, *Proposal to Study Feral Pigs in Kipahulu Valley Haleakala National Park*, Cooperative National Park Resources Studies Unit, Department of Botany, University of Hawaii, Honolulu, September 1977.

Springer, *op. cit.*

Tennessee Game and Fish Commission, *European Hog Research Project W-34* (Final Report), 1971.

R. L. Thompson, 'Feral hogs in national wildlife refuges', in G. Wood (ed.), *op. cit.*, pp. 11-15.

M. H. Tuff, *The Effect of the European Wild Boar* (Sus scrofa) *on the Woody Vegetation of Gray Beach Forest in the Great Smoky Mountains*, Management Report No. 18, U.S. Department of Interior, National Park Service, Gatlinburg, Tennessee, October 1977.

Wood and Barrett, *op. cit.*

23. Some of the problems of this evaluation are discussed in C. Tisdell, *On the Economics of Saving Wildlife from Extinction*, Research Report or Occasional Paper No. 48, Department of Economics, University of Newcastle, February 1979. ISBN 0 7259 0329 5.

24. For a copy of the survey form, see Appendix A in C. Tisdell, *Feral Pigs (and Other Feral or Introduced Mammals) in National Parks and Nature Reserves in New South Wales*, Department of Economics, University of Newcastle, N.S.W., September 1979. ISBN 0 7259 0355 4. For general information on New South Wales national parks, the reader may wish to consult A. Fairley, *A Field Guide to the National Parks of New South Wales*, Rigby, Adelaide, 1978.

25. Macquarie Marshes Nature Reserve at the time of the survey was under the administration of the Department of Lands, and grazing leases were in fact in force for the land, but it was expected to come under the control of the National Parks and Wildlife Service. Personal communication from J. R. Giles, National Parks and Wildlife Service, November 1979.

26. For estimates of the number of feral pigs in individual parks and nature reserves in New South Wales, see Tisdell, *op. cit.*, p. 9, Table 4. See also National Parks and Wildlife Service, *Control of Mammalian Pests*, Sydney, 1978.

27. For estimates of densities of feral pigs in individual parks and nature reserves, see Tisdell, *op. cit.*, p. 12, Table 5.

28. It has been suggested to me that the number of pigs indicated for Barrington Tops National Park may be an overestimate. There clearly is difficulty in obtaining an accurate estimate. Nevertheless feral pigs are present in large numbers in this park and pose a major threat.

29. For the allocation of National Parks and Wildlife Service Districts to regions, see Tisdell, *op. cit.*, Appendix B.

30. More detailed information on National Parks and Wildlife Service (N.S.W.) expenditure on feral pig control is available in Tisdell, *op. cit.*, pp. 33-43.
31. All the kill rates are well below the 70 per cent level considered desirable by Giles, Hone *et al.*, (*op. cit.*, p. 28) for effective control.
32. For some detailed observations on goats in New South Wales' parks, see Tisdell, *op. cit.*, pp. 50-54.
33. For further details about frequency with which feral or introduced mammals are considered sufficiently serious to warrant control in parks, see Tisdell, *op. cit.*, p. 46, Table 33.
34. Tisdell, *op. cit.*, Appendix C, Tables C4 to C7.
35. *Ibid.*, Appendix C, Table C8.
36. Co-operation is being achieved in a number of areas by the voluntary formation of Regional Wild Pig Advisory Councils, consisting of representatives of the National Parks and Wildlife Service, the Forestry Commission and Pastures Protection Boards.
37. The draft questionnaire is available as Appendix A to C. Tisdell, *A Report on a Survey of Feral Pigs in National Parks in Western Australia: Presence, Damages and Control Measures*, Department of Economics, University of Newcastle, N.S.W., October 1979. ISBN 0 7259 0359 7.
38. The density at Serpentine Falls National Park, W.A., of 21 ha/pig compares with 24 ha/pig at Willandra National Park (Griffith District), N.S.W., and 20 ha/pig at Warrumbungle National Park (Coonabarabran District), N.S.W., and is regarded as heavy.
39. An independent return covering Walyunga National Park and John Forrest National Park was received from the District Officer of the Metropolitan Area on 3 September 1979 through the Agricultural Protection Board. This return confirms the main points made by the rangers and amplifies some points. Pigs (in Walyunga) were said to be destroying plants in creek banks and soaks, and a few complaints had been received of pigs migrating from the park and of their causing damage to pastures and killing lambs.
40. See Frith, *op. cit.*, Ch. 1.
41. A copy of the survey form is included as Appendix 3 in Tisdell, *Feral Pigs in Queensland National Parks and Nature Reserves: Results of a Survey*, Department of Economics, University of Newcastle, 2308, Australia, November 1980. ISBN 0 7259 0396 1.
42. For details of estimated levels of feral pig populations and densities in individual Queensland parks, see Tisdell, *ibid.*, Table A1.1, pp. 25-27.
43. Tisdell, *ibid.*, Table A1.3, pp. 29-30.
44. Many of the noxious, introduced weeds that are spread by feral pigs in Australia

come from areas in other countries where wild pigs or similar species are naturally present.

45. Possibly those who advocate the eradication of feral pigs in Australia should start here, by excluding (by fencing) new colonies of pigs from one selected national park. If eradication proves unworkable there, what prospect is there for it being possible in Australia as a whole?
46. J. A. Sinden and A. C. Worrell, *Unpriced Values: Decisions without Market Prices*, Wiley, New York, 1979. See also Tisdell, *On the Economics of Saving Wildlife from Extinction*, previously cited.
47. Frith, *Wildlife Conservation*, previously cited.

CHAPTER 7. Pigs in Forests and as a Threat to Wildlife

1. For a discussion of different methods and problems involved in specifying objective functions, see J. A. Sinden and A. C. Worrell, *Unpriced Values: Decisions without Market Prices*, Wiley, New York, 1979.
2. H. Fradrich, 'Swine and peccaries', in B. Grizmek (ed.), *Grizmek's Animal Life Encyclopedia*, Van Nostrand Reinhold, New York, 1979, p. 85.
3. J. Giles, J. Hone *et al.*, *The Biology and Control of Feral Pigs*, Australian Meat Committee Final Report, June 1977, N.S.W. Department of Agriculture, Glenfield, 1977.
4. See, for example, R. C. Stegeman, 'The European wild boar in the Cherokee National Forest, Tennessee', *Journal of Mammalogy* 19 (1938): 3, 279-90.
5. V. G. Heptner, A. A. Nasimovic and A. G. Bannikov, *Die Säugetiere der Sowjetunion*, Vol. 1, Gustav Fischer, Jena, 1966, p. 82, as translated from the German by Dr M. E. Tisdell.
6. V. G. Henry and R. H. Conley, 'Fall foods of European wild hogs in the Southern Appalachians', *Journal of Wildlife Management* 36 (1972): 3, 854-60, p. 859.
7. G. W. Wood and T. E. Lynn, Jr, 'Wild hogs in Southern forests', in *Southern Journal of Applied Forestry* 1 (1977): 2.
8. Heptner *et al.*, *op. cit.*, p. 82.
9. *Ibid.*
10. *Ibid.*
11. S. Erl, 'Erfahrungen mit der Schwarzwildbewirtschaftung in einem Waldgatter', in R. König and R. R. Hofman (eds), *Schwarzwild-Symposion Giessen*, Ferdinand Enke Verlag, Stuttgart, 1980 (Special Issue 1, of *Schrifften des Arbeitskreises für Wildbiologie und Jagdwissenschaft and der Justus-Liebig-Universität Giessen)*, pp. 73 and 75, as translated by Dr M. E. Tisdell.
12. In the past (at least), feral pigs have caused considerable damage to exotic

forest plantations in New Zealand, especially in the early stages of growth. (K. A. Wodzicki, *Introduced Mammals of New Zealand: An Ecological and Economic Survey*, Department of Scientific and Industrial Research, Wellington, 1950, p. 233f.) Now that numbers have been reduced by hunting for commercial use, damage is less severe.

13. A. A. Sludskii, *The Wild Boar: Its Ecology and Economic Importance*, translated by P. Aukland from the original Russian edition (IZADATEL, Alma-Ata, 1956), CSIRO, Melbourne, 1974, p. 53.
14. S. P. Bratton, 'The effect of the European wild boar (*Sus scrofa*) on the high-elevation vernal flora in Great Smoky Mountains National Park', *Bulletin of the Torrey Botanical Club* 101 (1974): 4, 198-206, p. 198.
15. T. D. Howe and S. P. Bratton, 'Winter rooting activity of the European wild boar in the Great Smoky Mountains National Park', *Castanea* 41 (1976): 256-64, p. 263.
16. C. W. Smith and C. H. Diong, *Proposal to Study Feral Pigs in Kipahula Valley Haleakala National Park*, Technical Report No. 19, Cooperative National Park Resources Studies Unit, Department of Botany, University of Hawaii, Honolulu, September 1977.
17. Wodzicki, *Introduced Mammals of New Zealand*, previously cited, p. 233.
18. T. Kirk, 'The displacement of species in New Zealand', *Trans. N.Z. Inst.* 28 (1896): 1-27.
19. G. M. Thomson, *The Naturalisation of Animals and Plants in New Zealand*, Cambridge University Press, Cambridge, 1922, p. 39.
20. See, for example, J. C. Yaldwin (ed.), *Preliminary Results of the Auckland Islands Expedition 1972-73*, Department of Lands and Survey, Wellington, 1975; and C. N. Challies, 'Feral pigs (*Sus scrofa*) on Auckland Island: Status and effects on vegetation and nesting sea birds', *New Zealand Journal of Zoology* 2 (1975): 4, 479-90, and other references listed there.
21. Hone made some observations on pig rooting in an area of native forest adjacent to that area in which he studied the effect of pig rooting on pasture. See J. Hone, 'Effect of feral pig rooting on introduced and native pasture in north-eastern New South Wales', *Journal of the Australian Institute of Agricultural Science* 46 (1980): 2, 130-2. The forest was of a dry sclerophyll type, with the dominant tree types *Eucalyptus*, *Banksia*, *Casuarina* and *Acacia*. Although he was unable to estimate the effect of feral pig rooting on the forest, he suggests 'The litter of many Eucalyptus forests is in a continual state of disturbance by many large animals such as wombats, bandicoots and lyrebirds. Accordingly, pig rooting is not necessarily destructive unless its frequency is high.' (p. 132) But what is a high frequency, and can this

conclusion be applied to forests elsewhere in Australia?

22. Anon., 'New light on the origins of Australia's flora', *Ecos* 24 (1980): 3.
23. *Ibid.*
24. A. Fairley, *A Field Guide to the National Parks of New South Wales*, Rigby, Adelaide, 1978, p. 124.
25. Sphagnum bogs form the headwaters of many rivers and streams. They store large quantities of water soaked up like a sponge. They provide stability and continuity of water flow in rivers after months of no rainfall. They also provide high-quality water and reduce run-off considerably. Interesting plant species occur within the swamps. A rare frog found in such swamps is the corroboree frog *Pseudophryne corroboree*. In all probability it is eaten by wild pigs.
26. Charging a fee for hunting in forests is common in Europe and the U.S.A.
27. Wood and Lynn, *op. cit.*
28. Heptner *et al.*, *op. cit.* pp. 81 and 82. On the other hand, recent evidence by Baettig contradicts Heptner's view about pigs' usually preferring to remain in forests. In Waadt Canton, Switzerland, Baettig found that sows elected to move away from feeding in the forest, to feed themselves more easily and fully on agricultural crops, especially maize. The exception was in autumn-winter after oaks and birches had shed their seeds. M. Baettig, 'Das Schwarzwild im Kanton Waadt-Seine Jagd, Verbreitung und sein Konflikt mit der Landwirtschaft' in König and Hofman, *op. cit.*
29. Heptner *et al.*, *op. cit.*, p. 82.
30. *Year Book Australia 1979*, Vol. 63, Australian Bureau of Statistics, Canberra, p. 315.
31. J. C. McIlroy, 'The effects of forestry practices on wildlife in Australia: A review', *Australian Forestry* 41 (1978): 2, 78-94.
32. For a copy of the survey form, see Appendix C in C. Tisdell, *Results from a Survey of Feral Pigs in New South Wales State Forests*, Department of Economics, University of Newcastle, N.S.W., June 1980. ISBN 0 7259 0381 3.
33. For details of the statistical results of the survey, see Tisdell, *ibid.*
34. The numbers of wild pigs in forest districts is also listed in Tisdell, *ibid.*, Appendix A, Table 1, which also gives information about the distribution of feral goats. In general, more details are available in the report than are given in this chapter.
35. See Chapter 6.
36. For details of the averaging procedure of mean rank scores, see Tisdell, *op. cit.*
37. Bratton *op. cit.*
38. The impact on small ground marsupials and mammals of combined predation by cats, pigs and foxes could be serious.

39. See R. M. McDowall (ed.), *Freshwater Fishes of South-eastern Australia,* A. H. & A. W. Reed, Sydney, 1980. Australian freshwater fish feed on vegetation at the edge of billabongs or waterholes.
40. The effect on trout spawning by increased turbidity of streams was suggested to me by Mr R. Penfold, a pighunter and fisherman from Newcastle, N.S.W.
41. See the discussion of Queensland forests later in this chapter.
42. J. A. Springett, 'The underground movement: Soil', in H. F. Recher *et al.* (eds), *A Natural Legacy: Ecology in Australia,* Pergamon Press, Sydney, 1979, p. 54.
43. Robinson says of root-infecting fungi: 'The ectotrophic growth habit is almost certainly a response to a tropic stimulus from the host root, for it is well authenticated that fungi of this type only rarely make any vegetative growth through the soil in the absence of a living host. Thus during a non-parasitic phase these fungi are forced to remain in a state of semi-dormancy until stimulated into activity by the proximity of a potential host root.' R. K. Robinson, *Ecology of Fungi,* English Universities Press, London, 1967.
44. Deverall mentions that 'Three routes of entry for parasitic fungi are through wounds, natural openings (such as stomata) and directly through intact surfaces.' B. J. Deverall, *Fungal Parasitism,* Edward Arnold, London, 1969, p. 37.
45. This may become evident in Western Australia where *Phytophthora cinnamomi* has been accidentally introduced and is spread by foresters, machinery, etc., in forests. In this situation pigs also would help its spread.
46. For a copy of the Queensland survey form, see Appendix C, in C. Tisdell, *Results from a Survey of Feral Pigs in Queensland State Forests and Timber Reserves,* Department of Economics, University of Newcastle, N.S.W., August 1980. ISBN 0 7259 0384 8.
47. In this exposition, forest management groupings are referred to as 'sub-districts'. In practice, official sub-districts in Queensland each cover a greater area.
48. For further details about the distribution and other aspects of pigs in Queensland forests, see Tisdell, *Results From a Survey of Feral Pigs in Queensland State Forests and Timber Reserves.*
49. For further details on the impact of feral pigs in Queensland forests, see Tisdell, *ibid.*
50. See Chapter 6.
51. Forest Department circular.
52. Virus inoculation (possibly with swine fever) has been tried in Pakistan in an attempt to reduce the population of wild pigs (in this Islamic country, pigs are considered to be a pest). Correspondence from the National Council for Conservation of Wildlife in Pakistan, November 1980.

53. For some thoughts about the way in which feral pigs might assist the spread and occurrence of *Phytophthora cinnamomi*, see the earlier discussion about pigs in New South Wales forests.

CHAPTER 8. Health and Disease Risks from Wild Pigs

1. *Good News Bible*, The Bible Society in Australia, Canberra, 1976, p. 111.
2. A. Jones (gen. ed.), *The Jerusalem Bible*, Darton, Longman and Todd, London, 1966, pp. 702-3.
3. R. Burton, *All Colour World of Farm Animals*, Octopus Books, London, 1979, p. 51.
4. *Ibid.*
5. See, for example, R. A. Rappaport, *Pigs for the Ancestors: Ritual in the Ecology of the New Guinea People*, Yale University Press, New Haven, 1968. For a discussion of the role of pigs in the occurrence of human diseases in New Guinea see Barry Shaw, 'The children of the Kyaka Enga: Culture, diet, environment and health in Papua New Guinea Highlands Society 1950-1960', in L. Manderson (ed.), *Food, Culture and Society*, Cambridge University Press, Cambridge, forthcoming.
6. A. A. Sludskii, *The Wild Boar: Its Ecology and Economic Importance*, translated by P. Aukland from the original Russian edition (IZADATEL, Alma-Ata, 1956), CSIRO, Melbourne, 1974, pp. 165-6.
7. *Ibid.*, p. 166.
8. *Ibid.*, p. 166.
9. Center for Disease Control, *Trichinosis Surveillance Annual Summaries, 1973-1977*, 5 vols., U.S. Department of Health, Education and Welfare, Public Health Service, Atlanta, 1974-78.
10. G. W. Wood and R. H. Barrett, 'The status of wild pigs in the United States', *The Wildlife Society Bulletin* 7 (1979): 4, 237-46, p. 242.
11. G. W. Wood, J. B. Hendricks and D. E. Goodman, 'Brucellosis in feral swine', *Journal of Wildlife Diseases* 12 (1976): 579-82.
12. H. Becker *et al.*, 'Brucellosis in feral swine in Florida', *Journal of American Veterinaries Medical Association* 173 (1978): 1181-2.
13. J. Griffin, *Ecology of the Feral Pig on the Island of Hawaii*, Hawaii Department of Land and Natural Resources, Division of Fish and Game, Honolulu, 1978.
14. Wood and Barrett, *op. cit.*, p. 242.
15. *Ibid.*
16. R. C. Belden and M. R. Pelton, 'Wallows of the European wild hog in the mountains of East Tennessee', *Journal of the Tennessee Academy of Science* 51 (1976): 3 (July), 91-3, p. 91.
17. *Ibid.*, p. 92.

18. K. Hart, 'Feral pig problems on the South Coast', *Agriculture Gazette of New South Wales* 90 (1979): 6 (December), p. 1.
19. *Ibid.*, p. 2.
20. *Feral Animals in the Northern Territory*, Report of the Board of Inquiry established by the Northern Territory Legislative Council, Darwin, 1979, p. 96.
21. *Ibid.*, p. 93.
22. J. T. Martin, 'Wild pigs: Potential vectors of foot and mouth disease', *New Zealand Journal of Agriculture* 125: 6, 18-23, p. 19.
23. For general coverage of this topic (and the role of pigs in the spreading of endemic diseases), see E.M. Pullar, 'The wild (feral) pigs of Australia and their role in the spread of infectious diseases', *Australian Veterinary Journal* (1950): May, 99-110. See also M. D. Murray and W. A. Snowdon, 'The role of wild animals in the spread of exotic diseases in Australia', *Australian Veterinary Journal* 52 (1976): 547-54.
24. H. W. Dunne (ed.), *Disease of Swine*, 3rd edn, Iowa State University Press, Ames, Iowa, 1970.
25. *Ibid.*, pp. 309-36.
26. The National Farmer's Federation, *Farm Focus: The 80's*, Canberra, 1981, p. 64.
27. Australian marsupials have also been experimentally infected by the virus. See Murray and Snowdon, *op. cit.*, p. 533; and W. A. Snowdon, 'The susceptibility of some Australian fauna with foot and mouth disease', *Australian Journal of Experimental Biology and Medical Science* 46 (1968): 667-87. Snowdon concluded in his study 'that it would only be under exceptional conditions that the Australian fauna tested in this study would participate in the spread of foot and mouth disease under field conditions' (p. 667).
28. Martin, *op. cit.*
29. I. J. Roth, 'Diseases transmissible by feral pigs', roneoed paper delivered at Regional Wild Pig Advisory Council Seminar, Tamworth, N.S.W., March 1979. The 5 per cent figure is also given in O. Siegmund (ed.), *The Merck Veterinary Manual*, 2nd edn, Merck & Co., Rathway, N.J., 1961.
30. Pullar, *op. cit.*, p. 104.
31. For more details on the stability of the virus that causes foot-and-mouth disease, see Dunne, *op. cit.*, pp. 318-25.
32. Siegmund, *op. cit.*, p. 267.
33. A. P. Power and S. A. Harris, 'A cost-benefit evaluation of alternative control policies for foot-and-mouth disease in Great Britain', *Journal of Agricultural Economics* 24 (1973): 573-600. See also Ministry of Agriculture, Fisheries and Food (Northumberland Committee), *Report of the Committee of Inquiry on Foot-and-Mouth Disease*, 1968, Parts One and Two, Cnd. 3999 and 4225, H.M.S.O., London, 1969.
34. See Dunne, *op. cit.*, pp. 240-55; and Siegmund, *op. cit.*, pp. 312-14.

35. See Dunne, *op. cit.*, p. 241.
36. See Dunne, *op. cit.*, pp. 847-8.
37. See Dunne, *op. cit.*, pp. 337-55; and Siegmund, *op. cit.*, pp. 236-8.
38. See Dunne, *op. cit.*, p. 338.
39. *Ibid.*, p. 352.
40. See Siegmund, *op. cit.*, pp. 271-3.
41. See Dunne, *op. cit.*, pp. 177-239; and Siegmund, *op. cit.*, pp. 309-12.
42. Swine fever (or hog cholera) occurs in European wild boar, especially when their density is high. See Lutz Heck and Gunther Raschke, *Die Wildsauen*, Paul Parey, Hamburg, 1980, pp. 67-9.
43. See, for example, Dunne, *op. cit.*; and Siegmund, *op. cit.* See also F. V. Collins, *Meat Inspection*, Rigby, Adelaide, 1966. See also Pullar, *op. cit.*, p. 104.
44. See Dunne, *op. cit.*, p. 178.
45. See Dunne, *op. cit.*, pp. 225-6.
46. G. Edgar, L. Hart and J. T. Hayston, 'Studies in the viability of the virus of swine fever', *Report 14th International Veterinary Congress*, Vol. 2, p. 387.
47. R. R. Birch, 'Hog cholera transmission through infected pork', *Journal of the American Veterinarians Association* 4 (1917): p. 303.
48. See Dunne, *op. cit.*, pp. 370-88; and Siegmund, *op. cit.*, pp. 316-17.
49. See Dunne, *op. cit.*, pp. 270-91; and Siegmund, *op. cit.*, pp. 314-17.
50. Quoted in Dunne, *op. cit.*, p. 274.
51. See Dunne, *op. cit.*, pp. 292-308; and Siegmund, *op. cit.*, pp. 265-7.
52. Quoted in Dunne, *op. cit.*, p. 305.
53. See Dunne, *op. cit.*, pp. 702, 703.
54. *Feral Animals in the Northern Territory*, previously cited.
55. Quoted in Dunne, *op. cit.*, p. 702.
56. *Feral Animals in the Northern Territory*, previously cited, pp. 93, 95.
57. *Ibid.*, p. 95.
58. See Dunne, *op. cit.*, pp. 734-6; and Siegmund, *op. cit.*, pp. 898-900. A.G. Clegg and P.C. Clegg, *Man Against Disease*, Heinemann, London, 1973, pp. 233-5.
59. I. G. Kagan, 'Trichinosis in the United States', *Public Health Report* 74 (1959): p. 159.
60. W. J. Zimmermann, J. H. Steele and I. G. Kagan, 'Trichiniasis in the U.S. Population, 1966-1970: Prevalence, epidemiologic factors', *Public Health Report* 88 (1973): 606-23.
61. E. Barrett-Connor, C. F. Davis, R. N. Hamburger and I. Kagan, 'An epidemic of trichinosis after ingestion of wild pig in Hawaii', *Journal of Infectious Diseases* 133 (1976): 4, 473-7.
62. See Dunne, *op. cit.*, p. 178; Pullar, *op. cit.*; and Collins, *op. cit.*

See also N. K. Golding, 'Field and administrative aspects of the 1961 swine fever outbreak in New South Wales', *Australian Veterinary Journal* 38 (1962): p. 123;

and J. C. Keast and N. K. Golding, 'Further developments in relation to swine fever in New South Wales', *Australian Veterinary Journal* 40 (1964): p. 137.

63. See Dunne, *op. cit.*, pp. 416-32; Siegmund, *op. cit.*, pp. 373-84; and F. V. Collins, *op. cit.*, pp. 372-3.
64. Collins, *op. cit.*, p. 372.
65. J. C. Keast, L. R. Littlejohns, L. C. Rowan and J. S. Wannan, 'The role of the feral pig as a disease reservoir', *Australian Veterinary Journal* 39 (1963): p. 99.
66. *Ibid.*
67. See Dunne, *op. cit.*, pp. 457-66; Siegmund, *op. cit.*, pp. 346-54; and Collins, *op. cit.*, pp. 321-4.
68. Collins, *op. cit.*, p. 321.
69. See Dunne, *op. cit.*, pp. 508-62; Siegmund, *op. cit.*, pp. 400-8; and Collins, *op. cit.*, pp. 328-31.
70. R. Shuman and P. Wood, quoted in Dunne, *op. cit.*, p. 548.
71. See Dunne, *op. cit.*, pp. 433-66; and Siegmund, *op. cit.*, pp. 384-94.
72. Roth, *op. cit.*
73. Keast, Littlejohns *et al.*, *op. cit.*
74. *Feral Animals in the Northern Territory*, previously cited, p. 87.
75. J. H. Norton and A. D. Thomas, '*Brucella suis* in feral pigs', *Australian Veterinary Journal* 52 (1976): 293-4.
76. See Dunne, *op. cit.*, pp. 642-59; Siegmund, *op. cit.*, pp. 394-9; and Collins, *op. cit.*, pp. 350-2.
77. Roth, *op. cit.*
78. Dunne, *op. cit.*, p. 646.
79. *Feral Animals in the Northern Territory*, previously cited, p. 85.
80. *Ibid.*, p. 85.
81. *Ibid.*, p. 86.
82. G. P. Gard, J. R. Giles, R. J. Dwyer-Gray and G. M. Woodroofe, 'Serological evidence of inter-epidemic infection of feral pigs in New South Wales with Murray Valley encephalitis virus', *Australian Journal of Experimental Biology and Medical Science* 54 (1976): 297-302.
83. See E. M. Pullar and G. C. McLennan, 'Sparganosis in a Victorian pig', *Australian Veterinary Journal* (1949): December, 302-4;

    H. McL. Gordon, B. A. Forsyth and M. Robinson, 'Sparganosis in feral pigs in New South Wales', *Australian Veterinary Journal* (1954): May, 135-8;

Collins, *op. cit.*, pp. 441-2;

J. Drabble, *Textbook of Meat Inspection*, Halstead Press, Sydney, 1971, pp. 373-4;

P. L. Appleton and J. H. Norton, 'Sparganosis, a parasitic problem in feral pigs', *Queensland Agricultural Journal* 102 (1976): 4, 339-43.

84. Appleton and Norton, *op. cit.*, p. 339.
85. Appleton and Norton, *op. cit.*, p. 343.
86. Gordon, Forsyth and Robinson, *op. cit.*, p. 136.
87. See Collins, *op. cit.*, pp. 436-41; Drabble, *op. cit.*, pp. 371-3; and Clegg and Clegg, *op. cit.*, pp. 194-6.
88. See Clegg and Clegg, *op. cit.*, pp. 221-3.
89. Collins, *op. cit.*, p. 442.
90. G. A. Letts, 'Feral animals in the Northern Territory', *Australian Veterinary Journal* 40 (1964): March, p. 88. See also Pullar (1950), *op. cit.*, p. 104. L. J. Hone and J. Giles found various internal parasites in their sample of 1235 feral pigs in New South Wales between 1970 and 1975. These parasites included '*Physocephalus sexalatus*, a stomach worm causing erosion of the gastric mucosa in severe infections; *Macracanthorhyneus hirudinaceus*, a large intestinal tapeworm which imbeds deeply into the mucosa of the small intestine, the pleroceroid or spargana of *Spirometra erinacei*, found in the muscle fascia; *Metastrongylus* sp., lungworms; the cystercercoid of *Taenia hydatigena*, occasionally found attached to the diaphragm or liver; and a liver fluke, *Fasciola hepatica*.' L. J. Hone, J. R. Giles and others, *The Biology and Control of Feral Pigs*, Report to the Australian Meat Research Committee, N.S.W. Department of Agriculture, Glenfield, 1977, pp. 17, 19.
91. Collins, *op. cit.*, p. 416.
92. Collins, *op. cit.*, p. 415. Hone and Giles, *op. cit.*, reported that only four of the feral pigs inspected by them had mange mite.
93. Letts, *op. cit.*, p. 88. Hone and Giles, *op. cit.*, reported that in their sample lice were found on *all* feral pigs, and young pigs were heavily infested, especially during the winter.
94. Department of Prime Minister and Cabinet, *Review of Australian Quarantine Arrangements*, Australian Government Publishing Service, Canberra, 1977, p. 98.
95. *Ibid.*, p. 99.
96. *Ibid.*, pp. 49-55, 108-10. Also see this review for further details about the other control measures.
97. See, for example, Commonwealth Department of Health, Canberra, pamphlet, 'Foot and mouth disease: Early warning advice', undated.
98. Previously cited.
99. G. Child, P. Smith and W. von Richter, 'Tsetse control hunting as a measure of

large mammal population trends in the Okavango Delta, Botswana', *Journal of Mammalogy* 34 (1970): 34-75.

100. *Ibid.*, p. 73.
101. *Ibid.*, p. 73.
102. For further information on Australian policies regarding the control of exotic animal disease, see D. M. Flynn, *Report on Vertebrate Pests and Exotic Animal Diseases*, Australian Bureau of Animal Health, Canberra, June 1980.

CHAPTER 9. Methods and Economics of Control of Wild Pigs

1. K. Hart, 'Feral pig problems on the South Coast', *Agricultural Gazette of New South Wales* 90 (1979): 6 (December), p. 3.
2. J. Hone and J. O'Grady, *Feral Pigs and their Control*, Division of Animal Production Bulletin A 4.1.1, N.S.W. Department of Agriculture, 1980, p. 19.
3. *Ibid.*, for further details.
4. C. H. Diong, 'Studies of the Malayan wild pig in Perak and Johore', *Malayan Nature Journal* 26 (1973): 120-51, p. 145.
5. For a description and illustration of the silo trap, see Hone and O'Grady, *op. cit.*, p. 18.
6. Hone and O'Grady, *op. cit.*, p. 18.
7. Hart, *op. cit.*, p. 3.
8. Diong, *op. cit.*, p. 147.
9. G. H. Matschke, 'Trapping and handling of European wild hogs', *Transactions of the Annual Conference of the South East Association of the Fish and Game Commission* 18 (1962): 21-4. A box trap and other traps are also illustrated in J. R. Giles, 'Controlling feral pigs', *Agricultural Gazette of New South Wales* 84: January, 130-2. Further traps are described in J. R. Giles, 'Control of feral pigs', *Wool Technology and Sheep Breeding* (1977): July, 29-31.
10. M. S. Williamson and M. R. Pelton, 'New Design for large portable mammal trap', *Transactions of the Annual Conference of the South East Association of the Fish and Game Commission* 25 (1971): 315-22.
11. A. A. Sludskii, *The Wild Boar: Its Ecology and Economic Importance*, translated by P. Aukland from original Russian edition (IZADATEL, Alma-Ata, 1956) CSIRO, Melbourne, 1974, p. 261.
12. *Ibid.*
13. J. W. Plant, 'Electric fences will give feral pigs a shock', *Agricultural Gazette of New South Wales* 91 (1980): 2, 38-40.
14. For example, these are produced by Gallagher Electronics Ltd, P.O. Box 5324, Frankton, Hamilton, New Zealand.

15. Plant, *op. cit.*
16. *Ibid.*
17. D. M. Pharaoh, *Permanent Electric Fencing*, Information Bulletin 76.1, N.S.W. Department of Agriculture, Lismore, N.S.W., October 1976, p. 55.
18. C. L. Batcheler, '1080 is not the villain emotionalism paints it', *Avinews* (1978): October, 7-10. New Zealand Forest Service Reprint No. 1176.
19. *Agricultural Chemicals: A Synopsis of Toxicity and a Guide to Treatment*, N.S.W. Department of Public Health, Sydney, 1968, p. 34.
20. 'Compound 1080', *Rural Research in CSIRO* 23 (1958): 2-5, p. 2.
21. *Ibid.*, and see Batcheler, *op. cit.*
22. *Agricultural Chemicals*, previously cited, p. 34.
23. *Wild Pig Control Program*, Western Agricultural Region, N.S.W. Department of Agriculture, Dubbo, 1978, p. 1.
24. Batcheler, *op. cit.*, p. 9.
25. *Ibid.*
26. J. Giles, 'Compound 1080 - sodium monofluoroacetate', Regional Wild Pig Advisory Council Seminar on Wild Pigs, Summaries (roneoed), Tamworth, 1979, p. 3.
27. Batcheler, *op. cit.*, p. 9.
28. W. D. Hamilton, J. Wright and R. Kirby, 'Feral pig control trials on "Wynella", Dirrabandi', *Queensland Agricultural Journal* 102 (1976): 419-21, p. 420. Although the Queensland factory producing S.A.P. burned down, production of a substance with the same formula (C.S.S.P.) is reported to have commenced in New South Wales. See R. Benson, *Evaluating Feral Pig Damage and Costs of Control in Western New South Wales*, Research Workpaper No. 120, Division of Marketing and Economics, N.S.W. Department of Agriculture, Sydney, February 1980, p. 7.
29. For full details, see G. R. Mackintosh, 'Destruction of wild pigs by poisoned baits', *New Zealand Journal of Agriculture* 80 (1950): 3, 259-61.
30. Benson, *op. cit.*, p. 14.
31. Benson, *op. cit.*, p. 7.
32. J. A. Miller, 'Poisoning wild pigs with arsenic trioxide', *New Zealand Journal of Agriculture* 90 (1955): 5, p. 551.
33. Hone and O'Grady, *op. cit.*, p. 11.
34. C. W. Smith and C. H. Diong, *Proposal to Study Feral Pigs in Kipahula Valley Haleakala National Park*, Technical Report No. 19, Cooperative National Park Resources Studies Unit, Department of Botany, University of Hawaii, Honolulu, September 1977, p. 22.
35. Hone and O'Grady, *op. cit.*, p. 20.
36. J. Hone, J. O'Grady and H. Pedersen, *Decisions on the Control of Feral Pig Damage*, Ag. Bulletin 5, N.S.W. Department of Agriculture, Sydney, 1980, p. 17.

37. Sludskii, *op. cit.*, p. 155.

38. Sludskii, *op. cit.*, p. 152.

39. P. Le S. Milstein, 'The bushpig *Potamochoerus porcus* as a problem animal in South Africa', Proceedings Entomological Symposium, Pretoria, 27 September 1971, Transvaal Division of Nature Conservation, (roneoed), pp. 10-11.

40. Sludskii, *op. cit.*, p. 151.

41. Personal communication from National Council for Conservation of Wildlife in Pakistan, through Embassy of Pakistan in Australia, 25 November 1980.

42. Hone and O'Grady, *op. cit.*, p. 10.

43. Hone, O'Grady and Pedersen, *op. cit.*, p. 8.

44. *Ibid.*

45. See J. R. Giles, 'Control of mammalian pests', *Parks and Wildlife* 2 (1978): 2, p. 48; and J. R. Giles, 'Control of Feral Pigs', *Wool Technology and Sheep Breeding*, (1977): June/July, 29-31, esp. p. 29.

46. J. R. Giles, 'Control of mammalian pests', *Parks and Wildlife* 2 (1978): 2, 46-9. On page 48, Giles says, for example, 'The rather simplistic figures illustrate if one wishes to control a population of "pests" so that the damage caused by it is reduced, it is essential not to generate a high "potential sustained yield", or one will quickly end up with as many animals as before the control started.'

47. Hone and O'Grady, *op. cit.*, pp. 20-2.

48. J. C. Headley, 'Defining the economic threshold', in *Pest Control Strategies for the Future*, National Academy of Sciences, Washington, D.C., 1972, pp. 100-8.

49. *Ibid.*

50. Headley assumes that the productive unit is a farm but for some purposes it may be useful to assume that it is a region. The larger the region the more likely it is that the price of the product cannot be assumed to be constant and independent of the level of production. Nevertheless, as argued in Chapter 2, no price effects on produce are likely to arise from variations in wild pig numbers in Australia.

51. Headley, *op. cit.*, p. 104.

52. However, the implications of the Headley model do not depend on the special nature of this function. The implications of the model are the same provided that the total cost of controlling a pest increases with the attempted reduction in the population.

53. In this problem, there is one interior critical point, the one for which the marginal cost of control equals the marginal revenue from it, and this yields a relative maximum of profit. However, profit at the end-points $P = 0$ and $P = P_E$ have to be evaluated and compared with this value to determine the profit-maximizing strategy.

54. See, for example, C. A. Tisdell, *Microeconomics of Markets*, Wiley, Brisbane, 1982, Ch. 13.
55. Mathematically, the marginal revenue function can be obtained as follows: Letting λ represent the price of product, and as before assuming that x is its quantity of production, total revenue is

$$R = \lambda x$$
$$= \lambda(K - cbP_t^2 - A) \qquad \text{for } P_t > \hat{P}_t.$$

Hence, marginal revenue is

$$MR = \frac{dR}{dP} = -2\lambda cbP_t.$$

The parameters of this equation are λ, the price of the product, and cb, indicating the marginal impact on production of the pest population.
56. As discussed in Chapter 4, some graziers in the Hay region turned to greater commercial exploitation of feral pigs for amateur hunting when wool prices slumped.
57. The value of the marginal product of a factor of production is equal to the price of the product times the marginal physical product of the factor. See, for example, C. A. Tisdell, *Economics of Markets*, Wiley, Sydney, 1974, p. 245.
58. See Giles, 'Control of mammalian pests', previously cited, p. 48. See also Hone, O'Grady and Pedersen, *op. cit.*, pp. 14, 15.
59. Discounting allows for the fact that $1 in income available at a future date is worth less than this sum available now. This is because a return equal at least to the rate of interest can be earned on $1 if it is available now. Usually the rate of interest is used as a discount factor. For a full discussion of this matter, see Tisdell, *Microeconomics of Markets*, previously cited, Ch. 17.
60. Even weed pests can be considered to be very mobile, the mobility coming via the dispersal of their seed, etc. Indeed my earlier paper on *Simple Economic Models of Pest Control: Models with Possible Applications to Control of Feral Pigs and Other Wild Animals* (Research Report No. 33, Department of Economics, University of Newcastle, 1977) helped to stimulate some research into this aspect. See B. A. Auld, K. H. Menz and N. M. Monaghan, 'Dynamics of weed spread', *Protection Ecology* 1 (1978-79): 141-8; B. A. Auld, K. M. Menz and R. W. Medd, 'Bioeconomic model of weeds in pastures', *Agro-Ecosystems* 5 (1979): 69-84; K. M. Menz, B. G. Coote and B. A. Auld, 'Spatial aspects of weed control', *Agricultural Systems* 6 (1980-81): 67-75.
61. J. R. Giles, 'The ecology of the feral pig in New South Wales, II. The dynamics of a colony in the Macquarie Marshes, 1971-75', Noxious and Feral Research Centre of the N.S.W. Department of Agriculture, Glenfield, (roneoed), undated.
62. Hone, O'Grady and Pedersen, *op. cit.*, p. 10.
63. Sludskii, *op. cit.*, p. 194.

64. *Ibid.*, p. 193.

65. *Ibid.*, p. 189.

66. See, for example, G. W. Wood and R. H. Barrett, 'Status of wild pigs in the United States', *The Wildlife Society Bulletin* 7 (1979): 4, 237-46;
R. H. Barrett, 'The feral hog on the Dye Creek Ranch, California', *HILGARDIA* 46 (1978): 9, 283-355;
F. J. Singer, D. K. Otto, A. R. Tipton and C. P. Heble, *Home Ranges, Movements, and Habitat Use of European Wild Boar*, Report for the Superintendent of Great Smoky National Park, Department of Interior, Gatlinburg, Tennessee, May 1979.

67. See, for example, J. R. Martin, 'Movement of feral pigs in North Canterbury, New Zealand', *Journal of Mammalogy* 50 (1975): 914-15.

68. R. H. Barrett, 'The feral hog on the Dye Creek Ranch', previously cited, p. 305. Barrett gives many additional references (to those given here) to studies dealing with the mobility of wild pigs.

69. The problems involved are similar to those involved in the supply of pure public goods. See, for example, R. R. Russell and M. Wilkinson, *Microeconomics*, Wiley, New York, 1979, Ch. 18; and C. Tisdell, *Microeconomics: The Theory of Economic Allocation*, Wiley, Sydney, 1972, Ch. 21.

70. But more than co-ordination between landholders may be required to obtain a socially optimal level of control of a pest. Subsidies to encourage control effort, or fines or taxes on those failing to take socially adequate control measures may be needed. Ideally these should be of a magnitude that takes the spillover benefits of the pest control into account.

71. Auld, Menz and collaborators have made some progress along these lines (see note 60). Hone, O'Grady and Pedersen, *op. cit.*, p. 10, suggest that some spatial patterns of feral pig control may be superior to others.

72. Hone, O'Grady and Pedersen, *op. cit.*, p. 8.

73. H. B. Olaff, *Zur Biologie und Okologie des Wildschweines*, Paul Schopps, Franfurt a. Main, 1951.

74. See L. Heck and G. Raschke, *Die Wildsauen*, Paul Parey, Hamburg, 1980, p. 52.

75. V. G. Heptner, A. A. Nasimovic and A. G. Bannikov, *Die Säugetiere der Sowjetunion*, (V. G. Heptner and N. P. Naumov, eds), Vol. 1, Gustav Fischer, Jena, 1966, p. 53.

76. Wood and Barrett, *op. cit.*, p. 242.

77. *Ibid.*, p. 243.

78. Barrett, *op. cit.*, pp. 339-40.

79. It can be compatible with the suggestion by J. Giles that it may be advantageous to reduce the population of a pest *below* the level giving maximum sustainable yield.

80. Some of the types of allowance to be made for uncertainty are discussed in

Tisdell, *Simple Economic Models of Pest Control*, previously cited.

81. For further discussion of this aspect see Tisdell, *Some Economic Models of Pest Control*.
82. For a useful general discussion of this matter, see G. A. Norton, 'Economic aspects of pest control', (roneoed), Imperial College, London, undated.
83. This question is discussed to some extent by D. Hueth and U. Regev, 'Optimal agricultural pest management with increasing pest resistance', *American Journal of Agricultural Economics* 56 (1974): 543, 552.
84. If we take two landholders A and B able to use a pest control technique, their payoffs from different control strategies may be like those shown in the following matrix where the first entry in each cell is the payoff to A and the second that to B. For simplicity, the strategies available to the two landholders are divided into frequent and infrequent use of the pest control technique.

		B's strategies →	
		Infrequent use	Frequent use
A's strategies	Infrequent use	(5, 5)	(4, 6)
	Frequent use	(6, 4)	(2, 2)

Acting in their own selfish interest both landholders are likely to choose frequent use of the control technique and hence the gain of each is smaller than if infrequent use is made of the technique by the landholders. Some readers will recognize this as akin to the prisoner's dilemma model of game theory. See R. D. Luce and H. Raiffa, *Games and Decisions*, Wiley, New York, 1957, pp. 94-7.

85. See Heptner *et al.*, *op. cit.*, p. 82.
86. This is common in European countries. See, for example, Heck and Raschke, *Die Wildsauen*. For a discussion of analytical questions raised by such conflict, see C. Tisdell, 'Wildlife: A national asset or pest to be managed', in Department of Science and the Environment, *Environmental Economics*, Australian Government Publishing Service, Canberra, 1979, pp. 79-87.
87. J. E. Harting, *British Animals Extinct Within Historic Times*, Trübner & Co., Ludgate Hill, London, 1880, pp. 79-114 and esp. pp. 206-9.
88. *Ibid.*, p. 209.

CHAPTER 10. Concluding Observations

1. J. E. Harting, *British Animals Extinct Within Historic Times*, Trübner & Co., Ludgate Hill, London, 1880, p. 209.
2. They are declared to be noxious animals in New South Wales and this places a legal obligation on landholders to take action to rid his property of them.

They are also regarded as vermin in Queensland and Victoria.

3. Some readers may wish to refer to the following article which gives a critical evaluation of cost-benefit analysis and other references given there: J. A. Sinden, 'Pangloss, Pandora and Pareto for the aspiring benefit-cost analyst', *Review of Marketing and Agricultural Economics* 48 (1980): 2, 99-108.
4. For a discussion of compensation tests of social welfare, see, for example, I.M.D. Little, *A Critique Welfare Economics*, 2nd edn, Oxford University Press, London, 1957, Ch. 6; and D. M. Winch, *Analytical Welfare Economics*, Penguin, Harmondsworth, 1971, Ch. 3, pp. 143-50.
5. See, for example, E. J. Mishan, *Cost-Benefit Analysis: An Informal Introduction*, George Allen & Unwin, London, 1971, Ch. 48; and Sinden, *op. cit.*, p. 104.
6. See Mishan, *op. cit.*, Ch. 19.
7. Hoogvelt makes this point in relation to the application of cost-benefit analysis in developing countries. See Ankie M. M. Hoogvelt, *The Sociology of Developing Societies*, 2nd edn, Macmillan, London, 1978.
8. See, for example, C. J. Cicchetti and A. M. Freeman, 'Option demand and consumer surplus: Further comment', *Quarterly Journal of Economics* 85 (1971): 528-39; C. A. Tisdell, 'Provision of parks and the preservation of nature: Some economic factors', *Australian Economic Papers* 11 (1972): 154-62; C. Henry, 'Investment decision under certainty: The "irreversibility effect"', *American Economic Review* 64 (1974): 1006-12.
9. See H. J. Frith, *Wildlife Conservation*, Angus & Robertson, Sydney, 1973, Ch. 2; K. A. Wodzicki, *Introduced Mammals of New Zealand: An Ecological and Economic Survey*, Department of Scientific and Industrial Research, Wellington, 1950, Ch. 1.
10. A number of bidding or auctioning schemes for revealing preference and other techniques are reviewed in J. W. Bennett, Evaluation of Benefits of Natural Ecosystem Preservation with Special Reference to Existence Values, Ph.D. thesis, Australian National University, July 1981, Chs 2 and 3.
11. Some of the questions involved are covered in N. Myers, *The Sinking Ark: A New Look at the Problem of Disappearing Species*, Pergamon Press, Oxford, 1979; J. Passmore, *Man's Responsibility for Nature: Ecological Problems and Western Traditions*, Duckworth, London, 1974; C. A. Tisdell, *On the Economics of Saving Wildlife from Extinction*, Research Report or Occasional Paper No. 48, Department of Economics, University of Newcastle, 1979.
12. Bennett, *op. cit.*, pp. 195-6, and Ch. 8.
13. Bennett, *op. cit.*, covers many of these issues.
14. This point can be illustrated diagramatically.
15. See, for example, C. A. Tisdell, 'Wildlife: A national asset or pest to be

managed', in Department of Science and the Environment, *Environmental Economics*, Australian Government Publishing Service, Canberra, 1979, pp. 79-87.

16. The regulations appear not to be enforced because the authorities believe that a conviction is unlikely, given the 'loopholes' in the law.
17. For some information about bonuses paid in Australia in the past on wild pigs, see T. McKnight, *Friendly Vermin: A Survey of Feral Livestock in Australia*, University of California Press, Berkeley, 1976, pp. 49-51. McKnight argues that bonuses proved to be ineffective as a control policy. For similar information for New Zealand, see Wodzicki, *op. cit.*, pp. 236-7.
18. S. Whitehouse, 'Bounty Schemes in Vermin Control', *Western Australian Journal of Agriculture* 17 (1976): 3 (4th series), 85-9.
19. *Ibid.*, p. 86.
20. For further discussion of problems associated with bonus schemes, see Tisdell, ref. 15 (above).
21. The problem is similar to that often quoted in the literature of several companies drawing oil or gas from a common underground pool.
22. This research does not take account of collective good values of wildlife. See, for example, P. W. Barkley and D. W. Seckler, *Economic Growth and Environmental Decay: The Solution Becomes the Problem*, Harcourt Brace Jovanovich, New York, Ch. 9.
23. R. Allen, *How to Save the World: Strategy for World Conservation*, Kogan Page, London, 1980, p. 111.
24. Some of these issues are covered in Myers, *op. cit.*, and Passmore, *op. cit.*
25. However, diversity is not certain to create greater stability.
26. The Flynn Report recommends that an Australian national policy on feral animal control and eradication be formulated with first priority being given to feral pigs. It suggests that Commonwealth funding be provided for an initial pilot project designed to eradicate feral pigs in a location such as Kangaroo Island. See D. M. Flynn, *Report on Vertebrate Pests and Exotic Animal Diseases*, Australian Bureau of Animal Health, Department of Primary Industry, Canberra, June 1980, p. 1.

Index

Made in the USA
Monee, IL
13 September 2026

62977004R10249